Advances in

MICROBIAL PHYSIOLOGY

Advances in
MICROBIAL PHYSIOLOGY

Edited by
A. H. ROSE
School of Biological Sciences
Bath University
England

and

D. W. TEMPEST
Laboratorium voor Microbiologie
Universiteit van Amsterdam
Amsterdam-C
The Netherlands

VOLUME 12

1975

ACADEMIC PRESS
LONDON NEW YORK SAN FRANCISCO
A Subsidiary of Harcourt Brace Jovanovich, Publishers

ACADEMIC PRESS INC. (LONDON) LTD.
24/28 Oval Road
London NW1

United States Edition published by
ACADEMIC PRESS INC.
111 Fifth Avenue
New York, New York 10003

Library of Congress Catalog Card Number: 67-19850
ISBN: 0 12-027712-3

PRINTED IN GREAT BRITAIN BY
WILLIAM CLOWES AND SONS LIMITED
LONDON, COLCHESTER AND BECCLES

Contributors to Volume 12

W. A. HAMILTON, *Unit of Microbiology, Department of Biochemistry, Marischal College, University of Aberdeen, Scotland*

I. B. HOLLAND, *Department of Genetics, University of Leicester, Leicester LE1 7RH, England*

H. E. KUBITSCHEK, *Division of Biological and Medical Research, Argonne National Laboratory, Argonne, Illinois 60439, U.S.A.*

J. G. MORRIS, *Department of Botany and Microbiology, School of Biological Sciences, The University College of Wales, Aberystwyth, SY23 3DA, Wales*

TATSUO MATSUSHITA, *Division of Biological and Medical Research, Argonne National Laboratory, Argonne, Illinois 60439, U.S.A.*

N. SHAW, *Microbiological Chemistry Research Laboratory, School of Chemistry, The University of Newcastle upon Tyne, Newcastle upon Tyne, NE1 7RU, England*

Contents

Energy Coupling in Microbial Transport

W. A. HAMILTON

Physiology of Colicin Action

I. B. HOLLAND

Bacterial Glycolipids and Glycophospholipids

NORMAN SHAW

The Physiology of Obligate Anaerobiosis

J. G. MORRIS

DNA Replication in Bacteria

TATSUO MATSUSHITA AND HERBERT E. KUBITSCHEK

Energy Coupling in Microbial Transport

W. A. HAMILTON

Unit of Microbiology, Department of Biochemistry, Marischal College, University of Aberdeen, Aberdeen, Scotland

I. Introduction

During recent years the study of the transport of nutrients and ions across cell and organelle membranes has increasingly captured and held the attention of biologists, so that now it represents a major area of research effort. This contention is amply borne out by the rash of symposia and reviews that have appeared recently; one may cite, for example, the papers of Harold (1972), Kaback (1972), Kaback and Hong (1973) and Boos (1974). A feature of current work in the field of transport even more significant, however, than the dramatic increase in the number of papers published, is the change in emphasis of these papers.

Since the classic review of Cohen and Monod (1957), research into transport phenomena has been dominated by kinetic and genetic analyses. The mechanism of energy coupling, a prerequisite where a substrate is concentrated within the membrane-bounded volume, has been incompletely understood and only hinted at in the various models that have been put forward in an effort to describe the transport system. The particular excitement of current studies lies in our new, or at least developing, understanding of the molecular nature of the cellular mechanisms coupling the redox and chemical energies of metabolism to the transport and intracellular accumulation of nutrients. The present paper will confine its collation and discussion of transport data to this one aspect of the subject. Readers more concerned with questions of, for example, specificity, or who wish a more general review, are referred to the papers cited above, and to others introduced below.

II. The Permease Model

In *Escherichia coli* the transport of β-galactosides is characterized by the appearance of the chemically unmodified sugar within the cell. This was the system studied by the Paris school in the 1950s and reviewed by Cohen and Monod in 1957. Since then this, and closely related, sugar and amino-acid transport systems in bacteria have been examined and analysed by a large number of workers, employing the same considerations of specificity and kinetics (Koch, 1964; Winkler and Wilson, 1966; Scarborough *et al.*, 1968; Schachter and Mindlin, 1969; Kepes, 1971). In the model which has developed from these studies (Fig. 1), the diffusion of the polar sugar across the hydrophobic barrier of the membrane is facilitated by its affinity for and binding to a membrane component. The specificity and saturable character of this system, the protein nature of the membrane component, the fact that its synthesis can be both induced and repressed, and that it is coded for by a specific gene, suggested an enzyme-like mechanism. Accordingly, the membrane protein coded for

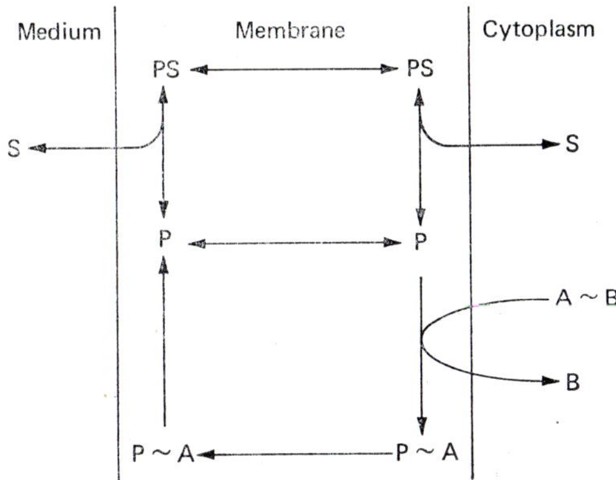

FIG. 1. The permease model. S indicates a sugar or other polar substrate; P the permease; A ~ B the theoretical high-energy compound; and P ~ A the activated form of the permease. Active transport results from the series of irreversible reactions P → P ~ A → P ~ A → P. After Kepes (1971).

by the *y* gene and controlled by the *i* gene in the *lac* operon in *Escherichia coli*, was given the name "permease". With the exception of a report by Koch (1971a), it is generally agreed that the permease system can function independently of a source of metabolic energy in catalysing both uptake to the point of transmembrane equilibrium, and exchange of extracellular and intracellular sugar. When, however, mutants which have lost the activity of the intracellular catabolic enzymes are used, or the substrate is replaced by a non-metabolized analogue, the phenomenon of active transport can be observed, with the accumulation of the still unmodified substrate to a concentration within the cell up to 1000-times greater than that in the extracellular medium. Under these conditions, a source of metabolic energy is obligatory, and the process is sensitive to the action of metabolic inhibitors such as iodoacetate or cyanide, and to uncouplers of oxidative phosphorylation. In the permease model the nature of the energy donor has variously been proposed as ATP itself, or as an unspecified high-energy compound, A ~ B. Values for K_m and V_{max} have been determined for entry and exit, under both energy-coupled and uncoupled conditions. From such analyses, some authors have concluded that the site of energy coupling is the binding of the sugar to the permease protein on the outer face of the membrane's osmotic barrier, thus affecting the process of entry; others claim that exit is effected through the action of energy coupling on the dissociation of the permease and sugar at the membrane's inner surface. In a recent paper describing their studies with a mutant, energy-uncoupled for lactose transport, Wilson and Kusch (1972) have re-inforced their earlier conclusion that energy coupling has no effect on entry, but decreases the exit rate of galactosides from the cell by lowering the affinity of the permease for the substrate at the inner border of the plasma membrane.

What is common to all derivatives of the permease model is the hypothesis that energy coupling is achieved by an interaction between ATP, or a related high-energy compound, and the permease, either free or in a complex with the transported substrate. This interaction chemically modifies the permease, possibly to a phosphorylated derivative, with a consequent alteration in either the value for K_m of association or of dissociation, or in the rate of translocation across the membrane of either the permease-substrate complex or the free permease. Komor *et al.* (1973a) have however carried out a detailed kinetic analysis of hexose transport in the alga *Chlorella vulgaris*. This system appears to be closely related to the β-galactoside transport in *E. coli*. The authors conclude that accumulation of the non-metabolized 6-deoxyglucose can only be explained kinetically on the basis of energy coupling affecting not either, but both, the affinity of the permease for the sugar, and the velocity constants for the flux of complexed and free permease.

From these conflicting views it would seem therefore that the permease model may be of only limited value with regard to consideration of the molecular mechanism of energy coupling and active transport.

III. Enzyme-Catalysed Reaction and Carrier-Mediated Transport

A. Kinetics versus Thermodynamics

Even to the choice of the name, the development of the permease model has been greatly influenced by our knowledge and understanding of enzyme action. The function of an enzyme is to catalyse a chemical reaction, i.e. to increase its rate. Clearly a kinetic analysis, with the determination of parameters such as K_m and V_{max}, is a valid method of studying such a system. However, any reaction, catalysed or uncatalysed, is also subject to the laws of thermodynamics, and will only proceed if accompanied by a decrease in the free energy of the system. Both the rate and the mechanism of the reaction are quite independent of this free energy change, and thus kinetic and energetic analyses of the reaction can proceed separately and are of little direct relevance to one another. When, however, one considers the mechanism of the reaction, and in particular the mechanism of catalysis, one sees that the rate of the reaction is controlled by the activation energy which must be present before the reactants will react, and so give rise to the products. The enzyme in a biological reaction increases the rate of that reaction by lowering the activation energy. Consequently, energetic conclusions, drawn from kinetic analyses of an enzyme reaction, will be relevant to the mechanism of that reaction, rather than to the position of final equilibrium.

B. Scalar versus Vectorial

When one turns from enzyme-catalysed reactions to carrier-mediated transport (and from now on I shall refer to the specific membrane component involved in facilitated diffusion by the less contentious and more generally applicable term "carrier"), one must note two critical differences. Other than in the special case of group translocation to be considered later, the substrate and product of the transport reaction are chemically identical. Secondly, unlike the scalar nature of an enzymic reaction taking place in solution, carrier-mediated transport is vectorial in that the substrate is translocated from one aqueous phase, across the hydrophobic barrier phase of the membrane, to a second aqueous phase. It will be a key point in this discussion that the full appreciation of these facets of transport phenomena is critical to our understanding of their mechanism, and in particular of the mechanism of energy coupling.

C. Group-Transfer Reactions: Active Transport and Oxidative Phosphorylation

In energy-independent facilitated diffusion, the substrate is transported into the cell to the point where the concentration on each side of the membrane is the same. Thereafter exchange can occur, but there will be no net flux in either direction. The substrate can only diffuse down its concentration gradient. It is to allow movement against the concentration gradient, and so produce intracellular accumulation, that energy coupling is required in active transport. Here it is the position of final equilibrium which is being altered and, as in enzyme-catalysed reactions, the transport must be coupled to another reaction, or translocation, in itself characterized by a decrease in free energy greater than the increase required for the intracellular accumulation of the translocated substrate. As discussed by Mitchell (1972), Lipmann's (1941) powerful concept of the coupling of group-transfer reactions requires the sharing of a common reactant. This can be illustrated by the coupling of the oxidation of AH_2 to the reduction of B through the shared linkage of the appropriate dehydrogenases to NAD^+. Similarly, 1,3-diphosphoglycerate couples oxidation of 3-phosphoglyceraldehyde to the phosphorylation of ADP. The study of energy coupling in active transport (and incidentally in oxidative phosphorylation) centres therefore on the search for the shared intermediate between the exergonic reactions of the cell's catabolism and the endergonic reactions of transport against a concentration gradient (or of ATP synthesis).

Such a comparison of active transport and oxidative phosphorylation cannot be considered as merely superficial. Both are obligatorily associated with membranes, and there is much evidence to suggest that the elusive shared intermediate may couple not only oxidation to phosphorylation but also the active transport of ions and nutrients to both these processes. As suggested by Harold (1972), in his excellent review of energy conservation and transformation by bacterial membranes, the chemical, conformational and chemiosmotic hypotheses of oxidative phosphorylation have their direct counterparts in the permease, respiration-linked, and ion-gradient models for active transport. In the first two, the coupling is direct, and the nature of the shared intermediate is chemical. In the last, the coupling is indirect and the shared intermediate, or high-energy state, takes the form of gradients of chemical and electrical potential across the coupling membrane.

We have already discussed the permease model and its development over the past 17 years. As a description of a component intimately concerned in the interaction between a cell and its environment, and a product of the function and control of the *lac* operon, its contribution to our understanding has been considerable. But, like the chemical hypothesis of phosphorylation, it has singularly failed in its attempt to describe the nature of energy coupling, and the identity of the high-energy intermediate remains as elusive as ever.

IV. The Redox Model

The respiration-linked or redox model of transport is very much the brain-child of Kaback and his coworkers. Their extensive work with subcellular membrane vesicle systems (Kaback and Stadtman, 1966) has been reviewed by Kaback (1972) and Kaback and Hong (1973). The basic tenet of this model is that the carrier responsible for solute translocation can also function as a redox carrier. Not only can it exist in either an oxidized or a reduced form, but these forms differ in their affinity for the transported substrate, and in their orientation within the membrane. The oxidized form, with an S–S disulphide bond, has a high affinity for the substrate at the outer membrane surface. On reduction, re-orientation of the carrier and a decrease in affinity results in release at the inner surface. The cyclic oxidation and reduction is dependent on the flow of reducing equivalents through the electron-transport chain. Originally it was suggested that the transport carrier was an integral component of this chain, but now, from work with mutants affecting the electron-transport and ATPase activities, the favoured hypothesis is that it lies on a shunt or loop, and that its response to respiratory activity is subject to some unspecified control (Hong and Kaback, 1972). Through this means of coupling, therefore, it is envisaged that the redox energy of respiration is transduced into the osmotic energy of the concentration gradient generated in active transport. It is suggested further that in its reduced form the carrier is capable of oscillation within or across the membrane, and so of catalysing energy-independent facilitated and exchange diffusions.

This model (Fig. 2) has been deduced from studies of the transport of a wide range of sugars and amino acids by subcellular membrane vesicles, which have been prepared from a large number of bacteria, including *E. coli*, *Staphylococcus aureus*, *Salmonella typhimurium*, *Bacillus subtilis* and *Azotobacter vinelandii*. Although a number of electron donors are capable of powering uptake, e.g. D-lactate, L-lactate, NADH, succinate, L-malate, D-α-glycerophosphate, and ascorbate plus phenazine methosulphate, they do not all do so with the same efficiency. It is claimed that all of the transport carriers in *E. coli* are integral components of the

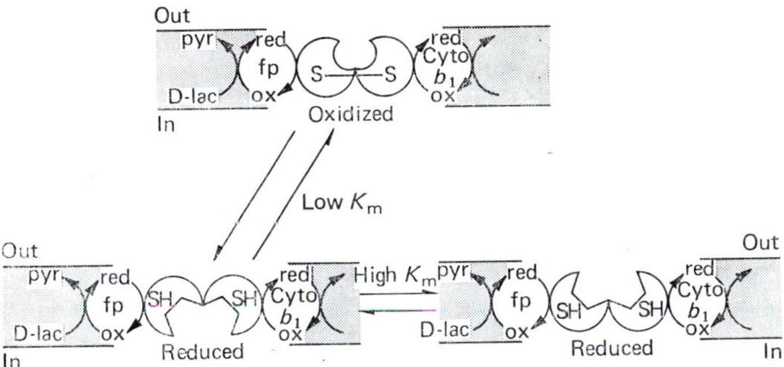

FIG. 2. The redox model. The transport carrier is located between the flavoprotein D-lactate dehydrogenase and cytochrome b_1. In its oxidized form, the binding site is exposed on the outer membrane surface and shows a high affinity for its substrate. On reduction, a conformational change brings the site to the inner surface and lowers its affinity. At higher internal concentrations of substrate, exchange diffusion results from oscillations within the membrane of the reduced form of the binding site. From Kaback and Barnes (1971).

D-lactate oxidase, and that they are sited between the flavoprotein dehydrogenase and cytochrome b_1 (Barnes and Kaback, 1971). In *Staph. aureus*, however, the role of primary electron donor is taken by D-α-glycerophosphate (Short *et al.*, 1972) and in *A. vinelandii* by L-malate (Barnes, 1972). Even in *E. coli*, mutants defective in D-lactate dehydrogenase have been shown to couple transport with increased efficiency to oxidation of succinate (Hong and Kaback, 1972). Moreover, Mitchell (1973) has pointed out that a more meaningful estimate of the efficiency of the coupling of oxidation of a particular electron donor to nutrient transport, e.g. of lactose by *E. coli*, is obtained by comparing the *ratio* of lactose uptake rate to the oxygen uptake rate with that for D-lactate-driven transport. From such an analysis it is seen that DL-α-hydroxybutyrate and L-lactate are apparently better energy sources than D-lactate; D-α-glycerophosphate, formate and succinate are about half as good, and only NADH is of minimal use.

A. Criticisms

1. Membrane Orientation

In his criticisms of the redox model, Harold (1972, 1974) discussed the importance of membrane orientation, and transport first of the electron donor to its dehydrogenase on the inner surface of the membrane. For example, the very limited permeability of bacterial membranes to NADH means that any oxidation of this compound by vesicle preparations will in all probability be carried out by vesicles in which the membrane orientation has been reversed, or by vesicles that have not completely sealed. As such vesicles would not be expected to be capable of active transport, the observed oxidation would therefore be quite irrelevant to considerations of transport. This point is exemplified by the work of Heppel *et al.* (1972). They showed that mutants with a defective transport of D-α-glycerophosphate were deficient also in uptake of proline driven by oxidation of this electron donor; D-lactate-driven transport, on the other hand, was the same as in the wild type. With relevance to the question of changing patterns of oxidation after lysis and resealing of the ghosts, they reported that, while there was no evidence for external oxidation of D-α-glycerophosphate or succinate when these compounds were added to a suspension of sphaeroplasts, there was with vesicles prepared from them. Clearly the vesicle population must contain a certain proportion of its number which are either inverted or, at the molecular level, have inadequately resealed. Alternatively, as recently proposed by Altendorf and Staehelin (1974), dislocation of membrane proteins may have taken place during the lysis, with an internally facing dehydrogenase in the sphaeroplast becoming an externally facing one in the vesicle.

From studies of oxidative phosphorylation in membrane preparations from *Micrococcus denitrificans* (John and Hamilton, 1970, 1971) and oxidative phosphorylation and amino-acid transport in preparations from *Mycobacterium phlei* (Hirata *et al.*, 1971; Hirata and Brodie, 1972; Asano *et al.*, 1973) another important consideration of membrane orientation is highlighted. Due to the impermeability of intact bacterial membranes to adenine nucleotides, oxidative phosphorylation can only be demonstrated either in particles with reversed membrane orientation, or in ghosts prepared with intravesicular ADP. Populations of membrane vesicles are liable to be mixed in terms of membrane orientation, in an undefined manner determined by the organism and the method of preparation. The apparent efficiency of coupling the energy of respiration to either phosphorylation or transport will be greatly influenced by this fact which, regrettably, is not always appreciated or allowed for.

2. *Obligatory Coupling to Respiration*

While it is clearly established that active transport in these systems can by powered by respiration, the conclusion that the carriers are themselves also redox carriers, and that they are primarily coupled to a particular dehydrogenase, seems very much less convincing. Furthermore, it is claimed that transport not only can be driven by respiration, but that it is obligatorily so coupled. The vesicles do not contain ATP, are incapable of oxidative phosphorylation, and transport is neither activated by added ATP nor inhibited by arsenate. While these findings could readily be explained in terms of membrane orientation and impermeability to ATP, Konings and Kaback (1973) also failed to obtain evidence of coupling of transport to ATP hydrolysis, even when ATP or an ATP-generating system was trapped within the vesicles during preparation. Even more significantly, these experiments were performed with anaerobically grown *E. coli* where ATP derived from glycolysis would be expected to be the cell's primary energy source. However, rather than conclude, as Kaback and his associates do from such data, that transport cannot be coupled to ATP hydrolysis other than through reversed electron transport, it might be more reasonable to consider that the relatively sensitive and complex ATPase system may have been damaged during the course of vesicle preparation. The development of the membrane vesicle has done much to simplify experimental procedures and to clarify interpretation of results in the transport field, but it must not be forgotten that the vesicle is none-the-less an experimental artefact. The absence of any ATP-stimulated activity in vesicles can hardly be taken as proof that ATP serves no useful function in whole cells either! To paraphrase Kaback (1972), even a negative hypothesis cannot be proved by negative results. Such polemics are, however, rendered superfluous by positive results achieved by Van Thienen and Postma (1973) when they were able to demonstrate stimulation of serine transport by ATP in membrane vesicles derived from *E. coli*.

3. *Uncouplers*

Probably, though, the greatest weakness of the redox hypothesis lies in its inability to offer a satisfactory explanation for the inhibition of active transport by uncouplers of oxidative phosphorylation. On a related matter, Lombardi *et al.* (1973) go against the enormous body of data derived from studies with bacterial, mitochondrial, chloroplast and artificial membranes, and claim that valinomycin does not make their vesicle membranes passively permeable to rubidium ions. In an idea very reminiscent of the earlier suggestion of Pressman (1965), they hypothesize that valinomycin facilitates the approach of the alkali metal ion to the active centre of its redox transport carrier, and that transport of rubidium creates, rather than responds to, a membrane potential. These conclusions rest on the absence of any exit of pre-accumulated rubidium when its uptake is inhibited by oxamate or *p*-chloromercuribenzene sulphonate, and on the inability of the vesicles to accumulate the lipid-soluble cation dibenzyldimethyl ammonium (DDA^+). Treatment of vesicles with DDA^+, however, did not cause either inhibition of valinomycin-induced rubidium uptake, nor of efflux of sodium ions. The lipid-soluble tetraphenylarsonium and triphenylmethylphosphonium ions did produce these effects, but they were not tested as indicators of membrane potential. Furthermore, Hirata *et al.* (1973) have pointed out that, under the conditions these experiments were carried out, DDA^+ is neither freely translocated across the membrane nor is it a reliable indicator of the magnitude of the membrane potential. We shall consider the membrane potential and the action of lipid-soluble ions in greater detail in a later section.

4. *Redox Potentials*

A final criticism of the redox model comes from Barnes (1973) himself. Through the use of L-malate and ascorbate as electron donors, and the study of their sensitivity to inhibitors of the branched electron-transport chains in *A. vinelandii*, he has demonstrated the existence of two sites for coupling of redox energy to transport. As the two sites are at very different redox potentials, it is extremely difficult to imagine how a single transport carrier with its redox-responsive sulphydryl group can be coupled to both sites.

It is clear then that, in the eyes of this reviewer, neither the permease nor the redox models of membrane transport can claim to offer a valid solution to the problem of the mechanism of energy coupling. But for a more detailed discussion of the various arguments for and against these hypotheses, the reader is referred to the articles of Kepes (1971), Kaback (1972), Kaback and Hong (1973), Harold (1972, 1974) and Boos (1974).

V. The Chemiosmotic Hypothesis of Energy Transduction

Attention has already been drawn to two fundamental features of carrier-mediated transport, namely that it does not involve chemical modification of the transported substrate, and that it has a vectorial character with the only difference between substrate and product being that of location. The term "substrate translocation" is in fact often used synonymously with carrier-mediated transport. If we revert again to our comparison with enzyme-catalysed reactions, it is now possible from consideration of these features to achieve some valuable insight into the mechanism of membrane transport, and in particular of energy coupling.

Considering the transport system in isolation, active transport represents an increase in free energy. Since the substrate is translocated rather than chemically altered, this increase in free energy results largely from a decrease in entropy, and is in the form of a trans-membrane gradient of chemical and, if the substrate is charged, electrical potential. In active transport, therefore, there must exist a mechanism for converting the chemical energy of metabolism into the osmotic energy of these trans-membrane gradients. In the permease model, this is achieved through the action of ATP on either binding or flux velocity constants, such that an asymmetry and unidirectional character is imposed on the mobile carrier. In the redox model, the energy transducer takes the form of a carrier which is capable of cyclic oxidation and reduction, accompanied by conformational changes in the protein affecting the location and affinity of the substrate-binding site. These two energy transducers are proposed on *a priori* grounds, and they are supported by the minimum of experimental evidence. Equally feasible, and more experimentally accessible, is the hypothesis that the energy coupled to active transport is already osmotic and in the form of a trans-membrane gradient.

Such an hypothesis suggests three questions that must be answered. How is this osmotic energy created from the exergonic reactions of metabolism? What chemical species constitute the components of the primary trans-membrane gradient(s)? How is this primary gradient coupled with substrate translocation to produce the secondary gradient which is the end product of active transport?

A. Vectorial Metabolism and Group Translocation

The binding of the substrate to form the enzyme-substrate complex is critical to the mechanism of enzyme action. The formation of this complex shows a high degree of specificity toward the substrate, and is accompanied by a conformational change at the enzyme's active site. After reaction, and an associated further change in the enzyme conformation, the product is released. At the molecular level on the enzyme surface, it is reasonable to suggest that the specificities toward the substrate and product might be concerned not only with the shape of the active site, but also with the route of approach to the active site. That is, the substrate might approach the active site from the left while, after reaction, the product might leave by another path to the right. For an enzyme in solution in the cytoplasm, such a process would have no measurable vectorial character. An enzyme which is, however, located within a membrane has the potentiality of demonstrating such a vectorial character as a macroscopic phenomenon, and one therefore subject to measurement and analysis. As an example, one may compare the action of the soluble enzymes hexokinase and pyruvate kinase coupled through ATP, with Enzymes I and II of the phosphotransferase system coupled through the system's unique heat-stable protein (HPr) (Fig. 3). This phosphotransferase system will receive considerable attention in a later section of this review (Section VII, page 40). Chemically the reactions are identical, but the effect of Enzyme II being membrane-bound is that, while glucose is extracellular, its product glucose 6-phosphate appears within the cell. Not glucose itself, but the "glucose-6" group has been translocated across the membrane by its reaction with the phosphoryl group donated from HPr. This powerful concept has been given the name "group translocation".

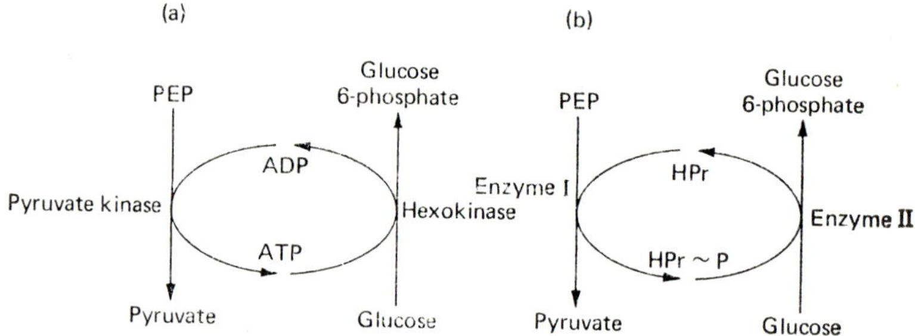

FIG. 3. Group transfer reactions. Transfer of phosphate-bond energy from phospho-enol pyruvate (PEP) to glucose by: (a) the coupling of pyruvate kinase and hexokinase through ATP; and (b) the coupling of Enzymes I and II of the phosphotransferase system through the protein HPr.

The prerequisites for an enzyme or system to demonstrate group translocation are that it should be situated within the osmotic barrier of a membrane, and that it should be anisotropic inasmuch as the reacting groups should not be equally accessible to the active site(s) from each of the aqueous phases on either side of the membrane. In Fig. 3, for example, the phosphotransferase system cannot convert intracellular glucose to glucose 6-phosphate.

B. THE PROTONMOTIVE FORCE

There is now a wealth of data demonstrating that both components of oxidative phosphorylation, namely electron-transport chain and ATPase, satisfy these two requirements for group translocation. The hydrogen- and electron-carrying components alternate in the redox chain and are arranged in loops. Associated with the passage of reducing equivalents down the chain is the separation of charge through the efflux of protons from the mitochondrial or bacterial membrane. In this manner of proton group-translocation, the redox energy is converted to the osmotic energy of the trans-membrane gradients of protons and of charge (Fig. 4), that is, to the protonmotive force (Δp) which is given by the relationship:

$$\Delta p = \Delta\psi - Z\Delta \text{ pH}$$

where $\Delta\psi$ is the membrane potential measured in mV, Δ pH is the trans-membrane pH gradient, and Z is the factor converting pH value into mV, which at 25°C equals almost 60. The ATPases of the mitochondrial and

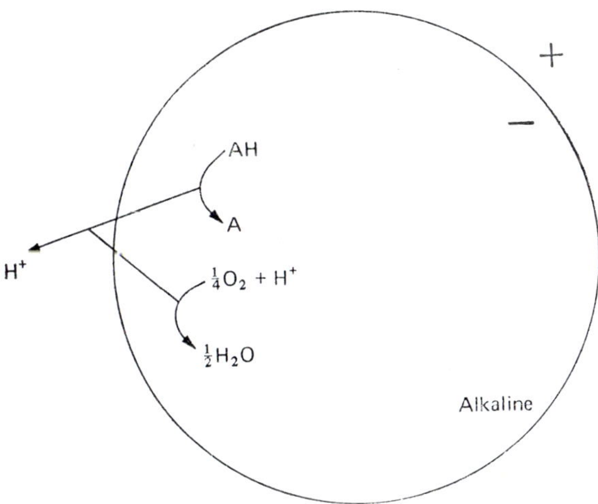

FIG. 4. The protonmotive force. A schematic representation of the generation from respiration of gradients of chemical, inside alkaline, and electrical potential, inside negative, as the result of proton extrusion.

bacterial membranes are also proton translocating, hydrolysis of intracellular ATP being coupled to the efflux of protons into the medium (Fig. 5). It is, of course, the cornerstone of the chemiosmotic hypothesis of oxidative phosphorylation that the protonmotive force generated by electron transport constitutes the mechanistic and energetic coupling which drives the ATPase in the direction of ATP synthesis (Mitchell, 1966; Greville, 1969; Harold, 1972).

C. Proton Symports

This protonmotive force, and its generation from either electron transport or ATP hydrolysis by proton group-translocation, thus fulfil the first two requirements of an ion-gradient hypothesis of active transport.

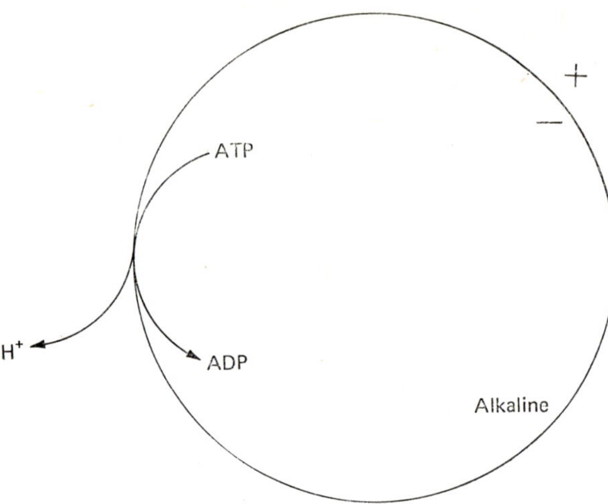

FIG. 5. Action of the proton-translocating ATPase. The reversible enzyme operates in the direction of ATP hydrolysis, with associated proton extrusion.

Mitchell (1973b) has in fact described how studies which led to the development of his chemiosmotic hypothesis had their origins in his search for a general theory of the mechanism of coupling between metabolism and transport (Mitchell, 1963). At that time the concept of group translocation was introduced, and its relevance to trans-membrane phenomena in general discussed. In the light of increasing knowledge of these

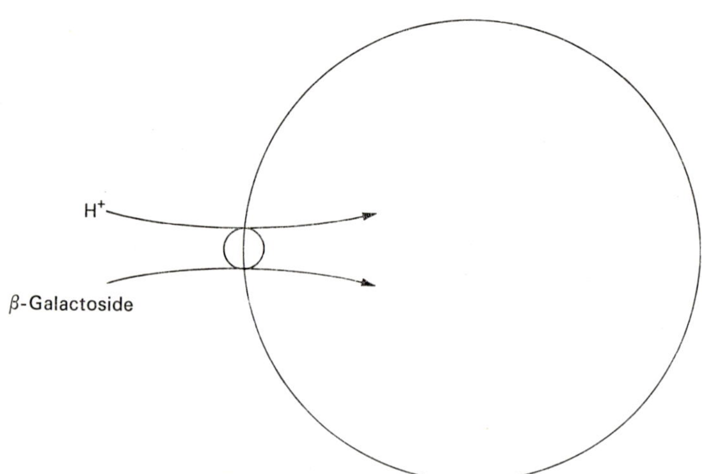

FIG. 6. β-Galactoside proton symport.

systems, a more detailed and less hypothetical account was given (Mitchell, 1970). In these two symposia, Mitchell (1963, 1970) also proposed that the β-galactoside permease and other substrate-specific nutrient transport carriers might function as proton symports. That is to say, the carriers are bifunctional, having binding sites for both the substrate to be transported and for protons (Fig. 6). Should such a carrier exist within a bacterial or mitochondrial membrane, for example, across which there are gradients of chemical and electrical potential in the form of the protonmotive force, then clearly the flow of protons down their energy gradient into the cell, or organelle, could be coupled to, and drive, the active transport and accumulation of the substrate. As will be discussed in some detail in Section VI (page 21), there is now a wealth of experimental data supporting the existence and function of these proton-linked transport mechanisms across bacterial, yeast, fungal, algal and mitochondrial membranes.

D. Uncouplers and Ionophores

Before leaving this general statement of the ion-gradient or chemiosmotic model of transport, I should like to deal with the action of uncouplers and ion-translocating antibiotics, or ionophores as they have become known.

In the presence of uncouplers, sensitive cells lose their ability to couple energy to active transport, although facilitated diffusion is unaffected. The action of uncouplers in dissociating oxidation from phosphorylation, so that the rate of respiration is increased but without the concomitant synthesis of ATP, is explained in the chemical hypothesis by the proposed hydrolysis of the high-energy intermediate X ~ I which is the first reactant common to all three coupling sites. The permease model accepts this explanation and simply predicts a decrease in the effective concentration of the high-energy compound which reacts with the carrier. Unfortunately for this view, it is clearly established that active transport remains sensitive to the action of uncouplers under anaerobic conditions, and that what is more, ATP levels and ATP-dependent reactions are unaffected by uncouplers (Pavlasova and Harold, 1969). On the basis of a comparison with data from phosphorylating systems, the redox model of transport would tend to predict an increase in transport rather than the observed inhibition. The complete failure of this model to deal with uncoupler action is freely admitted (Kaback, 1972; Kaback and Hong, 1973). Once again the most satisfactory explanation, and its supporting experimental evidence, come from the chemiosmotic model. One of the four basic postulates of this hypothesis of energy transduction in a membrane system is that the membrane itself should have a low conductivity to protons (Mitchell, 1966). Even in his preliminary statement of the hypothesis, Mitchell (1961a) proposed that uncouplers dissolve in the membrane and act as circulating carriers conducting protons across the membrane. This action has since been confirmed in mitochondria (Mitchell and Moyle, 1967), bacteria (Harold and Baarda, 1968b) and synthetic lipid bilayers (Hopfer *et al.*, 1968). Clearly such an action, uncoupling oxidation from phosphorylation by "short circuiting" the proton current back across the membrane, will equally abolish the driving force for nutrient translocation mechanisms dependent on the chemical and electrical potentials of the protonmotive force.

In increasing membrane permeability specifically to protons, uncouplers such as 2,4-dinitrophenol (DNP), tetrachlorosalicylanilide (TCS), carbonylcyanide *m*-chlorophenylhydrazone (CCCP) and the fluoro derivative (FCCP), collapse both the pH gradient and the membrane potential components of the total protonmotive force. A flux of a single species in this manner is said to be catalysed by a uniport mechanism; where the species is charged, the flux is classed as electrogenic. A similar electrogenic flux of alkali-metal cations, in particular potassium, is catalysed by the peptide antibiotics valinomycin and the enniatins, and by the macrotetralide actins. The gramicidins are less specific toward potassium, affecting also sodium-ion permeability and, at higher concentrations, that of protons. Apart from this last-mentioned effect, these ionophores can only cause uncoupling under certain conditions. Consider, as an example, valinomycin added to a suspension of respiring mitochondria in potassium-containing medium. An electrogenic influx of potassium ions will occur in response to the metabolically derived membrane potential, inside negative. The pH gradient is not however affected, and so the protonmotive force is not completely dissipated, i.e. the mitochondria are not uncoupled. They do, however, attempt to compensate for the decreased membrane potential by further proton efflux, with consequent increase in the pH gradient. It is the magnitude of this pH gradient and the extent of the internal alkalinization which eventually produces the secondary uncoupling effect characteristic of valinomycin. Alternatively, in the presence of a permeant anion, the electrogenic influx of potassium ions may be compensated by an anion flux in the same direction. This in turn leads to an accumulation of electrically neutral but osmotically active material within the organelle. The consequent mitochondrial swelling and stretching of the membrane again lead to an uncoupling type of action. It should be noted, though, that the direction of flux of potassium ions will depend on the relative magnitudes of the membrane potential and the concentration gradient of ions across the membrane. In media lacking or containing low concentrations of potassium ions, the concentration gradient directed outward may be greater than or equal to the potential gradient of the protonmotive force directed inward. Thus the mitochondrial uncoupling action of valinomycin is dependent on the presence of a sufficient concentration of potassium ions in the medium. Similarly, the inhibition by valinomycin of amino-acid uptake by *Staph. aureus* requires the presence of potassium in the medium (Gale and Llewellin, 1972; Niven and Hamilton, 1972). On the other hand, Gale and Llewellin (1972) have shown that valinomycin inhibition of amino-acid transport in *Streptoccocus faecalis* is maximal at low extracellular concentrations of potassium ions. Also, Harold and Baarda (1967) demonstrated that the bacteriostatic effect of valinomycin and gramicidin on this organism resulted from loss of potassium ions from the cells, and that it could be reversed by increasing the concentration of the ion in the medium. It is possible that these interspecies differences may result from *Strep. faecalis* normally having a lower metabolic membrane potential than *Staph. aureus* or mammalian mitochondria. The data at present available are preliminary and inconclusive on this point. Harold and Papineau (1972a) claim a value for the membrane potential in *Strep. faecalis* of between 150 and 200 m*V*. The estimate varies with the method of assay, and furthermore is only demonstrable in potassium-deficient cells, a condition that might be expected to stimulate the potential in an effort to accumulate potassium within the cells. The estimate of 120 mV for the potential in *Staph. aureus* (Jeacocke *et al.*, 1972) is very much a minimal value; most probably it is a good deal closer to the figure of 180 mV characteristic of mitochondria.

Antibiotics of the carboxylic polyether class, e.g. nigericin and monensin, catalyse an electroneutral exchange of protons for alkali–metal cations. Nigericin shows a degree of specificity as a K^+/H^+ antiport, monensin as a Na^+/H^+ antiport. These antibiotics are therefore capable of decreasing the pH gradient across a membrane, but only with a resultant increase in the magnitude of the membrane potential. That is to say, they do not act as uncouplers in mitochondria or in bacteria. Harold and Baarda (1968a) described how the bacteriostatic action of nigericin on *Strep. faecalis* is associated with the loss of potassium ions from the cells and internal acidification. Submitochondrial particles, which have been prepared by sonication, have their membrane orientation reversed so that the respiration-linked proton flux is directed inwards, and the protonmotive force is composed of a membrane potential, inside positive, and a pH gradient, inside acid. With such structures, neither valinomycin nor nigericin alone can cause uncoupling. In combination, however, they do uncouple. Nigericin catalyses the exchange of internal H^+ for K^+, which then effluxes again through the action of valinomycin. Uncoupling in this case results from a short-circuiting through the two linked proton and potassium cycles (Fig. 7). This effect has also been noted in phosphorylating particles prepared from *M. denitrificans* (John and Hamilton, 1971), and was taken as evidence for the reversal of membrane orientation in these vesicles also.

A more extended consideration of uncouplers and ionophores can be found in articles by Harold (1970), Henderson (1971), Gale *et al.* (1972) and Hamilton (1974). These compounds must not be thought of only as ecological or experimental curiosities. Despite the primary extrusion of protons, the protonmotive force in mitochondria and bacteria exists largely in the form of a membrane potential expressed through gradients

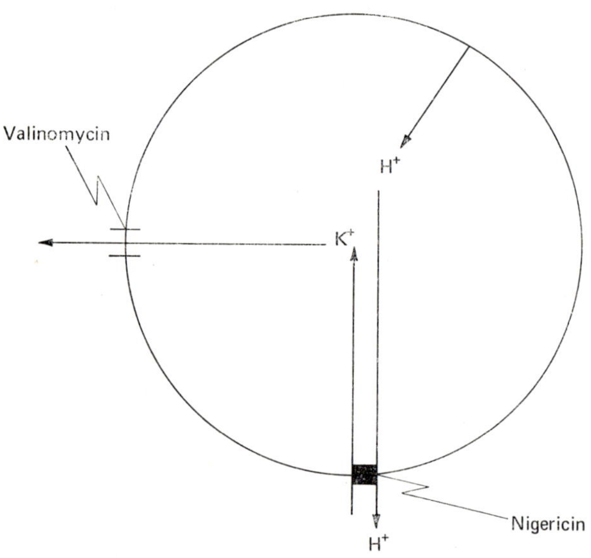

FIG. 7. The uncoupling action of the ionophores nigericin and valinomycin in submitochondrial particles, with their reversed membrane orientation and influx of protons associated with respiratory or ATPase activity.

of ions other than protons. Even in the absence of the $[Na^+ + K^+]$-stimulated ATPase, bacterial, algal and fungal cells have high intracellular concentrations of potassium and low concentrations of sodium, in parallel with mammalian cells. The existence of "natural" ionophores of the valinomycin and nigericin types represents the most reasonable explanation of these phenomena.

One further point of very considerable significance emerges from the consideration of the ionophores and their mechanism of action. As evidenced most strikingly when they function in creating or responding to potentials across artificial lipid bilayers, the ionophores are entirely passive in an energetic sense. They catalyse, or facilitate, the diffusion of an ion or ions across the membrane, in a direction and to an extent predetermined by the trans-membrane gradients of chemical and electrical potential already in existence. As with my earlier discussion of enzyme-catalysed reactions, the catalytic and energetic functions are quite separate. A process of "facilitated diffusion", of potassium ions for example, is converted to one of "active transport", not by modification of the "carrier" itself, but by the coupling either to a larger gradient of protons in the opposite direction through the nigericin K^+/H^+ electroneutral antiport, or to a gradient of potential in the same direction through the valinomycin/K^+ electrogenic uniport. These of course represent a direct experimental verification of the principles of the chemiosmotic model for energy coupling in transport. It is worth stressing again that the permease and redox models differ fundamentally from this view, inasmuch as they both require a direct interaction between the carrier and the energy source, with a consequent alteration in the conformation or redox state of the protein carrier. Manifestly such postulated mechanisms can have no relevance to the ionophore-catalysed translocation of ions across biological, and even less across artificial lipid bilayer membranes. None the less, such translocations mirror, in terms of both energetics and control, the characteristics of nutrient transport across cell and organelle membranes. Studies with uncouplers and ionophores, therefore, through their relevance to the *a priori* development of the concepts and through the experimental verification they afford of the model's predictions, represent a cornerstone of the chemiosmotic hypothesis of energy-transducing mechanisms in the biological membrane.

E. Lipid-Soluble Ions

Closely related to these considerations of ionophore-facilitated ion translocations are the extensive studies with the lipid-soluble synthetic ions such as dibenzyldimethyl ammonium, triphenylmethylphosphonium, phenyldicarbaundecaborane and tetraphenyl boron (Liberman and Skulachev, 1970; Skulachev, 1971, 1972a; Griniuviene *et al.*, 1974). By virtue of their lipid solubility these ions are capable of carrier-free electrogenic translocation across membranes. They thus represent the simplest and least ambiguous model possible for the study of the role of the membrane potential in transport, and the testing of the predictions of the chemiosmotic model. Additionally, these ions have found an important application in their use to gain a measure of membrane potential in cells that are too small to allow use of micro-electrodes. The distribution of any freely permeable ion across a membrane is related to the membrane potential by the Nernst equation:

$$\Delta\psi = \frac{RT}{nF} \ln \frac{(DDA^+)_i}{(DDA^+)_o}$$

where R is the gas constant, T the absolute temperature, n the valency of the ion, F the faraday, and $(DDA^+)_i$ is the activity of the dibenzyldimethylammonium ion in the internal phase. Converting to log to the base 10, RT/nF becomes Z, as in the equation for the protonmotive force with a value at 25°C of approximately 60 (see page 13). The use of tritiated dibenzyldimethylammonium ions therefore allows one to assay the trans-membrane distribution of the ion, and hence to measure the membrane potential. This technique has been used successfully with whole cells of *Strep. faecalis* (Harold and Papineau, 1972a) and membrane vesicles of *E. coli* (Hirata *et al.*, 1973). Equivalent assays of membrane potential in whole cells of *Staph. aureus* (Jeacocke *et al.*, 1972) and *Strep. lactis* (Kashket and Wilson, 1973) have been carried out from the estimate of distribution of potassium ions in the presence of valinomycin.

F. Protonmotive Force in Micro-Organisms

These considerations of membrane potential and its measurement lead us on to the data relevant to the existence of a protonmotive force across the microbial cell membrane, and its generation from proton group-translocation associated with electron transport and/or ATP hydrolysis.

Respiration-driven electrogenic extrusion of protons has now been recorded for whole cells of *E. coli* (Lawford and Haddock, 1973; West and Mitchell, 1972; Griniuviene *et al.*, 1974), *M. denitrificans* (Scholes and Mitchell, 1970), *Staph. aureus* (Jeacocke *et al.*, 1972), a range of organisms including *B. subtilis* and *E. coli* (Meyer and Jones, 1973), the yeasts *Candida utilis* and *Saccharomyces carlsbergensis* (Garland *et al.*, 1972), and for membrane vesicles prepared from *E. coli* (Reeves, 1971), *M. denitrificans* (John and Hamilton, 1971), *M. lysodeikticus* (Tikhonova, 1974). Hirata *et al.* (1973) report that respiration of vesicles from *E. coli* on D-lactate can generate a membrane potential of 100 mV, inside negative. In *Staph. aureus* (Jeacocke *et al.*, 1972) and *Strep. faecalis* (Harold and Papineau, 1972a, b) electrogenic proton extrusion occurs under anaerobic conditions, presumably through the action of the membrane ATPase on ATP derived from glycolysis. Estimates of the protonmotive force in these experiments, and also with respiring cells of *Staph. aureus*, were of the order of 200 mV, interior alkaline and negative. There is mounting evidence therefore that, in micro-organisms as in mitochondria, the redox and hydrolysis energies of the cell's metabolism can be transduced into the form of trans-membrane gradients of chemical and electrical potential.

The mechanisms of group translocation thus giving rise to the proton-motive force fall strictly outwith the area of consideration of this review. They are discussed in the articles of Mitchell (1966), Greville (1969) and Harold, (1972). Of especial interest, though, are the experiments on energy coupling in reconstituted respiratory systems in synthetic lipid bilayers (see Hinkle, 1973; Skulachev, 1972b; Kayushiv and Skulachev, 1974; Drachev *et al.*, 1974), and Mitchell's (1973c) thought-provoking article on a postulated mechanism of the proton-translocating ATPase of the mitochondrial and prokaryotic cell membrane. The [$Na^+ + K^+$]-stimulated ATPase is also considered in this model, and contrasted with the more classic enzymological scheme devised by Skou (1972).

VI. The Chemiosmotic Model of Transport

My examination of the ion-gradient or chemiosmotic model of energy coupling must now turn to the evidence supporting its predictions with regard to membrane transport. These are that:

(a) neutral substrates, such as sugars, will be translocated on a proton symport. This will involve the net flux of both protons and charge, and hence will be dependent on the gradients of both pH value and potential, and
(b) anions, such as phosphate, will also be translocated on a proton symport. This translocation will however be electroneutral, and so influenced only by the pH gradient; and
(c) cations, such as potassium, will be translocated on a uniport. This flux being electrogenic and not involving protons, will be driven solely by the membrane potential.

A. Proton Symports

1. Sugars

The first experimental verification of this model came from the work of West (1970). He showed that the flux of lactose down a concentration gradient into resting *E. coli* was accompanied by alkalinization of the medium. The fluxes of lactose and of protons were found to be strictly coupled with a stoicheiometry of 1:1 (West and Mitchell, 1973). The second prediction, that this transport should be electrogenic, was also verified by the demonstration that the influx of β-galactoside and protons was accompanied by an equal and opposite efflux of potassium ions (West and Mitchell, 1972). In this paper the authors also demonstrated the equivalence of their symport to the *M* protein (Fox and Kennedy,

1965) of the β-galactoside permease; uptake of galactoside driven by either a pH gradient derived from a small respiratory pulse or a diffusion potential, inside negative, consequent upon addition of the permeant anion thiocyanate (Fig. 8); and the lack of evidence that the symport uses sodium ions rather than protons as cosubstrate.

Henderson (1974) found proton fluxes associated also with the transport of the sugars galactose and arabinose by *E. coli*. In their induction and specificity, these two systems are quite distinct from each other, and from the galactoside permease, or proton symport. They both

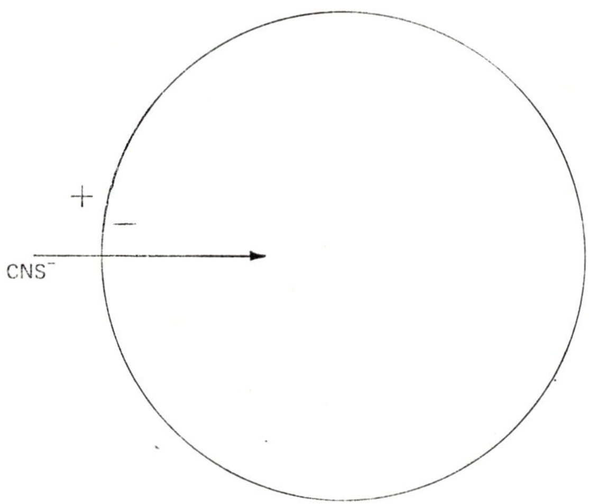

Fig. 8. The generation of a diffusion potential, inside negative, with the permeant anion thiocyanate.

demonstrate approximate stoicheiometries of 1 : 1 for protons fluxing per molecule of sugar translocated. Again, extending the findings of West and Mitchell (1972) with the β-galactoside system, Henderson (1974) has shown that the rate of decay of the transient acidification of the medium, consequent upon the addition of a pulse of oxygen to an anaerobic suspension of arabinose-grown cells, is accelerated by addition of D-fucose, a substrate for the arabinose uptake system; in other words, the uptake of sugar causes an increase in the rate of proton influx.

2. *Amino Acids in Bacteria and Yeasts*

Associated proton movement has also been reported in studies of amino-acid uptake in a number of organisms. Gale and Llewellin (1972) found proton-to-amino acid stoicheiometries of 0·62 and 0·91, respectively, for uptake of aspartate and glutamate by *Staph. aureus*. Their method was quite different from the pulse method employed by West and by Henderson, in that Gale and Llewellin suspended their resting cells at pH 5·5 and observed the rate of drift of acid into the cells, and the effect thereon of the presence of the amino acid. This steady-state method has also been used by Eddy and his coworkers in their studies of amino-acid transport in yeasts (Eddy *et al.*, 1970a; Eddy and Nowacki, 1971; Seaston *et al.*, 1973). The majority of their work has been done with *Saccharomyces carlsbergensis*, with more recently an extension to include *Sacch. cerevisiae* and *Sacch. fragilis*. They have found that, at a pH value of around 4·5, amino acids such as glycine may be accumulated up to 200-fold, and that this accumulation is accompanied by an influx of protons and an efflux of potassium ions. These effects are demonstrated in cells which have been inhibited with 2-deoxyglucose and antimycin, and are thus independent of energy metabolism. The stoicheiometries of these fluxes vary with experimental conditions. For example, uptake of glycine or phenylalanine is accompanied by two equivalents of protons after exposure for 2–4 min to the metabolic inhibitors; after 20 min, however, this value drops to 1·2 (Eddy and Nowacki, 1971). With starved cells in the absence of both inhibitors and any added substrate, again the uptake of glycine, phenylalanine, leucine and lysine was found to be associated with an influx of 2·1 equivalents of protons and a quantitatively similar efflux of potassium ions. In yeast cells containing sodium rather than potassium ions as the principal intracellular cation, the flux of protons increased to about three. In *Sacch. cerevisiae* the rapid absorption of glycine, citrulline and methionine through the general amino-acid transport system occurs with an uptake of approximately two extra equivalents of protons, whereas the slower absorption of methionine, proline and, possibly, arginine through their specific systems is associated with only one equivalent (Seaston *et al.*, 1973). These authors also showed that transport of sugars such as maltose, α-methylglucoside, sucrose and lactose by *Sacch. carlsbergensis* and *Sacch. fragilis* caused an accelerated rate of proton uptake.

In discussing these data, Eddy and his colleagues have drawn an analogy with sodium- and potassium-linked transport systems of the mammalian cell membrane (Schultz and Curran, 1970). According to this ion-gradient hypothesis, the three components of the driving force for translocations are the sodium-ion gradient, directed inwards, the potassium-ion gradient, directed outwards, and the membrane potential, inside negative, resulting from the 3:2 sodium:potassium-ion stoicheiometry of the [$Na^+ + K^+$]-stimulated ATPase. Eddy suggests that, in the proton-linked transport reactions he has demonstrated in yeasts, the potassium-ion gradient may be an integral component of the driving force for the transport and accumulation of amino acids and sugars. The strictly chemiosmotic interpretation of these findings would, however, state that the total driving force is composed of the chemical and electrical potentials of the protonmotive force only, and that any associated potassium-ion movements are purely secondary and the result of the need to preserve electroneutrality across the membrane. In support of this interpretation one may quote the findings of Eddy *et al.* (1970a) that efflux of potassium ions from starved cells accumulating glycine was abolished by addition of glucose as a metabolizable substrate. In the first

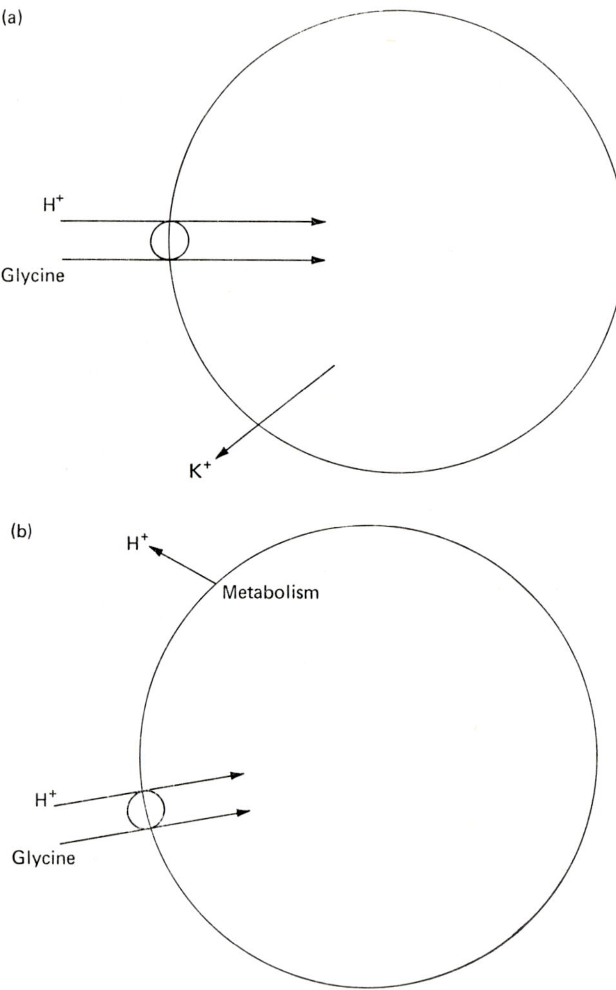

FIG. 9. Maintenance of electroneutrality during transport. (a) In the absence of metabolism, the electrogenic glycine proton symport is compensated for by K^+ efflux. (b) During metabolism, proton efflux generates a potential, inside negative, to which the glycine proton symport responds.

case, translocation of the amino acid by a proton symport in response to the pH gradient across the membrane will result in the net transfer of one positive charge, which is compensated for by efflux of a potassium ion. The action of the metabolically driven proton pump however is to create not only a pH gradient but also a potential, inside negative. Transfer of the positive charge during transport therefore re-establishes rather than disturbs electroneutrality across the membrane, and so no compensatory flux of potassium is observed (Fig. 9).

Further to these points of interpretation, the varying and generally high proton-to-amino acid stoicheiometries observed by Eddy's group are at variance with those recorded by other workers, mostly however with bacterial systems. It remains to be seen whether these differences reflect the different method employed by Eddy and his coworkers in determining the magnitude of the proton flux, or are indicative of a real quantitative, and possibly qualitative, difference between the transport systems of yeasts and bacteria.

3. *Other Transport Systems in Fungi*

Only a comparatively small amount of work has been done on energy-coupling mechanisms in other yeast and fungal transport systems. However, Hunter and Segel (1973) suggest, from their findings with the uncoupling action of a number of weak acids at or below their pK_a values, that a proton or charge gradient may be involved in energizing membrane transport in *Penicillium chrysogenum*. An active transport of glucose by *Neurospora crassa* has been reported by Scarborough (1970), but a mechanism of energy coupling was not suggested. Brown and Romano (1969) and Mark and Romano (1971) claim that accumulation of sugars by *Aspergillus nidulans* is not the result of phosphorylation and substrate modification. On the other hand, Van Steveninck (1970) claims that, in the yeast *Sacch. cerevisiae*, α-methylglucoside is accumulated by an inducible maltose transport mechanism, and that the sugar is phosphorylated in the process of translocation, the intracellular α-methyl-glucoside phosphate being the precursor of the intracellular α-methylglucoside.

4. *Organic Acids*

Reverting to the more fully characterized bacterial translocations, and in particular to the consideration of electroneutrality raised in connection with Eddy's work to which reference has already been made, further evidence for the presence of proton symports is afforded by the study of anion transport. Ghei and Kay (1973) studied the uptake of the dicarboxylic acids succinate, fumarate and malate by an inducible transport system in *B. subtilis*. The transport is sensitive to uncouplers and, from the fact that at least one free carboxyl group is required for activity, it is suggested that the acids may be transported as anions by a proton symport. In their studies of gluconate transport by *E. coli*, Robin and Kepes (1973) monitored shifts in pH value in the medium, and concluded that electroneutrality must be maintained through the functioning of either a gluconic acid uniport, or a gluconate proton-symport, or a gluconate hydroxyl-antiport. As pointed out earlier (West and Mitchell, 1972), these processes are thermodynamically equivalent and cannot readily be distinguished. Winkler (1973) has analysed the hexose-phosphate transport system in *E. coli*. He noted the sensitivity to uncouplers and concluded that, unlike the β-galactoside system he had studied earlier with Wilson (Winkler and Wilson, 1966), energy coupling was achieved by raising the affinity of influx. Although this conclusion was reached by the same experimental techniques and analysis used in the earlier study, a number of significant points are raised in the discussion of their data. It is noted, for example, that the lipid-soluble triphenylmethylphosphonium ion did not inhibit the transport, and this was taken as evidence that the membrane potential was probably not involved as a driving force for the translocation. Finally the point was made that, at neutral pH values, the sugar phosphates are anionic and their transport must therefore be associated with either cation or proton influx, or with anion efflux. Although no evidence is put forward, it is proposed that this point may bear directly on energy coupling. Clearly a proton symport responsive only to the pH gradient would satisfactorily explain the experimental and theoretical observations.

5. *Hexoses in Chlorella*

Other extensions of earlier data and models to consider the possible relevance of proton fluxes are evident in the papers of Komor (1973) and West and Wilson (1973). The work of Tanner and his colleagues on hexose transport in the alga *Chlorella vulgaris* has already been referred to in connection with the impossibility of determining on purely kinetic grounds whether energy coupling is associated with an altered affinity of the carrier for its substrate, or with altered rates of diffusion of the free and complexed carrier (Komor *et al.*, 1973a; see page 4). This interesting system has been analysed in a series of papers (Komor *et al.*, 1972, 1973b; Decker and Tanner, 1972). Transport and accumulation of the glucose analogues 3-O-methylglucose and 6-deoxyglucose have been shown to be dependent on energy derived from either respiration or photosynthesis. The steady-state influx occurs at twice the rate of the initial influx, although both increase the basal respiratory rate to the same extent, and, whereas anaerobiosis in the dark inhibits initial influx completely, the steady-state influx is only 50% decreased. Uncouplers completely inhibit both net and steady-state influx, but do not initiate efflux. In fact the steady-state efflux is also completely inhibited by uncouplers. These data are built into a model which is largely derived from the classical permease model of β-galactoside transport in *E. coli*. Energy coupling is seen as speeding up diffusion inward of the carrier-sugar complex, and diffusion outward of the free carrier. It is suggested that this may be achieved by a chemical modification of the carrier, e.g. by phosphorylation. Net efflux would be dependent on the significant diffusion inward of the free carrier, a reaction which is claimed to be energy-dependent in view of the sensitivity of efflux to the presence of uncouplers. It is further hypothesized that the steady-state efflux by outward diffusion of the carrier-sugar complex is an energy-generating reaction, and that this is the basis of the observed positive transmembrane effects. The authors eliminate the possibilities of the transport being dependent on phosphorylation of the sugar, or there being any cotransport of sodium or potassium ions. In his most recent paper, however, Komor (1973) has shown that uptake of 6-deoxyglucose, driven by either respiration or photosynthesis, is accompanied by alkalinization of the medium. An approximate stoicheiometry of 1:1 was established for proton and sugar uptakes.

6. *Mutants Uncoupled for Lactose Transport*

Wilson and Kusch (1972) have isolated two mutants of *E. coli*, X_{71-54} from the X_{71} strain of K_{12}, and ML_{308-22} from ML_{308}. They class these mutants as energy-uncoupled for lactose transport. Carrier function is unaltered, as evidenced by assays of *o*-nitrophenol β-galactoside hydrolysis, counterflow with thiomethyl β-galactoside-loaded cells in the presence of azide, and the initial rate of entry of this methylgalactoside. Only the ability to accumulate galactosides against a concentration gradient is affected. That the lesion is in the carrier, or *M* protein, was supported by the increased sensitivity of the mutant to sulphydryl inhibitors in facilitated diffusion assays, and the fact that the β-galactosidase and transacetylase showed equal activity in wild-type and mutant strains. In agreement with the earlier model (Winkler and Wilson, 1966), these data were taken as supporting the mechanism of energy coupling as being a reduction in the affinity of the carrier for the sugar at the point of exit. More recently, however, the same two mutants have been studied by West and Wilson (1973) in a series of experiments inspired by West's work on the β-galactoside proton symport in *E. coli*. The mutants showed undiminished activity when compared with the parent organisms with regard to the ability to produce a standard acid pulse and exponential decay on addition of a small pulse of air-saturated medium to anaerobic cells. However, alkalinization of the medium associated with the addition of thiomethylgalactoside to an anaerobic suspension of cells was greatly diminished in the mutants. The values for ML_{308} and ML_{308-22}

were, respectively, 42 and 1·8 ng ions H^+/mg cell dry weight per minute, and for X_{71} and X_{71-54} 39·6 and 18·9 ng ions H^+/mg cell dry weight per minute. The authors concluded therefore that uncoupling of transport and energy was in fact an uncoupling of β-galactoside transport from proton transport, and that the altered phenotype is due to a modified β-galactoside carrier. The interesting study of Wong and MacLennan (1973) indicates that the lesion affects the lipid moiety of the *M* lipoprotein.

Various studies with ATPase-deficient mutants and their relevance to transport will be discussed later in this article (page 38), but at this point in the development of our argument the work of Rosen (1973a, b) must be mentioned. The mutant NR_{70}, derived from *E. coli* K_{12} strain 7, is lacking in the Mg^{2+}-stimulated ATPase, and consequently incapable of oxidative phosphorylation; additionally the cells lose the ability to accumulate sugars and amino acids driven by either respiration or ATP hydrolysis. Rosen has shown that the mutant has a high proton permeability, comparable to the wild type in the presence of the proton uncoupler carbonyl cyanide *m*-chlorophenylhydrazone (CCCP), and that this permeability can be decreased in the presence of the ATPase inhibitor N,N′-dicyclohexylcarbodiimide (DCCD). This finding exactly parallels the similar effects of oligomycin and DCCD on the proton permeability of F_1-deficient mitochondrial and chloroplast membranes (Mitchell, 1973c). Further, the respiration-driven transport of thiomethylgalactoside in the mutant was greatly increased by DCCD under conditions which had minimal effect on the parent organism. Whereas, in the mutants studied by West and Wilson the lesion in transport resulted from the inability of the carrier to respond to the proton gradient, in Rosen's ATPase mutant the lesion is more general and results from the inability of the cell to establish and maintain a proton gradient. These studies therefore constitute very convincing evidence in favour of the protonmotive force as the primary form of the "high-energy state" relevant to active transport, and to energy-transducing mechanisms in general. Evidence from very different studies of the action of colicins A, E and K and staphylococcin 1580 on *E. coli* and *Staph. aureus*, respectively, leads to the same general conclusion (Fields and Luria, 1969; Jetten and Vogels, 1973).

B. Involvement of Ions Other Than Protons

Before considering in some more detail a number of important studies which have attempted to extend our detailed understanding of the chemiosmotic model of substrate translocations, it is relevant at this point to discuss briefly some papers which report the involvement of ions other than protons in bacterial nutrient-transport systems. Although West and Mitchell (1972), Komor *et al.* (1972) and Asghar *et al.* (1973) have discounted the involvement of sodium ions as cosubstrate in sugar translocations in *E. coli*, *Chlorella vulgaris* and *Strep. faecalis*, Stock and Roseman (1971) have reported a sodium-dependent thiomethylgalactoside uptake and a thiomethylgalactoside-dependent sodium uptake in *Sal. typhimurium*, and claimed that this demonstrates a sodium-sugar symport. In fact, lithium gives a greater activation of the sugar uptake, and the sugar-dependent sodium uptake is extremely difficult to measure accurately due to the activity, the authors claim, of a separate sodium pump causing rapid efflux of the ion. A paper by Shiio *et al.* (1973) reports the sodium-dependent uptake of threonine by a number of Gram-positive and negative organisms. Cotransport of the ion was not demonstrated, however, nor was uptake of the amino acid one of active transport. The sodium dependence is therefore more likely to be concerned with an activation of the carrier than as a component of the driving force. A similar comment seems justified in respect of the potassium-dependent transport of citric acid by *Aerobacter aerogenes*, reported by Eagon and Wilkerson (1972). The data of Willecke *et al.* (1973) concerning cotransport of magnesium and citrate in *B. subtilis* is, however, very much more convincing. The authors suggest that a citrate-magnesium complex, carrying a net negative charge, might be transported on, for example, a proton symport. In agreement with the earlier work of Thompson and MacLeod (1971) and Sprott and MacLeod (1972) on uptake of amino acids in a marine pseudomonad, Halpern *et al.* (1973) claim that sodium is required for transport of glutamate by *E. coli*, and potassium for its accumulation. More recently, however, Thompson and MacLeod (1973) have specifically ruled out the transmembrane gradients of either sodium or potassium as being components of the driving force for amino-acid uptake in their organism.

C. Transport Driven by Artificially Induced Gradients of pH Value and Potential

Within the last twelve months there have been published a number of papers which tested one of the most striking predictions of the chemiosmotic model, namely that it should be possible to drive, or "energize", nutrient transport by induced gradients of pH and electrical potential in metabolically resting cells. Verification of this prediction must surely be one of the major contributions in raising the status of chemiosmosis from hypothesis to theory.

1. *Whole Cells of* Streptococcus faecalis

The earlier findings of Harold and Papineau (1972a, b) established the generation of an ATPase-dependent extrusion of protons in *Strep.*

faecalis with the establishment of a protonmotive force, inside alkaline and negative. Asghar *et al.* (1973) have now extended the study of amino-acid transport in this organism. They have characterized a common transport system for the neutral acids glycine, alanine, serine and threonine, and shown that it can catalyse both an energy-independent exchange and an energy-dependent accumulation of amino acid up to 400 times the concentration in the external medium. This active transport is sensitive to the ATPase inhibitor DCCD and to the proton translocating uncouplers carbonylcyanide CCCP. Additionally the antibiotics nigericin and valinomycin can cause inhibition of uptake, and efflux of previously accumulated amino acid. In catalysing the electrogenic flux of protons back into the cell, the uncouplers act by collapsing both the chemical and electrical components of the protonmotive force. Nigericin causes an electroneutral exchange of protons for potassium ions, and so will collapse the pH gradient while maintaining the membrane potential. The action of valinomycin is complementary in that it affects only the membrane potential as a result of the increase in the electrogenic movement of potassium ions. As discussed by the authors and already alluded to in this article (see page 16), the direction of flux of the potassium ions will depend on the relative magnitudes of the metabolic potential, causing influx, and the chemical concentration gradient, usually causing efflux. The fact that in these experiments valinomycin addition resulted in dissipation of the membrane potential and consequent inhibition of active transport, only in the presence of increased concentrations of potassium ion in the external medium, is indicative of a relatively low value for the metabolically derived potential in this organism.

Although these data indicate that both the pH gradient and the membrane potential are functional in driving the accumulation of the neutral amino acids being assayed, the authors choose to suggest that the maintenance of an alkaline pH value in the cytoplasm is more important than the pH gradient *per se*. The evidence for this view seems rather tenuous, however, and it should be noted that maximal inhibition, comparable to that obtained with uncouplers, is only obtained with addition of both nigericin and valinomycin. Furthermore, when one considers the proposed functioning of a proton symport with a neutral amino acid, or sugar, one sees that formation of the proton-substrate-carrier complex should be favoured by a pH gradient, inside alkaline. Once formed, movement of this entity bearing a net positive charge should be dependent on the potential, inside negative. In the case of the sugars, this response to both pH value and potential gradients has, of course, been confirmed (West and Mitchell, 1972).

A development of this argument helps to explain an apparently serious discrepancy between the findings of Asghar *et al.* (1973) and Gale and Llewellin (1972) with regard to the influence of potassium-ion concentration on inhibition by valinomycin of the transport of amino acids by *Strep. faecalis*. As already discussed (page 17), inhibition of the uptake of glycine and threonine by the antibiotic is only evident at concentrations of potassium ion high enough to allow influx of the ion against the decreased outwardly directed concentration gradient, with consequent collapse of the membrane potential. Gale and Llewellin (1972) however show that valinomycin-inhibition of aspartate transport does not occur at high external concentrations of potassium ion, and is in fact maximal at 1 m*M*. The simplest explanation of this difference rests on the recognition that aspartate will carry a net negative charge at neutral pH values, and consequently the proton-substrate-carrier complex will be uncharged and therefore unresponsive to the membrane potential. As an explanation of the inhibition found with low concentrations of potassium ion in the medium, one can suggest that the potassium diffusion potential, outside positive, developed on addition of valinomycin might be sufficiently large to suppress the metabolic proton extrusion, and thus decrease the magnitude of the pH gradient which is the driving force for uptake of negatively charged substrates such as aspartate.

The entirely opposite dependence on potassium-ion concentrations for inhibition of aspartate uptake by valinomycin in *Staph. aureus*, which Gale and Llewellin (1972) have also reported, can be explained if one assumes a higher value for the metabolically derived membrane potential in this organism. At low external concentrations of potassium ion, the potential driving ion influx, and the concentration gradient driving efflux, will be essentially in balance with consequently little or no net flux or effect on either pH gradient or membrane potential. At higher concentrations, however, net influx will occur with a resultant decrease in the membrane potential. As in the case of the uncoupling effects of valinomycin with mitochondria, the cell responds by increasing the extent of proton efflux. This attempt to maintain the membrane potential exclusively in the form of hydrogen ions leads to an excessive pH gradient and alkalinization of the cytoplasm with a breakdown of normal metabolic processes.

Clearly the validity of such explanations depends on the measured values for the pH value, membrane and potassium diffusion potentials. When values were reported earlier in this article for the membrane potentials in *Strep. faecalis* (Harold and Papineau, 1972a) and *Staph. aureus* (Jeacocke *et al.*, 1972; see page 17), it was pointed out that at this stage these results could only be considered as preliminary. Hirata *et al.* (1973) discussed some of the possible reasons for discrepancies in such measurements, citing their own disagreement with the findings of

Lombardi *et al.* (1973). Apart from the direct relevance of such adequate quantitation to individual experiments, it lies at the heart of considerations of the nature of the driving force, and hence of the mechanism of translocation and the nature of the carrier-substrate interaction.

In the absence of any demonstrable metabolism or ATP synthesis in starving cells, Asghar *et al.* (1973) have generated potassium-diffusion potentials and gradients of pH value. The former is obtained by adding valinomycin to a resting cell suspension in a medium containing a low concentration of potassium ions; the induced electrogenic efflux of the ion creates a potential, inside negative (Fig. 10). When cells which have been equilibrated to a pH value of 9 are pulsed with acid, a pH gradient

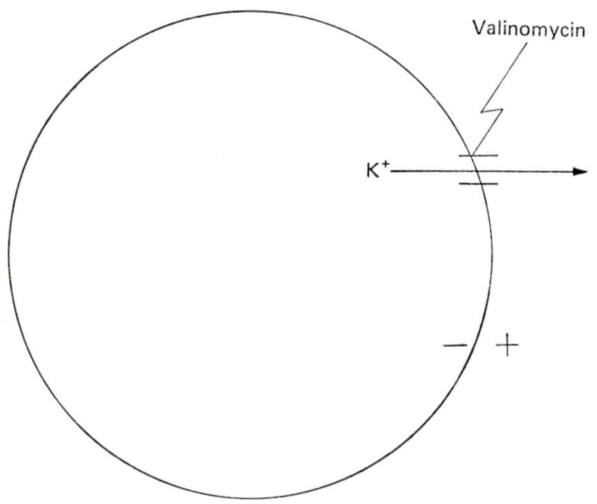

FIG. 10. Generation of a diffusion potential, inside negative, from the valinomycin-induced efflux of K^+.

is established, inside alkaline. These two induced gradients of electrical and chemical potential are thus in the same direction as their metabolically produced counterparts. Both cause accumulation of threonine by an uncoupler-sensitive process which is however insensitive to DCCD. Gradients of electrical or chemical potential, however produced, are therefore adequate driving forces for uptake of neutral amino acids in *Strep. faecalis*.

2. *Whole Cells of* Streptococcus lactis

A very similar series of experiments was carried out by Kashket and Wilson (1973) with resting cells of another anaerobic bacterium, *Strep. lactis*. They showed that the carrier for sugar transport was active in catalysing exchange diffusion in the absence of added fermentable substrate, while accumulation required input of metabolic energy. As with

E. coli, however, transport of thiomethylgalactoside down a concentration gradient into the cells is accompanied by an influx of protons. As with Asghar *et al.* (1973) the creation of a pH gradient, inside alkaline, by exposing the cells to pH 6, caused an accumulation within the cells to a concentration 20 times that in the external medium. Again, the development of a potential, inside negative, from valinomycin-induced potassium efflux can also drive uptake of thiomethylgalactoside. In these experiments, an influx of protons was also noted. For each experimental point it was possible to measure the magnitudes of: (a) the membrane potential from application of the Nernst equation,

$$\Delta\psi = 59 \log \frac{[K^+]_i}{[K^+]_o}$$

to the concentrations of freely diffusible potassium; (b) the pH gradient from the same treatment of intracellular and extracellular [^{14}C]-methylamine (Rottenberg *et al.*, 1972); and (c) the corresponding value for distribution of [^{14}C]-thiomethylgalactoside across the membrane. Values for the membrane potential and pH gradient are combined to give a measure of the total protonmotive force, which may then be compared with the concentration gradient of thiomethylgalactoside, here also expressed in electrical units. As predicted by the chemiosmotic hypothesis, a straight-line relationship is obtained, but one which does not pass through the origin. Kashket and Wilson (1973) propose that this discrepancy (at zero accumulation of thiomethylgalactoside the line extrapolates to a protonmotive force of 25 mV) may arise from a high estimate of the membrane potential due to use of concentrations rather than activities in solving the Nernst equation.

3. *Vesicles of* Escherichia coli

Harold and his colleagues (Hirata *et al.*, 1973) have turned their attention to the classic experimental system of Kaback's redox model, namely membrane vesicles derived from *E. coli*. They have sought to verify the two basic predictions of the chemiosmotic hypothesis as applied to membrane translocations, namely that oxidation of D-lactate by the vesicles can generate a membrane potential, inside negative, and that development of such a potential, even in the absence of metabolism, is in itself sufficient to drive active transport. Although they assay transport and uptake of the neutral amino acid proline, again the possible role of the pH gradient is not considered and the authors concentrate on membrane potential, which they measure from the trans-membrane distribution of the lipid-soluble cation dibenzyldimethyl ammonium in the presence of of trace amounts of the anion triphenyl boron.

It was found that vesicles respiring on D-lactate do indeed generate a potential, inside negative, of about 100 mV. This dibenzyldimethyl ammonium-measured potential can be decreased by the increase in electrogenic ion fluxes resulting from the addition of carbonylcyanide *m*-chlorophenylhydrazone (CCCP) or valinomycin to the system, but it is unaffected by nigericin with its capacity only for electroneutral H^+/K^+ exchange. In the absence of respiratory substrate, the vesicles can develop a potassium diffusion potential of the order of 60 mV on addition of valinomycin. These vesicles can couple this potential to accumulation of proline. This transport is insensitive to the respiratory inhibitor 2-heptyl-4-hydroxyquinoline-N-oxide (HOQNO) and to the ATPase inhibitor DCCD, and thus appears to be quite independent of metabolic processes. It does however depend on a functional carrier, as evidenced by a sensitivity equal to the wild type to *p*-chloromercuribenzoate. Also, a mutant deficient in normal proline transport does not demonstrate potassium efflux-linked uptake in vesicles, although glycine and lysine are accumulated. As predicted by the chemiosmotic model, the transport is sensitive to the proton uncouplers.

4. Whole Cells of Staphylococcus aureus

The work of my own laboratory had its beginnings in a study of the mechanism of the bacteriostatic action of tetrachlorosalicylanilide (TCS) against *Staph. aureus* (Hamilton, 1968). This antibacterial compound has been shown to increase proton permeability and to act as a classical uncoupler in both microbial and mitochondrial systems; along with other agents and antibiotics such as CCCP and valinomycin, TCS has become established as one of the standard experimental tools in studies of energy-transduction mechanisms. In our initial work, a most significant observation was that, while accumulation of glutamate was completely eliminated by TCS, that of lysine was only partially affected. This lack of sensitivity of lysine transport to uncouplers had in fact also been noted by Gale (1954) some 14 years earlier, when he suggested that lysine might be translocated as a cation in response to a Donnan equilibrium.

Our knowledge of chemiosmotic phenomena associated with membranes now allows us to extend this observation and appreciate its full significance. While the proton-translocating property of TCS (Gale used 2,4-dinitrophenol) will collapse both the pH and the potential gradients derived from metabolism, the cells will retain a Donnan potential. In medium containing a low concentration of potassium ions, provided the cells have a significant permeability to the ion, this Donnan potential will be inside negative, and largely in the form of a potassium diffusion potential. Through the use of valinomycin, therefore, and manipulation of potassium concentrations both in the cells and in the medium, it should be possible to verify directly the proposed dependence of lysine transport solely on a membrane potential. These experiments have been carried out with resting suspensions of *Staph. aureus* which have been endogenously depleted and treated with TCS as well as with valinomycin (Niven *et al.*, 1973). It was shown that the extent of lysine uptake could be varied by altering the concentration of potassium in the medium, and hence the magnitude of the concentration gradient. This effect was quantitated by substituting in the Nernst equation the respective values for the measured concentration gradients of potassium and of the amino acid. Under conditions of a potassium equilibrium potential of 98 mV, the calculated value for the lysine equilibrium potential was 100 mV. Manipulation of the experimental conditions to give a higher potassium potential of 148 mV resulted in increased lysine uptake and a potential of 140 mV.

In the presence of TCS and valinomycin, the magnitude of the potassium equilibrium potential can also by altered by addition of acid or alkali to the medium. Hence, addition of alkali (a pH 5 to 7 transition, for example) will cause a certain efflux of protons. This being an electrogenic flux, it will be balanced by an equal influx of the other permeant ion in the system, namely potassium. That is, addition of alkali will effectively increase the intracellular concentration of potassium, and so also increase the magnitude of the potassium diffusion potential (Fig. 11). As predicted by the model, such an alkali addition is indeed accompanied by an increased rate of uptake of lysine. This effect is reversible, and acid addition slows the transport of the amino acid. A significant point about this second experimental design is that conditions which lower the magnitude of the pH gradient, inside alkaline, simultaneously increase the membrane potential, inside negative, and *vice versa*. It is clear therefore that transport of the basic amino acid lysine in *Staph. aureus* can be driven solely by a membrane potential, derived either from metabolism or from the position of Donnan equilibrium in non-metabolizing cells. In terms of the chemiosmotic model of transport, the lysine carrier functions as a uniport catalysing translocation of lysine cation.

While the major form of lysine at neutral pH values is the cation with a single positive charge, neutral amino acids such as glycine and isoleucine are uncharged, and the acidic glutamate and aspartate are anions with a single negative charge. As with the sugars, the chemiosmotic model of transport proposes that isoleucine, say, will be translocated on a proton symport, and that the net charge on the proton-isoleucine-carrier complex ensures that the extent of uptake is dependent on the membrane potential as well as on the pH gradient. Since the proton glutamate carrier complex carries no net charge, transport of the

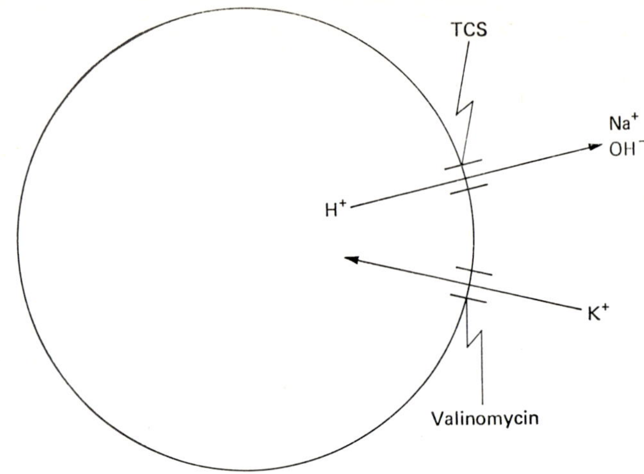

FIG. 11. The effect on diffusion potential following addition of alkali. Addition of alkali to tetrachlorosalicylanilide- and valinomycin-treated cells causes some efflux of protons; this efflux is electrogenic and is compensated for by a K^+ influx, thus increasing the intracellular concentration of K^+ and so also the potassium diffusion potential.

acidic amino acids should be driven only by the pH gradient. This model is summarized in the scheme:

Amino acid	*Charge at pH 7*	*Mechanism of transport*	*Charge on transported species*	*Driving force*
Lysine	+	Uniport	+	$\Delta\psi$
Isoleucine	0	H^+-Symport	+	$\Delta p = \Delta\psi - Z\,\Delta\,\text{pH}$
Glutamate	−	H^+-Symport	0	$-Z\,\Delta\,\text{pH}$

We have now further tested and verified the predictions of this scheme (Niven and Hamilton, 1973, 1974). Under conditions comparable to those used in demonstrating the increased uptake of lysine on pulsing the cells with alkali, exactly the opposite effect was noted with glutamate. Again the effect was reversible, and glutamate transport was markedly stimulated by addition of acid to suspensions of non-metabolizing cells.

In these experiments with cells which have relatively high permeabilities to protons and potassium ions (either endogenous or induced by TCS and valinomycin), the position of Donnan equilibrium is reached with the balancing of potassium efflux by proton influx. As compared with metabolizing cells, the direction of the pH gradient is reversed to inside acid, and at equilibrium $\Delta p = 0$ and $\Delta\psi = Z\Delta$ pH. The effect, therefore, of acid or alkali addition to such cells is to disturb the equilibrium and alter, at least transiently, all three parameters, the protonmotive force, the membrane potential and the pH gradient.

An unequivocal demonstration of the response of glutamate transport solely to the pH gradient therefore depends on an experimental manipulation of this parameter without affecting the potential or total protonmotive force. When cells treated with TCS and valinomycin have been

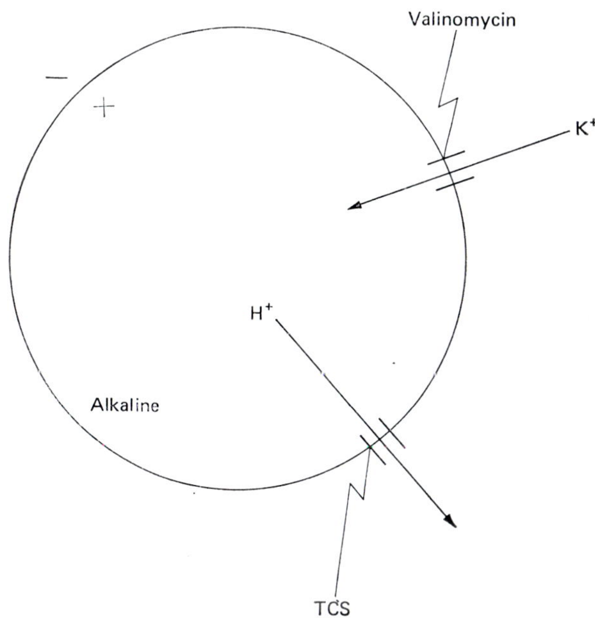

Fig. 12. Generation of a pH gradient from a suspension of tetrachlorosalicylanilide- and valinomycin-treated potassium-depleted cells in a high potassium medium, showing the electrogenic influx of K^+, causing proton efflux, and the establishment of the pH gradient, inside alkaline.

depleted of intracellular potassium prior to being suspended in a medium containing a high concentration of potassium ions, the Donnan equilibrium will be reached by potassium influx and proton efflux, and will be characterized by a potential, inside positive, and a pH gradient, inside alkaline (Fig. 12). *At equilibrium* these cells accumulate glutamate in response to this pH gradient which now has the same sign as that developed during metabolism.

As indicated above, addition of acid to cells in a medium low in potassium results in a transient protonmotive force, which dissipates as the permeant ions flux to the new position of equilibrium. A similar transient protonmotive force can be developed on adding valinomycin to a suspension of cells with a diminished permeability to potassium ions. This latter condition is the same as that used by Cockrell *et al.* (1967) and Reid (1970) in demonstrating ATP synthesis in mitochondria, and by Asghar *et al.* (1973), Kashket and Wilson (1973) and Hirata *et al.* (1973) in studying transport of neutral sugars and amino acids in bacterial systems. By both mechanisms we have also shown transient accumulation of glycine and isoleucine in response to a transient and non-metabolic protonmotive force.

This study therefore constitutes an unequivocal qualitative demonstration of the application of the chemiosmotic model to translocation of amino acids by a microbial cell; the basic lysine is transported by a uniport in response to the membrane potential, the acidic glutamate by a proton symport in response to the pH gradient, and the neutral glycine and isoleucine by a proton symport in response to the total protonmotive force. The metabolic generation of these gradients of chemical and electrical potential has already been demonstrated for *Staph. aureus* (Jeacocke *et al.*, 1972), and what remains to be done is: (a) to demonstrate the proton flux associated with uptake of glutamate and glycine; and (b) to obtain a measure of the magnitudes of the induced pH gradient and protonmotive force in these experiments for comparison with the amounts of amino-acid accumulation. These problems are currently under study but, as pointed out by Asghar *et al.* (1973), the magnitude and rate of these amino-acid fluxes are very much less than those found with sugars such as thiomethylgalactoside, and the associated proton fluxes are therefore very much more difficult to measure.

D. Adenosine Triphosphatase and Electron-Transport Mutants

One of the most powerful weapons in the armoury available to the microbiologist, the use of mutants, has now been brought to bear on the problem of energy coupling in transport. Schairer and Haddock (1972) used their ATPase-deficient mutant A103c, derived from *E. coli* K_{12} strain A1002, to demonstrate that thiomethylgalactoside accumulation can be driven either by respiration or by hydrolysis of ATP produced glycolytically. Another mutant (Unc 253) isolated by the same group (Schairer and Gruber, 1973) is very closely linked with A103c from the results of transduction studies. Although it is also deficient in oxidative phosphorylation and in ATP-driven transhydrogenase and thiomethylgalactoside accumulation, the nature of the lesion in the ATPase complex must be different from A103c since Unc 253 demonstrates normal ATPase activity.

Prezioso *et al.* (1973) conclude that the ATPase complex and its activity have no role in respiration-driven transport of thiomethylgalactoside, amino acids and rubidium by whole cells and vesicles of *E. coli*. The mutant they used (AN120) derived by Butlin *et al.* (1971) from K_{12} AN180, is deficient in ATPase activity and yet is unaltered in these transport reactions. Simoni and Shallenberger (1972) take the diametrically opposite view from their work with mutants derived from *E. coli* ML308-225. Both S2-21, deficient in electron transport, and DL-13 which is D-lactate dehydrogenase negative, show normal uptake of alanine and proline in whole cells. In vesicles they find enough residual respiration of D-lactate in S2-21 to power maximal uptake, which in any case is only 2% of that found in the whole cells. Mutant DL-13 however shows no D-lactate-driven transport in vesicles. Only in particular situations therefore is transport obligatorily coupled to respiration in general, and to oxidation of D-lactate in particular. These authors also isolated an ATPase-deficient mutant (DL-54) which showed some degree of loss of respiration-driven transport in whole cells, and a very marked loss in vesicles. As with the two ATPase mutants, A103c and Unc 253 isolated by Schairer and his coworkers, the nature of the lesions in AN120 of Prezioso *et al.* (1973) and DL-54 must be more fully characterized before they can hope to add enlightenment rather than confusion to our understanding of the energy-coupling mechanisms involved.

Other studies on the isolation and initial characterization of respiration and ATPase mutants are those of Butlin *et al.* (1971), Kanner and Gutnick (1972), Hong and Kaback (1972), Nieuwenhuis *et al.* (1973), Van Thienen and Postma (1973), Gibson and Cox (1973) and Yamamoto *et al.* (1973).

The findings of Rosen (1973a, b) can be repeated here. His ATPase mutant (NR70) lacks the enzyme protein, has increased proton permeability, cannot couple either respiration or ATP hydrolysis to transport, and can regain proton impermeability and coupled transport functions on treatment with the ATPase inhibitor DCCD. Van Thienen and Postma (1973) also found that DCCD stimulates respiration-driven uptake of serine into vesicles derived from their ATPase mutants N_{144} and K_{11} which in turn had come from strain $K_{12}A_{120}$. The papers of Simoni and Shallenberger (1972), Bragg and Hou (1973) and Berger (1973) constitute a very similar and most significant study of the ATPase mutant DL-54. The mutant, which has lost about 95% of its ATPase activity, has been shown by Bragg and Hou (1973) to have a modified ATPase enzyme protein which is more readily lost than in the wild type when producing vesicles. Although transport activity is diminished in cells, the effect is considerably more pronounced in vesicles (Simoni and Shallenberger, 1972). Berger (1973) has shown that this loss of transport activity in vesicles of DL-54 can be reversed by titrating with DCCD. It is concluded that the ATPase, in addition to its enzymic role, has some structural role in respect of membrane function, and in particular of the energy-transducing reactions. It remains to be demonstrated whether with mutants DL-54, N_{144} and K_{11}, as with NR70, this function can be equated with the permeability of the membranes to protons.

E. Conclusions

In concluding this major part of the present review, one may summarize the findings and conclusions that have been discussed. "Active transport" is the term used for the accumulation within a cell, or organelle, of a chemically unmodified substrate to a chemical potential higher than that in the extracellular medium. As such it shows a requirement for metabolic energy, which, depending on the organism and the conditions of growth and assay, may be derived from photosynthesis, respiration or hydrolysis of ATP produced by, for example, glycolysis. In addition to a sensitivity to inhibitors of the appropriate energy-producing mechanism, all active transport processes are sensitive to the proton-translocating uncouplers. Of the three principal models for the coupling of energy to the translocating carrier function, the overwhelming mass of the evidence supports that based on the chemiosmotic hypothesis. Additionally this model is the simplest in terms of its assumptions, and the one most readily tested experimentally. However opposing the final conclusions might be, no research into energy-coupling mechanisms in active transport can now be taken seriously unless it considers and tests the predictions of the chemiosmotic hypothesis relevant to the system under study.

VII. Group Translocation and Transport

A. The Phosphotransferase System

Substrate translocations which are linked to and driven by transmembrane ion gradients are secondary translocations in terms of their coupling to metabolic energy. The primary translocation is that of the protons, or of the sodium and potassium ions in mammalian cells. There is no reason *a priori* why mechanisms should not exist for the primary coupling of substrate translocation to an exergonic metabolic reaction. Since this might be most readily achieved through the chemical modification of the substrate by a membrane-bound enzyme, the process would most accurately be classified as a group translocation. Such a mechanism was first proposed by Mitchell (1961b). The phosphoenolpyruvate phosphotransferase (PT) system for group translocation of a number of sugars by certain bacteria was discovered by Kundig *et al.* (1964). It has been the subject of a number of review articles (Kaback, 1970; Harold, 1972; Roseman, 1972; Kornberg, 1973a), and most recently of a brilliant series of papers by Roseman and his coworkers on lactose transport in *Staph. aureus* (Simoni *et al.*, 1973a, b; Hay *et al.*, 1973; Simoni and Roseman, 1973).

Figure 3 (page 12) gave a very schematic representation of the PT system, coupling the phosphate-bond energy of phosphoenolpyruvate (PEP) to conversion of a sugar to its phosphate derivative. Where it is found, this series of reactions is now generally held to account for both the transport of sugars and the initial step in their metabolism. The PT system appears to be confined to anaerobic and facultative bacteria and to be absent from aerobic species (Romano *et al.*, 1970); there is no evidence for its occurrence in the fungi. Whereas all sugars are transported by the system in *Staph. aureus*, only some, for instance glucose, enter the cell by this mechanism in *E. coli*. The small heat-stable protein (HPr), and Enzyme I which catalyses its phosphorylation from PEP, are constitutive and soluble, and do not show sugar specificity. Enzyme II, which catalyses transfer of the phosphate group from HPr to the sugar, is sugar-specific and is an inducible membrane-bound enzyme. In fact the Enzyme II is now known to be considerably more complex than a single protein component. In *E. coli* and *Sal. typhimurium* it has been found (Kundig and Roseman, 1971a, b) that the functional complex requires two proteins, II-A and II-B, and phosphatidylglycerol. Proteins II-A and II-B are both membrane bound, but, whereas II-B appears to be constitutive, separate II-As are induced for transport of glucose, mannose and fructose. The system in *Staph. aureus* is different, as depicted in Fig. 13. The two sugar-specific components are Enzyme II, which is

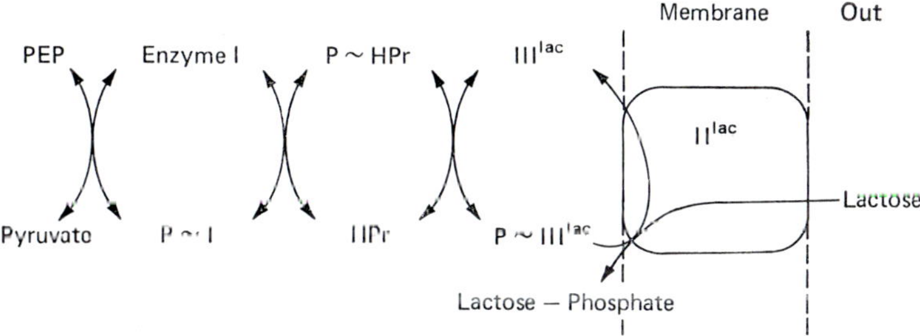

FIG. 13. The phosphotransferase system. Schematic representation of the reactions involved in the transfer of phosphate from phosphoenolpyruvate to lactose in *Staphylococcus aureus*. From Simoni *et al.* (1973).

membrane-bound, and Factor III which is soluble. The lactose PT system transports lactose and galactose, the inducer being intracellular galactose 6-phosphate. Separate Enzyme II and Factor III have been demonstrated for mannitol transport, and Factors III for transport of sorbitol and fructose. The specific components of the glucose system have not yet been unequivocally identified. The heat-stable proteins (HPr) in *E. coli* and *Sal. typhimurium* appear to be identical. In *Staph. aureus*, the HPr also has a molecular weight of 9600 daltons but the amino-acid composition varies from that in the HPr of the Gram-negative organisms. The protein from *Staph. aureus* has one histidine residue and two tyrosines, while that from *E. coli* has two histidines and no tyrosine residues.

In the lactose PT system in *Staph. aureus*, the three soluble components, Enzyme I, HPr and Factor III, are all phosphorylated during sugar translocation. With the intermediate production of a phosphoenzyme, Enzyme I catalyses transfer of a single "high energy" phosphate from PEP to the N-1 of a histidine residue in HPr. The reversible transfer of the phosphate to Factor III is a self-catalysing reaction and involves no other protein component. Factor III turns out to consist of three identical subunits of molecular weight approximately 12,000 daltons, each of which can bind one phosphate to the N-3 of a histidine residue. A ternary complex is then formed with lactose, Factor III carrying up to three phosphates, and Enzyme II, which catalyses the transfer of the phosphate to the galactoside-6 position, without itself being phosphorylated.

The standard free energy of hydrolysis of PEP is about −13·5 kcal per mol and, from the measured values of the equilibrium constants for the transfer reactions, it is possible to arrive at an approximate value of −12 kcal per mol for the standard free energies of hydrolysis of phospho-HPr and phospho-Factor III. The corresponding values for ATP and a sugar-phosphate, respectively, are −7 and −3·5 kcal per mol. From these values we can see that: (a) only PEP, and not ATP, could drive this mechanism; (b) all of the phospho-derivatives must be 'high energy'; and (c) the drop in free energy occurs at the stage of the final Enzyme II-catalysed reactions. Presumably, therefore, this last stage is also the site of translocation, possibly associated with a conformational change in the ternary complex.

It is claimed that such a translocation step does not take place in the absence of energy in the form of phospho-Factor III, and that the PT system is not therefore capable of facilitated diffusion. Some doubt remains on this point, however, and Gachelin (1970) has taken the opposite viewpoint.

The PT system therefore appears to be a completely different transport mechanism from the substrate translocations described in the previous sections. Pavlosova and Harold (1969), for instance, have clearly differentiated between the uncoupler-sensitive thiomethylgalactoside accumulation in anaerobic *E. coli* by the β-galactoside permease, and the uncoupler-resistant PT intracellular accumulation of glucose 6-phosphate from glucose. In *Arthrobacter pyridinolis*, fructose can be transported either by a specific PT system with the product being, unusually, fructose 1-phosphate, or accumulated by a standard cyanide- and uncoupler-sensitive mechanism driven by respiration (Krulwich *et al.*, 1973). There are, however, some remarkable points of overlap, with regard to function and control, which will be considered at the end of this article. At this point, though, one can mention the ideas put forward by Simoni and Roseman (1973) and by Mitchell (1973a). Simoni and Roseman (1973) draw a parallel with the permease model, and make the point that active transport might reasonably be expected to occur by phosphorylation of the sugar-specific carrier, but without transfer of the phosphoryl group to the sugar itself. Mitchell (1973a) stresses that, depending on the identity of the group which is actually translocated, e.g. glucose-6^{+} or glucose-6-O^{-}, the possibility exists for the PT system also to have associated proton movement, and so be responsive to the protonmotive force.

B. Other Group Translocations

A number of other systems have been reported in which transport appears to occur by primary group translocation. Klein *et al.* (1971) studied uptake of fatty acids by *E. coli*. The mechanism is inducible and shows the same specificity as the acyl-CoA synthetase, which is the first enzyme in degradation of the fatty acids. Mutants deficient in the acyl-CoA synthetase show similar lesions of fatty-acid transport which the authors conclude takes place by a process of vectorial acylation. Transport of purine bases and nucleosides by *E. coli* has been extensively studied in Hochstadt-Ozer's laboratory (Hochstadt-Ozer and Stadtman, 1971a, b, c; Hochstadt-Ozer, 1972). It was demonstrated that nucleosides were first hydrolysed to the free base, which was then subjected to group translocation by a phosphoribosyltransferase with intracellular accumulation of the purine nucleoside monophosphate and pyrophosphate. This system differs significantly from the PT system in that the phosphoribosyltransferase has a periplasmic location and is largely lost from the cells following an osmotic shock treatment. Also, the phosphoribosyl pyrophosphate can stimulate uptake when added exogenously.

Although it is not microbial in origin, one can mention here the beautiful demonstration by Storelli *et al.* (1972) of sucrose transport by a group translocation mechanism, and conversion to glucose and fructose by a sucrase-isomaltase complex incorporated into an artificial lipid bilayer. Apart from the interest in the transport system itself in intestinal tissues, this important paper is the first demonstration of a reconstituted transport system in an artificial membrane and as such is of very considerable importance.

VIII. Other Transport Mechanisms

Needless to say, not all microbial translocations fit easily into either of the two main mechanisms so far considered in this article. For example, the extremely interesting and unique mechanism for the mycobactin-dependent uptake of iron by *Mycobacterium smegmatis* is driven by the enzyme ferrimycobactin reductase (Ratledge and Marshall, 1972). Whereas, in their studies, Postma *et al.* (1973) conclude that the energy-requiring transport of Krebs-cycle intermediates by *A. vinelandii* is directly coupled to ATP. They observed that oxidation of exogenous intermediates of the cycle was inhibited by the ATPase inhibitors DCCD, Dio-9 and oligomycin. Also, similar coupling between influx and efflux was noted to that reported by Tanner's group in their studies of hexose transport in *Chlorella vulgaris* (Komor *et al.*, 1972; Decker and Tanner, 1972).

A. Periplasmic Binding Proteins

In this context the paper by Berger (1973) is highly relevant. Active transport of proline in *E. coli* is clearly shown to be dependent on electron transport under aerobic conditions, and on hydrolysis of ATP derived from glycolysis under anaerobic conditions. Uncouplers prevent energy coupling under both conditions. These results confirm those of Klein and Boyer (1972), and represent the classic pattern for an active transport, most probably driven by the protonmotive force. In a similar study, however, Berger (1973) showed that the mechanism of energy coupling in accumulation of glutamine was quite different. Aerobically it showed sensitivity to cyanide and uncouplers, but additionally required an active ATPase. Anaerobically the transport was resistant to both cyanide and uncouplers, and did not require a functional ATPase. Arsenate inhibited the transport under both sets of experimental conditions. It was concluded that ATP is directly coupled to transport of glutamine. An interesting, and perhaps most significant, point is that the mechanism of glutamine transport involves a specific binding protein (Weiner *et al.*, 1971; Weiner and Heppel, 1971; Heppel *et al.*, 1972; Boos, 1974).

Transport of certain sugars and amino acids in a number of Gram-negative bacteria has been found to be sensitive to osmotic-shock treatments. This sensitivity is associated with the loss from the periplasmic space of proteins which can bind specifically the substrate whose transport activity has been affected. Berger (1973) claims to have evidence that, in *E. coli*, other amino-acid transport systems, which like that for glutamine involve a periplasmic binding protein, are also driven by ATP itself. They also show very much lower sensitivity to the sulphydryl reagent N-ethylmaleimide as compared with translocations that are sensitive to osmotic shock such as those for proline and thiomethylgalactoside.

In their work on transport and accumulation of β-methylgalactosides by *E. coli*, Parnes and Boos (1973a, b) noted a number of important differences from the lactose system. Counterflow could not be demonstrated in energy-poisoned cells; trans-membrane stimulation was found for exit but not for entry; energy coupling affected entry rather than exit. The total transport system appears to require a periplasmic galactose-binding protein in addition to the membrane-bound carrier. Since it is claimed that the binding protein is involved only in entry of the sugar, exit being by another pathway, it is tempting to suggest that the binding protein may play a role in energy coupling. Unlike Berger's data with glutamine, however, β-methylgalactoside transport is absent from the ATPase mutant AN120 under anaerobic conditions. It is stimulated by oxidation of D-lactate and most probably does not involve ATP directly.

The exact role of the periplasmic-binding proteins in transport and other membrane-associated phenomena (Adler *et al.*, 1973) remains somewhat obscure at present. In view of the experimental observations already described, it will be extremely interesting to observe if their function in energy-coupling mechanisms can be confirmed and further elucidated.

IX. Transport as a Site of Cellular Control

A possible interaction between the group-translocating PT system and the substrate translocation mechanism(s) of active transport has already been alluded to. This is particularly evident when we turn to consider control of these processes and their integration with the overall metabolism of the cell. Although the subject of control might seem to be somewhat outside the strict scope of this review, clearly the energetics and control of a biological process must be intimately linked, functionally even if not structurally. The extremely interesting, although still somewhat confusing, data that are presently coming forward make a most valid reason for concluding this article with what must inevitably be a rather fleeting and even speculative look at transport as a site of cellular control.

The phenomenon of diauxic growth has been recognized for many years, and characterized with such systems as *E. coli* growing in a carbon–ammonium-salts medium with glucose and lactose, or *A. aerogenes* growing on glucose and citrate. Repression and induction of the catabolic pathway for the second carbon source is known to involve also a permease or transport mechanism. This holds true for the unusual diauxie in *Pseudomonas aeruginosa* where Krebs-cycle intermediates are used preferentially to glucose (Hamilton and Dawes 1959, 1960, 1961). The inducible glucose transport has been shown to be both repressed and inhibited by, for example, citrate and the products of its metabolism (Midgley and Dawes, 1973; Mukkada *et al.*, 1973).

Catabolite repression (Paigen and Williams, 1970) represents the "coarse control" giving rise to diauxic growth. In cells where the rate-limiting process is anabolism rather than catabolism, as may arise with a rapidly catabolized carbon source or under conditions of a diminished rate of protein synthesis, synthesis of further catabolic enzymes is repressed. The effect can be overcome by addition of cyclic AMP; rapid rates of catabolism are found to be associated with lower levels of intracellular cyclic AMP. It appears that the binding of RNA polymerase to the promotor gene of the *lac* operon requires the presence of cyclic-AMP and its specific binding protein. Catabolite repression therefore arises from a fault in the attachment of RNA polymerase, and hence an absence of the synthesis of the appropriate mRNA.

The complementary "fine control" of catabolite inhibition has been discovered by McGinnis and Paigen (1969, 1973). They showed that intracellular glucose (derived from exogenous lactose as a nutrient) could cause catabolite repression of the enzymes of mannose catabolism in *E. coli*. Exogenous glucose was required for the demonstration of catabolite inhibition. The authors concluded that the cause of the inhibition was glucose transport, and that the effect was most probably mediated through an inhibition of mannose transport. While glucose and mannose are both transported in *E. coli* by the PT system, the majority of reports of this effect have involved inhibition, by the transport of a PT sugar, of transport of a second sugar which does not enter the cell by group translocation.

Winkler and Wilson (1967) reported inhibition by glucose of β-galactoside transport in *E. coli*. Koch (1971b) analysed this effect further, and obtained evidence for what he described as a direct interaction indicative of a shared component, and an indirect interaction indicative of a decrease in the cell's energy pool.

In their studies of inhibition by the analogue 3-deoxy-3-fluoro-D-glucose of utilization of lactose by *E. coli*, Miles and Pirt (1973) found evidence for both catabolite repression and inhibition. The evidence for repression was indirect and based on the reversal by cyclic-AMP of a long-term repressive effect of the analogue on lactose utilization. Mutants resistant to the action of the glucose analogue were found to be deficient in Enzyme II of the PT system, which is responsible for transport of 3-deoxy-3-fluoro-D-glucose and its accumulation within the cell as the 6-phospho derivative.

In two recent review articles, Kornberg (1973a, b) discussed catabolite inhibition of transport of galactose, lactose, maltose, xylose, arabinose, glycerol (all non-PT sugars) and fructose (a PT sugar) by rapid transport of glucose into *E. coli*. In mutants defective in Enzyme II, the inhibition has been noted with glucose 6-phosphate rather than with glucose. Even in the absence of glucose, mutants deficient in Enzyme I of the PT system grew poorly on media containing as carbon source lactose, galactose, melibiose, maltose, glycerol or succinate. It was found that the cells were unable to transport sufficient inducer for synthesis of the transport and catabolic enzymes required. This lack could be overcome by addition of cyclic AMP. It was concluded that the intracellular inducers for these non-PT transport systems required an active PT system for their own initial accumulation within the cell. In this connection it is interesting to note the report by Kusch and Wilson (1973) that their *E. coli* mutant which is energy-uncoupled for lactose transport showed severely depressed rates of induction with lactose or low concentrations of a gratuitous inducer. The ability to accumulate the inducer, in this case by the action of the proton symport, appears to be required for maximum rates of induction.

The interactions and control of sugar transport mechanisms in the Gram-negative organisms *E. coli* and *Sal. typhimurium* have also been studied by Roseman and his colleagues (Roseman, 1972; Saier and Roseman, 1972). They have isolated a series of so-called "tight" and "leaky" mutants deficient in either Enzyme I or HPr of the PT system. When these mutants were examined for their ability to induce the non-PT transport and catabolic enzyme systems for lactose in *E. coli* and for melibiose, glycerol and maltose in *Sal. typhimurium*, it was found that the tight mutants could not be induced; whereas the leaky mutants could be induced, they were readily repressed by any sugar in the medium which was transported by the PT system. Although Makman and Sutherland (1965) had previously shown that addition of glucose to a culture of *E. coli* caused an efflux of cyclic-AMP, Saier and Roseman (1972) found that this did not occur in all strains of *E. coli*, and not at all in *Sal. typhimurium*. It cannot therefore be generally applicable as the mechanism of the repression of induction found in these mutants. A functional Enzyme II is required for the effect to be evident, and it appears that the repression is a direct consequence of inducer exclusion by catabolite inhibition of the inducer transport mechanism.

X. Concluding Remarks

Even in this very brief description of an exciting area of research that is just beginning its development, the role and importance of membrane translocation is very evident. Clearly one can also see something of a pattern forming out of the interactions between the various transport systems, and their function and regulation. In the "simpler" study of energy-coupling mechanisms, it has been this attempt to trace the outlines of a logical and integrated pattern that has directed the choice of papers and the form of discussion in this article. As such, no pretence can be made of a comprehensive coverage, and perhaps a personal bias may be all too obvious to some readers. It is the earnest hope of the author, however, that any such criticisms may be more than fully compensated for by an increased level of comprehension of the facts and their relevance, in an area of study that has too often in the past owed more to the imagination of the model builder than to the realism of the experimenter.

XI. Acknowledgements

Figure 2 is reproduced with the permission of H. R. Kaback and the *Journal of Biological Chemistry*, and Fig. 13 with the permission of S. Roseman and the *Journal of Biological Chemistry*. I am grateful to K. H. Altendorf, W. Boos, L. Grinius, F. M. Harold, P. J. F. Henderson and R. C. Valentine for the chance to see their manuscripts prior to publication and to W. Boos for his thoughtful and constructive comments on this manuscript.

My very great debt to Peter Mitchell and to Frank Harold must be obvious to everyone reading this article. I should also like to thank my colleagues who, in ways large and small, direct and devious, have contributed so much to my own understanding of the subject, and to the pleasure I have derived from it: Steve Collins, Tim Horne, Robin Jeacocke, Philip John, Ray Lindsay, Wright Nichols, Donald Niven, Dave Rowley, Michael Schedel.

References

Adler, J., Hazelbauer, G. L. and Dahl, M. M. (1973). *Journal of Bacteriology* **115**, 824.

Altendorf, K. H. and Staehelin, L. A. (1974). *Journal of Bacteriology* **117**, 888.

Asano, A., Cohen, N. S., Baker, R. F. and Brodie, A. F. (1973). *Journal of Biological Chemistry* **248**, 3386.

Asghar, S. S., Levin, E. and Harold, F. M. (1973). *Journal of Biological Chemistry* **248**, 5225.
Barnes, E. M. (1972). *Archives of Biochemistry and Biophysics* **152**, 795.
Barnes, E. M. (1973). *Journal of Biological Chemistry* **248**, 8120.
Barnes, E. M. and Kaback, H. R. (1971). *Journal of Biological Chemistry* **246**, 5518.
Berger, E. A. (1973). *Proceedings of the National Academy of Sciences of the United States of America* **70**, 1514.
Boos, W. (1974). *Annual Review of Microbiology* (in press).
Bragg, P. D. and Hou, C. (1973). *Biochemical and Biophysical Research Communications* **50**, 729.
Brown, C. E. and Romano, A. H. (1969). *Journal of Bacteriology* **100**, 1198.
Butlin, J. B., Cox, G. B. and Gibson, F. (1971). *Biochemical Journal* **124**, 75.
Cockrell, R. S., Harris, E. J. and Pressman, B. C. (1967). *Nature, London* **215**, 1487.
Cohen, G. N. and Monod, J. (1957). *Bacteriological Reviews* **21**, 169.
Decker, M. and Tanner, W. (1972). *Biochimica et Biophysica Acta* **266**, 661.
Drachev, L. A., Kaulen, A. D., Ostroumov, S. A. and Skulachev, V. P. (1974). *Federation of European Biochemical Societies Letters* **39**, 43.
Eagon, R. G. and Wilkerson, L. S. (1972). *Biochemical and Biophysical Research Communications* **46**, 1944.
Eddy, A. A., Indge, K. J., Bracken, K. and Nowacki, J. A. (1970a). *Biochemical Journal* **120**, 845.
Eddy, A. A., Bracken, K. and Watson, G. (1970b). *Biochemical Journal* **120**, 853.
Eddy, A. A. and Nowacki, J. A. (1971). *Biochemical Journal* **122**, 701.
Fields, K. L. and Luria, S. E. (1969). *Journal of Bacteriology* **97**, 57.
Fox, C. F. and Kennedy, E. P. (1965). *Proceedings of the National Academy of Sciences of the United States of America* **54**, 891.
Gachelin, G. (1970). *European Journal of Biochemistry* **16**, 342.
Gale, E. F. (1954). *Symposium of the Society for Experimental Biology* **8**, 242.
Gale, E. F., Cundliffe, F., Reynolds, P. E., Richmond, M. H. and Waring, M. J. (1972). "The Molecular Basis of Antibiotic Action", p. 121. Wiley–Interscience Publications.
Gale, E. F. and Llewellin, J. M. (1972). *Biochimica et Biophysica Acta* **266**, 182.
Garland, P. B., Clegg, R. A., Downie, J. A., Gray, T. A., Lawford, H. G. and Skyrme, J. (1972). *Federation of European Biochemical Societies Symposium* **28**, 105.
Ghei, O. K. and Kay, W. W. (1973). *Journal of Bacteriology* **114**, 65.
Gibson, F. and Cox, G. B. (1973). *Essays in Biochemistry* **9**, 1.
Greville, G. D. (1969). *Current Topics in Bioenergetics* **3**, 1.
Griniuviene, B., Chmieliauskaite, V. and Grinius, L. (1974). *Biochemical and Biophysical Research Communications* **56**, 206.
Halpern, Y. S., Dover, H. B. S. and Druck, K. (1973). *Journal of Bacteriology* **114**, 53.
Hamilton, W. A. (1968). *Journal of General Microbiology* **50**, 441.
Hamilton, W. A. (1974). *In* "Industrial Aspects of Biochemistry", (B. Spencer, ed.), p. 607. North-Holland Publishing Company, Amsterdam.
Hamilton, W. A. and Dawes, E. A. (1959). *Biochemical Journal* **71**, 25P.
Hamilton, W. A. and Dawes, E. A. (1960). *Biochemical Journal* **76**, 70P.
Hamilton, W. A. and Dawes, E. A. (1961). *Biochemical Journal* **79**, 25P.
Harold, F. M. (1970). *Advances in Microbial Physiology* **4**, 45.
Harold, F. M. (1972). *Bacteriological Reviews* **36**, 172.
Harold, F. M. (1974). *Annals of the New York Academy of Sciences* **227**, 297.
Harold, F. M. and Baarda, J. R. (1967). *Journal of Bacteriology* **94**, 53.

Harold, F. M. and Baarda, J. R. (1968a). *Journal of Bacteriology* **95**, 816.
Harold, F. M. and Baarda, J. R. (1968b). *Journal of Bacteriology* **96**, 2025.
Harold, F. M. and Papineau, D. (1972a). *Journal of Membrane Biology* **8**, 27.
Harold, F. M. and Papineau, D. (1972b). *Journal of Membrane Biology* **8**, 45.
Hay, J. B., Simoni, R. D. and Roseman, S. (1973). *Journal of Biological Chemistry* **248**, 941.
Henderson, P. J. F. (1971). *Annual Review of Microbiology* **25**, 394.
Henderson, P. J. F. (1974). *Proceedings of the Fifth International Conference on Biological Membranes* (in press).
Heppel, L. A., Rosen, B. P., Friedberg, I., Berger, E. A. and Weiner, J. H. (1972). "Molecular Basis of Membrane Transport", (J. F. Woessner and F. Huijing, eds.), p. 133. Academic Press, New York.
Hinkle, P. C. (1973). *Federation Proceedings. Federation of American Societies for Experimental Biology* **32**, 1988.
Hirata, H. and Brodie, A. F. (1972). *Biochemical and Biophysical Research Communications* **47**, 633.
Hirata, H., Asano, A. and Brodie, A. F. (1971). *Biochemical and Biophysical Research Communications* **44**, 368.
Hirata, H., Altendorf, K. and Harold, F. M. (1973). *Proceedings of the National Academy of Sciences of the United States of America* **70**, 1804.
Hochstadt-Ozer, J. (1972). *Journal of Biological Chemistry* **247**, 2419.
Hochstadt-Ozer, J. and Stadtman, E. R. (1971a). *Journal of Biological Chemistry* **246**, 5294.
Hochstadt-Ozer, J. and Stadtman, E. R. (1971b). *Journal of Biological Chemistry* **246**, 5304.
Hochstadt-Ozer, J. and Stadtman, E. R. (1971c). *Journal of Biological Chemistry* **246**, 5312.
Hong, J.-S. and Kaback, H. R. (1972). *Proceedings of the National Academy of Sciences of the United States of America* **69**, 3336.
Hopfer, U., Lehninger, A. L. and Thompson, T. E. (1968). *Proceedings of the National Academy of Sciences of the United States of America* **59**, 484.
Hunter, D. R. and Segel, I. H. (1973). *Journal of Bacteriology* **113**, 1184.
Jeacocke, R. E., Niven, D. F. and Hamilton, W. A. (1972). *Biochemical Journal* **127**, 57P.
Jetten, A. M. and Vogels, G. D. (1973). *Biochimica et Biophysica Acta* **311**, 483.
John, P. and Hamilton, W. A. (1970). *Federation of European Biochemical Societies Letters* **10**, 246.
John, P. and Hamilton, W. A. (1971). *European Journal of Biochemistry* **23**, 528.
Kaback, H. R. (1970). *Annual Review of Biochemistry* **39**, 561.
Kaback, H. R. (1972). *Biochimica et Biophysica Acta* **265**, 367.
Kaback, H. R. and Barnes, E. M. (1971). *Journal of Biological Chemistry* **246**, 5523.
Kaback, H. R. and Hong, J- S. (1973). *Critical Reviews in Microbiology* **2**, 333.
Kaback, H. R. and Stadtman, E. R. (1966). *Proceedings of the National Academy of Sciences of the United States of America* **55**, 920.
Kanner, B. I. and Gutnick, D. L. (1972). *Journal of Bacteriology* **111**, 287.
Kashket, E. R. and Wilson, T. H. (1973). *Proceedings of the National Academy of Sciences of the United States of America* **70**, 2866.
Kayushiv, L. P. and Skulachev, V. P. (1974). *Federation of European Biochemical Societies Letters* **39**, 39.
Kepes, A. (1971). *Journal of Membrane Biology* **4**, 87.

Klein, K., Steinberg, R. Fiethen, B. and Overath, P. (1971). *European Journal of Biochemistry* **19**, 442.
Klein, W. L. and Boyer, P. D. (1972). *Journal of Biological Chemistry* **247**, 7257.
Koch, A. L. (1964). *Biochimica et Biophysica Acta* **79**, 177.
Koch, A. L. (1971a). *Journal of Molecular Biology* **59**, 447.
Koch, A. L. (1971b). *Biochimica et Biophysica Acta* **249**, 197.
Komor, E. (1973). *Federation of European Biochemical Societies Letters* **38**, 16.
Komor, E., Haass, D. and Tanner, W. (1972). *Biochimica et Biophysica Acta* **266**, 649.
Komor, E., Haass, D., Komor, B. and Tanner, W. (1973a). *European Journal of Biochemistry* **39**, 193.
Komor, E., Loos, E. and Tanner, W. (1973b). *Journal of Membrane Biology* **12**, 89.
Konings, W. N. and Kaback, H. R. (1973). *Proceedings of the National Academy of Sciences of the United States of America* **70**, 3376.
Kornberg, H. L. (1973a). *Proceedings of the Royal Society, London, Series B* **183**, 105.
Kornberg, H. L. (1973b). *Symposium of the Society of Experimental Biology* **27**, 175.
Krulwich, T. A., Sobel, M. E. and Wolfson, E. B. (1973). *Biochemical and Biophysical Research Communications* **53**, 258.
Kundig, W., Ghosh, S. and Roseman, S. (1964). *Proceedings of the National Academy of Sciences of the United States of America* **52**, 1067.
Kundig, W. and Roseman, S. (1971a). *Journal of Biological Chemistry* **246**, 1393.
Kundig, W. and Roseman, S. (1971b). *Journal of Biological Chemistry* **246**, 1407.
Kusch, M. and Wilson, T. H. (1973). *Biochimica et Biophysica Acta* **311**, 109.
Lawford, H. G. and Haddock, B. A. (1973). *Biochemical Journal* **136**, 217.
Liberman, E. A. and Skulachev, V. P. (1970). *Biochimica et Biophysica Acta* **216**, 30.
Lipmann, F. (1941). *Advances in Enzymology* **1**, 99.
Lombardi, F. J., Reeves, J. P. and Kaback, H. R. (1973). *Journal of Biological Chemistry* **248**, 3551.
McGinnis, J. F. and Paigen, K. (1969). *Journal of Bacteriology* **100**, 902.
McGinnis, J. F. and Paigen, K. (1973). *Journal of Bacteriology* **114**, 885.
Makman, R. S. and Sutherland, E. W. (1965). *Journal of Biological Chemistry* **240**, 1309.
Mark, C. G. and Romano, A. H. (1971). *Biochimica et Biophysica Acta* **249**, 216.
Meyer, D. J. and Jones, C. W. (1973). *European Journal of Biochemistry* **36**, 144.
Midgley, M. and Dawes, E. A. (1973). *Biochemical Journal* **132**, 141.
Miles, R. J. and Pirt, S. J. (1973). *Journal of General Microbiology* **76**, 305.
Mitchell, P. (1961a). *Nature, London* **191**, 144.
Mitchell, P. (1961b). "Membrane Transport and Metabolism", (A. Kleinzeller and A. Kotyk, eds.), p. 22, Academic Press, New York.
Mitchell, P. (1963). *Biochemical Society Symposium* **22**, 142.
Mitchell, P. (1966). *Biological Reviews, Cambridge* **41**, 445.
Mitchell, P. (1970). *Symposium of the Society for General Microbiology* **20**, 121.
Mitchell, P. (1972). *Journal of Bioenergetics* **3**, 5.
Mitchell, P. (1973a). *Journal of Bioenergetics* **4**, 63.
Mitchell, P. (1973b). "Mechanisms in Bioenergetics", (F. G. Azzone, L. Ernster, S. Papa, E. Quagliariello and N. Siliprandi, eds.), p. 177. Academic Press, New York.
Mitchell, P. (1973c). *Federation of European Biochemical Societies Letters* **33**, 267.
Mitchell, P. and Moyle, J. (1967). *Biochemical Journal* **104**, 588.

Mukkada, A. J., Long, G. L. and Romano, A. H. (1973). *Biochemical Journal* **132**, 155.

Nieuwenhuis, F. J. R. M., Kanner, B. I., Gutnick, D. L., Postma, P. W. and Van Dam, K. (1973). *Biochimica et Biophysica Acta* **325**, 62.

Niven, D. F. and Hamilton, W. A. (1972). *Biochemical Journal* **127**, 58P.

Niven, D. F., Jeacocke, R. E. and Hamilton, W. A. (1973). *Federation of European Biochemical Societies Letters* **29**, 248.

Niven, D. F. and Hamilton, W. A. (1973). *Federation of European Biochemical Societies Letters* **37**, 244.

Niven, D. F. and Hamilton, W. A. (1974). *European Journal of Biochemistry* **44**, 517.

Paigen, K. and Williams, B. (1970). *Advances in Microbial Physiology* **4**, 251.

Parnes, J. R. and Boos, W. (1973a). *Journal of Biological Chemistry* **248**, 4429.

Parnes, J. R. and Boos, W. (1973b). *Journal of Biological Chemistry* **248**, 4436.

Pavlasova, E. and Harold, F. M. (1969). *Journal of Bacteriology* **98**, 198.

Postma, P. W., Cools, A. and Van Dam, K. (1973). *Biochimica et Biophysica Acta* **318**, 91.

Pressman, B. C. (1965). *Proceedings of the National Academy of Sciences of the United States of America* **53**, 1076.

Prezioso, G., Hong, J.-S., Kerwar, G. K. and Kaback, H. R. (1973). *Archives of Biochemistry and Biophysics* **154**, 575.

Ratledge, C. and Marshall, B. J. (1972). *Biochimica et Biophysica Acta* **279**, 58.

Reeves, J. P. (1971). *Biochemical and Biophysical Research Communications* **45**, 931.

Reid, R. A. (1970). *Biochemical Journal* **116**, 12P.

Robin, A. and Kepes, A. (1973). *Federation of Biochemical Societies Letters* **36**, 133.

Romano, A. H., Eberhard, S. J., Dingle, S. L. and McDowell, T. D. (1970). *Journal of Bacteriology* **104**, 808.

Roseman, S. (1972). "The Molecular Basis of Biological Transport", (J. F. Woessner and F. Huijing, eds.), p. 181. Academic Press, New York.

Rosen, B. P. (1973a). *Biochemical and Biophysical Research Communications* **53**, 1289.

Rosen, B. P. (1973b). *Journal of Bacteriology* **116**, 1124.

Rottenberg, H., Grunwald, T. and Avron, M. (1972). *European Journal of Biochemistry* **25**, 54.

Saier, M. H. and Roseman, S. (1972). *Journal of Biological Chemistry* **247**, 972.

Scarborough, G. A. (1970). *Journal of Biological Chemistry* **245**, 3985.

Scarborough, G. A., Rumley, N. K. and Kennedy, D. P. (1968). *Proceedings of the National Academy of Sciences of the United States of America* **60**, 951.

Schacter, D. and Mindlin, A. J. (1969). *Journal of Biological Chemistry* **244**, 1808.

Schairer, H. U. and Gruber, D. (1973). *European Journal of Biochemistry* **37**, 282.

Schairer, H. U. and Haddock, B. A. (1972). *Biochemical and Biophysical Research Communications* **48**, 544.

Scholes, P. and Mitchell, P. (1970). *Journal of Bioenergetics* **1**, 309.

Schultz, S. G. and Curran, P. F. (1970). *Physiological Reviews* **50**, 637.

Seaston, A., Inkson, C. and Eddy, A. A. (1973). *Biochemical Journal* **134**, 1031.

Shiio, I., Miyajima, R. and Kashima, N. (1973). *Journal of Biochemistry, Tokyo* **73**, 1185.

Short, S. A., White, D. C. and Kaback, H. R. (1972). *Journal of Biological Chemistry* **247**, 298.

Simoni, R. D. and Shallenberger, M. K. (1972). *Proceedings of the National Academy of Sciences of the United States of America* **69**, 2663.
Simoni, R. D., Nakagawa, T., Hays, J. B. and Roseman, S. (1973a). *Journal of Biological Chemistry* **248**, 932.
Simoni, R. D., Hays, J. B. Nakagawa, T. and Roseman, S. (1973b). *Journal of Biological Chemistry* **248**, 957.
Simoni, R. D. and Roseman, S. (1973). *Journal of Biological Chemistry* **248**, 966.
Skou, J. C. (1972). *Federation of European Biochemical Societies Symposium* **28**, 339.
Skulachev, V. P. (1971). *Current Topics in Bioenergetics* **4**, 127.
Skulachev, V. P. (1972a). *Journal of Bioenergetics* **3**, 25.
Skulachev, V. P. (1972b). *Federation of European Biochemical Societies Symposium* **28**, 371.
Sprott, G. D. and MacLeod, R. A. (1972). *Biochemical and Biophysical Research Communications* **47**, 838.
Stock, J. and Roseman, S. (1971). *Biochemical and Biophysical Research Communications* **44**, 132.
Storelli, C., Vogeli, H. and Semenza, G. (1972). *Federation of European Biochemical Societies Letters* **24**, 287.
Tikhonova, G. V. (1974). *Biochemical Society Transactions* **2**, 466.
Thompson, J. and MacLeod, R. A. (1971). *Journal of Biological Chemistry* **246**, 4066.
Thompson, J. and MacLeod, R. A. (1973). *Journal of Biological Chemistry* **248**, 7106.
Van Steveninck, J. (1970). *Biochimica et Biophysica Acta* **203**, 376.
Van Thienen, G. and Postma, P. W. (1973). *Biochimica et Biophysica Acta* **323**, 429.
Weiner, J. H., Furlong, C. E. and Heppel, L. A. (1971). *Archives of Biochemistry and Biophysics* **142**, 715.
Weiner, J. H. and Heppel, L. A. (1971). *Journal of Biological Chemistry* **246**, 6933.
West, I. C. (1970). *Biochemical and Biophysical Research Communications* **41**, 655.
West, I. C. and Mitchell, P. (1972). *Journal of Bioenergetics* **3**, 445.
West, I. C. and Mitchell, P. (1973). *Biochemical Journal* **132**, 587.
West, I. C. and Wilson, T. H. (1973). *Biochemical and Biophysical Research Communications* **50**, 551.
Willecke, K., Gries, E.-M. and Oehr, P. (1973). *Journal of Biological Chemistry* **248**, 807.
Wilson, T. H. and Kusch, M. (1972). *Biochimica et Biophysica Acta* **255**, 786.
Winkler, H. H. (1973). *Journal of Bacteriology* **116**, 203.
Winkler, H. H. and Wilson, T. H. (1966). *Journal of Biological Chemistry* **241**, 2200.
Winkler, H. H. and Wilson, T. H. (1967). *Biochimica et Biophysica Acta* **135**, 1030.
Wong, P. T. S. and MacLennan, D. H. (1973). *Canadian Journal of Biochemistry* **51**, 538.
Yamamoto, T. H., Mevel-Ninio, M. and Valentine, R. C. (1973). *Biochimica et Biophysica Acta* **314**, 267.

Note added in proof.

The *E. coli* ATPase mutant DL-54 (see page 40) has now been further studied by K. H. Altendorf, F. M. Harold and R. D. Simoni (*Journal of Biological Chemistry* (1974) **249**, 4587). They have shown that vesicles show an increased permeability to protons and that this effect can be reversed by treatment with DCCD.

Physiology of Colicin Action

I. B. Holland

Department of Genetics, University of Leicester, Leicester LE1 7RH, England

I. Introduction

Although colicins have been recognized and studied intermittently for more than 40 years, it is only recently that definitive information concerning their nature and mechanism of action has begun to emerge. For so long dwarfed by studies of the bacterial viruses, with which they share an ecological niche, colicins and their extrachromosomal determinants are now being actively studied in relation to bacterial physiology and DNA replication, respectively. In particular, colicins are seen as probes of the structural and functional organization of the bacterial membrane, yielding results with important implications for the nature of any biological process in which high molecular-weight proteins impinge upon the surface of a responsive cell. In consequence, particular emphasis will be placed in this review on that aspect of colicin studies which involves penetration of the cell envelope.

Since the interaction of colicins with bacterial cells is so intimately concerned with the structure of the surface envelope, it is desirable to commence with a brief outline of the basic organization of this structure. This description relies heavily upon the studies of several workers; in particular, De Petris (1967), Braun and Rehn (1969), Braun *et al.* (1973), and Schnaitman (1971a, b), and reference should be made to the original papers for detailed information. The outermost region of the Gram-negative cell envelope is usually composed of lipopolysaccharide units whose terminal lipid base is embedded in a lipoprotein layer and which together constitute the outer membrane. This membrane appears to completely cover the underlying peptidoglycan network which constitutes the rigid skeleton of the envelope (see Fig. 6, p. 129). The outer membrane is apparently permeable to many small molecules, but acts as a protective barrier to several antibiotics and possibly to other potentially harmful compounds (see Normark and Westling, 1971). Treatment of cells with ethylenediamine tetra-acetic acid (EDTA) leads to the disintegration of the outer membrane (Leive, 1968) although in electron-microscope sections it has a typical unit membrane appearance. This suggests that the outer membrane may not be a fully continuous lipoprotein layer, but may frequently be completely penetrated by lipopolysaccharide units, perhaps held in stable association with the lipoprotein by Mg^{2+} ions. At the internal face, the outer membrane is covalently linked to the underlying peptidoglycan by molecules of a specific lipoprotein distributed every 100 Å, or so, throughout the surface. In consequence of this tight coupling, the outer membrane and peptidoglycan layer are frequently considered together as constituting the "cell wall". Beneath the cell wall, and largely separated from it by the periplasmic space, is the cytoplasmic membrane, the main permeability barrier of the cell and the site of numerous cellular functions. Studies by Bayer (1968a, b) indicate that the inner, or cytoplasmic, membrane is in contact with the peptidoglycan, and perhaps even with the outer membrane layer itself, by means of 200–400 tubular structures per cell; these appear in electron micrographs as continuations of the cytoplasmic membrane when cells are briefly plasmolysed. Many different bacteriophage binding sites have been localized specifically in the region of the tubules, and entry of phage DNA may be accomplished via these surface channels. Many of the obstacles to the understanding of the mechanism of colicin penetration of the cell surface would be removed if it could be shown that these channels are also involved in colicin action.

It is now well established that colicins are antibacterial agents, protein in nature, whose production is determined by structural genes exclusively localized on extrachromosomal DNA molecules called Col factors. Col factors are found in the majority of naturally occurring strains of the Enterobacteriaceae, although no satisfactory account has yet been advanced to explain the significance of this. The activity spectra of colicins is usually limited to bacterial species closely related to the colicin-producing strains, and extensive studies by Fredericq and his collaborators have established that the basis of this specificity is determined by receptors, analogous to bacteriophage receptors, present in the surface layers of sensitive bacteria. Fredericq's studies also demonstrated that a great variety of colicins exist which may be distinguished on the basis of their different activity spectra. Much of this early work was reviewed previously by Fredericq (1957), and more recently by Reeves (1973), and will not be considered further here. This review will, in fact, concentrate upon the major recent developments in the study of colicin action which have occurred since the reviews of Reeves (1965b) and Nomura (1967). Firstly, however, two earlier sets of studies concerning the biochemistry of colicin action will be considered briefly, as these constitute the basis of all our current knowledge. Jacob *et al.* (1952) demonstrated that colicin E1 promptly arrested the growth of sensitive cells, although respiration proceeded at the pretreatment rate. More significant, perhaps, was the finding that colicin E1 killing, like bacteriophage killing, was a one hit process, indicating that a single "adsorbed particle" could kill a sensitive cell without cooperation of other "particles". Single-hit killing by other colicins has subsequently been confirmed by many workers and Reeves (1973) has reviewed their findings in detail. As might be anticipated, although a single hit is sufficient to kill a bacterial cell, it is frequently observed that many molecules of colicin have to be adsorbed before a lethal hit occurs. In other words, many adsorbed colicin molecules may never initiate any biochemical change. Thus, although the probability of a single colicin molecule promoting a single hit may be as low as one in a thousand under some conditions, it may approach unity under other conditions, providing direct evidence that a single colicin molecule can be sufficient to cause cell death. This feature is extremely important and must be fully taken into account in considering any model of the molecular basis of colicin action.

In the first of a continuing series of experiments, Nomura (1963) demonstrated that different groups of colicins could be readily distinguished on the basis of their early effects upon macromolecular synthesis in treated cells. Colicins K and E1 were shown to arrest cell growth completely, whilst colicins E2 and E3 produced specific effects upon DNA metabolism and protein synthesis, respectively. Moreover in this, and in an earlier study, Nomura and Nakamura (1962), made the important observation that the viability of cells, apparently killed by colicin K, could be restored by treatment with trypsin. This finding provided strong evidence that colicin K promotes its lethal effect from its site on the cell surface, where it is still accessible to trypsin. Moreover, this result clearly demonstrated that the action of colicin K, at least under the experimental conditions employed, is reversible. On the basis of his findings, Nomura (1964) formulated a model of colicin action which in general outline, if not in detail, still describes the stepwise interaction of all varieties of colicins with sensitive bacteria.

The model defines an initial *receptor* site in the cell surface, and a *biochemical target* specific for each colicin whose modification, as a result of colicin action, may itself be lethal or lead to the disruption of a *killing target*. The model specifically proposes that colicin action upon the biochemical target is indirect, and depends upon the mediation of a specific *transmission* system which connects colicin and target along the membrane. As will be discussed at length in the course of this review, colicin action is still seen as a two-stage process; binding to the surface receptor, followed, with a certain probability, by a second step which results in the alteration of the biochemical target and cell death. However, this second step appears likely to involve direct interaction between colicin and target without the intervention of a transmission step. Even so, the mechanism required to bring together a colicin molecule, which initially binds to the outer membrane, and target sites, which are located either in the inner membrane or in the cell interior, is still not clear. This modified view of colicin action implies that colicins of the E1 type, although penetrating the outer membrane, may remain largely extracellular whilst acting upon targets in the cytoplasmic membrane. Colicin E3, in contrast, must also penetrate the inner membrane (at least partially, if not wholly) in order to interact with its ribosomal target. How penetration of the envelope is achieved is now the central problem of colicin action. In consequence, there are two aspects of colicin studies which are of fundamental importance: (i) elucidation of the structure of a protein molecule, which like a prototype bacteriophage is uniquely programmed to penetrate the bacterial envelope before inactivating specific functional sites, and (ii) use of this probe to reveal, during its passage, both the functional organization of the cell membrane and the workings of the target sites themselves.

Allusion has already been made to the superficial similarity between colicins and bacteriophages with regard to the penetration of sensitive cells (see also Fields (1969) for a more detailed comparative analysis of the killing effects of colicin E1 and phage), and to the localization of colicin genes on extrachromosomal plasmids which, in many respects, resemble temperate bacteriophage genomes. The analogy between bacteriophage and colicins may be drawn closer since colicinogenic bacteria, like lysogenic bacteria, are immune to the homologous colicin or virus particle, respectively (Fredericq, 1957; 1958; Nomura, 1963). Furthermore, immunity, at least in the case of colicin E3, appears to be expressed through the direct interaction of colicin with an immunity protein (Bowman *et al.*, 1973), reminiscent of the interaction between prophage repressor and superinfecting viral genome.

Considerable progress has been made in the last few years in the isolation and characterization of various colicins and their specific receptor complexes. Studies suggesting that colicins are probably highly asymmetric and elongated molecules, in solution, will therefore be reviewed first. The possible mechanisms which may lead to the formation of an active colicin target complex (Complex II), in contrast to the apparently inert colicin–receptor complex (Complex I), will then be considered. The many factors, both physiological and genetical, known to affect this transition will also be reviewed, followed by an examination of the primary biochemical changes induced by the three major groups of colicin (E2, E3 and the E1 types). This aspect will include a survey of studies carried out both *in vivo* with whole cells and *in vitro* with subcellular systems. Studies of the latter kind have, in the case of colicin E3, led to the fundamental finding that this colicin is able to promote by direct interaction the modification of its biochemical target, the 30S ribosomal subunit. Finally, the genetics and physiology of those colicin-tolerant mutants that have so far been isolated will be discussed. These mutants, which are easily obtained, appear to adsorb normal amounts of colicin and are therefore presumably blocked in some step leading to Complex II formation. However, the majority of these mutations are relatively non-specific, and the analysis of their properties has not yet revealed much precise information concerning the mechanism of Complex II formation. On the other hand, such mutants, including several which are conditional lethal mutants, are proving to be extremely valuable in the study of the synthesis of membrane proteins and of the organization of these proteins within the membrane.

Reeves (1965b) has previously pointed out the need to identify precisely specific colicins by a notation which includes the strain of origin of the corresponding Col$^+$ factor. This recommendation has considerable merit but since most workers in the field now use standard colicin E1 (ML), E2 (P9), E3 (CA38), Ia (CA53), Ib (P9) and K (235)-producing strains, this form of notation will be omitted in the text for the sake of convenience.

II. Colicin Production and Col Factors

The synthesis of colicins is determined by genes encoded on bacterial plasmids. Such plasmids, or colicinogenic factors, are widely distributed amongst the Enterobacteriaceae and vary considerably in size and genetic complexity. Small plasmids of the Col E1 type contain an amount of DNA equivalent to between five and ten cistrons, whilst plasmids of the Col I type may be up to 12-times larger. Col I factors correspondingly carry many more genetic determinants which function, for example, to promote conjugal transfer of the Col factor, and any other small plasmids present in the same cell. In addition, Col I factors mobilize the transfer of chromosomal genes under certain conditions. For an extensive review of the properties of Col factors and other plasmids the reader is referred to Clowes (1972) and Willetts (1972).

Colicin synthesis is largely, if not completely, repressed in the majority of Col E2 cells, but upon induction most cells produce colicin (Ozeki *et al.*, 1959). Induction may be achieved by treatment of colicinogenic cultures with ultraviolet radiation or mitomycin C (Reeves, 1963; Herschmann and Helinski, 1967a, b), by temperature shock (Senior, 1968) or chloramphenicol (Ben-Gurion, 1970; Kennedy, 1971). In some cases, induction is accompanied by extensive replication of the Col$^+$ factor (Amati, 1964; De Witt and Helinski, 1965) but this is not essential for colicin production

(Kennedy *et al.*, 1973). Similarly, although induced colicin synthesis is often accompanied by death of the cell (Ozeki *et al.*, 1959), exceptions to this have now been observed (Hausmann and Clowes, 1971; Herschmann and Helinski, 1967b). Furthermore, under conditions where induction and release of colicin E2 does result in cell death, this is not apparently due to the bacteriocidal effects of the colicin itself (Margolin and Kennedy, 1973). The expression of colicin genes is not, therefore, necessarily a lethal biosynthesis, as has been frequently suggested.

III. Nature of Colicins

Induced Col^+ cultures synthesize up to 3×10^5 molecules of colicin per cell (see Isaacson and Konisky, 1973), which is equivalent to several per cent of the total cellular protein. Colicin synthesis is not always accompanied by lysis of the bacteria (Schwartz and Helinski, 1971) and, in this case, the method of release from the cells is unknown. After induction, and providing the producing cells carry colicin receptors, large amounts of colicin remain bound to the cell surface (Isaacson and Konisky, 1973); consequently purification is often greatly facilitated by salt extraction of colicin from the surface of induced cells, without prior disruption of the bacteria (Herschmann and Helinski, 1967a). Standard enzyme purification procedures may then be used to obtain homogeneous preparations of colicin. In this way highly purified preparations of colicin K (Jesaitis, 1970; Goeble, 1973), colicin E1 (Schwartz and Helinski, 1971), colicins E2 and E3 (Herschmann and Helinski, 1967a), colicin D (Timmis, 1972) and colicins Ia and Ib (Konisky and Richards, 1970) have been obtained and shown to be simple proteins. Several colicins, in particular, V, A and K (Hutton and Goebel, 1962; Barry *et al.*, 1965; Goeble, 1962) have, however, been isolated in a complex form with the colicin protein tightly bound to the bacterial O-antigen. This is found when, for example, colicin K is extracted and purified from the culture filtrate of non-induced cells, where it is present in low levels. The reason for these different forms of colicin K are not known, but the bacteriocidal and immunological properties of the two forms appear identical (Jesaitis, 1970).

A. Chemistry

Colicin K, which is negatively charged at pH 8·3, has a molecular weight of 45,000 daltons and appears to lack cysteine residues (Jesaitis, 1970). Colicin E1, in contrast, is a basic protein with a molecular weight of 56,000 daltons and, like colicins E2 and E3, appears to contain a single cysteine residue (Schwartz and Helinski, 1971). The total amino-acid composition of colicin E1 is, however, quite distinct from that of the other

E colicins and resembles more those of colicins I and K. Nevertheless, colicin E1 does not appear to have any antigenic homology with these colicins (Konisky, 1973). Colicins E2 and E3 are both relatively neutral proteins, with molecular weights close to 60,000 daltons. The amino-acid compositions of colicins E2 and E3 are very similar, and immunochemical analysis suggests that the two colicins probably have an identical region of their polypeptide in common (Herschmann and Helinski, 1967a). This region probably constitutes that part of the molecule which interacts with the bacterial receptor, since this receptor appears to be the same for colicins E2 and E3. Colicins Ia and Ib, whose activities may only be distinguished on the basis of the immunity of the homologous Col I producing strain, are basic proteins with molecular weights close to 80,000 daltons. The I colicins also appear to lack cysteine, and the comparison of their amino-acid fingerprints, along with immunochemical analyses, indicate that the primary structure of the two colicins is very similar (Konisky, 1972; Isaacson and Konisky, 1972). Colicin D is composed of a single polypeptide chain with a molecular weight of 92,000 daltons and, in contrast to other colicins, it contains six cysteine residues. In amino-acid composition, this colicin somewhat resembles colicins E2 and E3; nevertheless, no antigenic homology is detectable between colicin D and these two colicins (Timmis, 1972).

Many colicins, including E2 (Herschmann and Helinski, 1967a), E3 (Glick *et al.*, 1972), K (Jesaitis, 1970) and colicin I (Konisky and Richards, 1970), exist in purified preparations as multiple forms which differ in net charge. From a study of the properties of two biologically active forms of colicin E2, Herschmann and Helinski (1967a) concluded that, although the two species differed in net charge, their primary structure was probably identical. Consequently, it was suggested that different forms of colicin E2 are "conformers", differing only in tertiary structure.

Purified colicins, which are stable for many months if kept lyophilized, have specific activities of 10^3 arbitrary units/μg protein (Holland, 1968; Ringrose, 1970). Sensitive bacteria are inhibited at concentrations of 0·1 to 0·6 ng of colicin/ml (Ringrose, 1970; Jesaitis, 1970), which corresponds to molecular multiplicities as low as 25–50 molecules per cell (Holland and Holland, 1970). Since colicins irreversibly adsorb to bacterial cells, and display single-hit killing kinetics, colicin concentrations can be calculated from the Poisson distribution: $N/No = e^{-m}$, where N/No is the surviving fraction of bacteria after a standard time, and m is the colicin multiplicity per cell in killing units (KU). Although single-hit killing is always observed with colicin preparations, the probability of a single adsorbed molecule achieving a lethal hit appears to vary widely, depending on the specific activity of the colicin preparation, the cultural conditions, and the genotype of the sensitive bacteria. Consequently, examination of the literature reveals a wide range of quoted titres for purified colicins, including 3×10^9–2×10^{10} KU/μg protein for colicin E1 (Feingold, 1970; Seto *et al.*, 1973), 3×10^9–10^{11} KU/μg protein for colicins E2 and E3 (Beppu and Arima, 1967; Maeda and Nomura, 1966) and 2–4 $\times 10^{11}$ KU/μg protein for colicin I (calculated from Konisky and Cowell, 1972). Since the molecular weights of the colicins are known, the number of colicin molecules per killing unit can be calculated and this varies from about 50 to 3000 on the basis of the above data, with 50–100 molecules per KU being the most frequently reported values. As shown

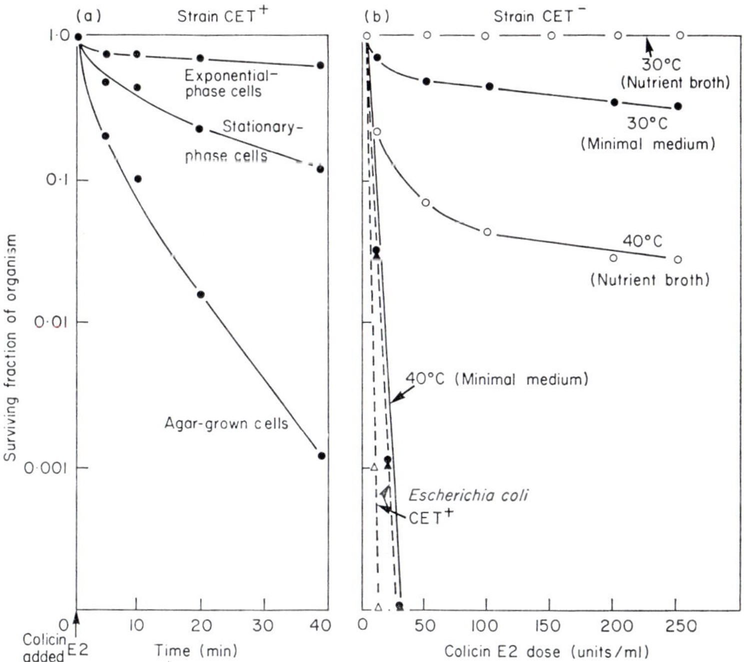

FIG. 1. Sensitivity of *Escherichia coli* strains to colicin E2 under various conditions. (a) Nutrient-broth suspensions (10^8 bacteria/ml) from liquid cultures in exponential- or stationary-phase of growth, and from nutrient-agar plates, were each treated at 37°C with two units of colicin E2/ml, and viable counts determined at intervals. In each experiment, from the number of cells killed, the number of killing units (KU) was calculated as indicated in the text (p. 62). Assuming a molecular weight of 60,000 daltons for colicin E2, the number of molecules/KU was also calculated and found to be 800, 200 and 55 for exponential-phase, stationary-phase, and agar-grown cells, respectively. (b) Suspensions of ASH 101, a mutant specifically tolerant to colicin E2, grown in either M9-minimal salts medium, or nutrient broth, were treated with different concentrations of colicin E2 for 30 minutes at either 37°C or 40°C, and finally plated out to determine the surviving fraction. For comparison, the sensitivity of the Cet+ parent strain under each régime was included.

in Fig. 1a, the probability of killing, per molecule of colicin E2, varies between 0·12%–2%, depending upon the cultural conditions and, presumably, upon the physiological state of the bacteria. In contrast, Mitusi and Mizuno (1969) reported that the activity of purified colicin E2 was markedly stabilized by bovine serum albumin such that the specific activity increased from $1{\cdot}8 \times 10^9$ KU/μg to 2×10^{12} KU/μg protein in its presence. In the latter case, one killing unit corresponds to only five molecules of colicin E2, and the probability of one molecule achieving a lethal hit is therefore about 20%. Reeves (1965a) has described preparations of colicin E2 with similar high specific activities. These findings are extremely important as they provide direct support for the conclusion, drawn from bacterial survival studies, that one, or a very few, molecules are capable of killing a single cell. Any mechanism proposed to explain, for example, the inhibition of cell division, or the promotion of DNA degradation, by colicin E2 must take this into account.

B. Molecular Architecture

One of the most intriguing aspects of recent studies of colicins has been the accumulation of information concerning their molecular shape. In the case of colicins K, Ia, Ib, and the three E colicins, calculation of molecular weights based upon sedimentation coefficients, and assuming that colicin is a spherical unhydrated molecule, yields values considerably lower than estimates from either gel filtration or from sodium dodecylsulphate (SDS)-polyacrylamide gel electrophoresis. Furthermore, sedimentation equilibrium studies, or molecular weight determinations based upon a combination of sedimentation coefficient, partial specific volume and diffusion coefficient, all confirm the higher molecular weight values. Finally, calculation of the average frictional coefficient (f), from diffusion and sedimentation coefficients, reveals that the frictional ratio, f/f_0 (where f_0 is the frictional coefficient of the corresponding unhydrated sphere) is about 1·8 for colicin I, 2·0 for colicin E1, 1·7 for colicin K, 1·45 for colicin E2, 1·4 for colicin E3 and 1·58 for colicin D (see summary of data; Konisky, 1973). Therefore, the anomalous centrifugation data of these colicins, and the unusually high viscosity of colicin K solutions (Jesaitis, 1970) can in all probability be ascribed to the unusual asymmetry of the molecules. The alternative explanation, that colicins are spherical molecules associated with enormously large amounts of bound water, appears, as reasoned by Konisky (1973), to be extremely unlikely. Assuming average values of 0·3 gram water per gram of protein, Konisky has also estimated the axial ratios of various colicins from the frictional ratio. For oblate, or plate-shaped molecules, the calculated values range from 8·8 for colicin E3 to 20 for colicin E1, indicating that the minor axis for the latter may be 10–20 Å, perhaps only a single α-helix in thickness. Additional evidence for the elongated or asymmetric form of colicin molecules has been provided by Konisky (1973), who has demonstrated that colicin Ia molecules appear in electron micrographs as large rounded particles with diameters of about 200 Å, which is quite consistent with an oblate form. Finally, although other configurations, including prolate or even relatively spherical forms cannot yet be ruled out, Konisky has argued (see Fisher, 1964) that, in view of their high content of polar amino-acids, colicins must of necessity be elongated molecules in order to maximize the interaction of these amino acids with the aqueous environment (Konisky, 1974, and personal communication).

C. Function and Structure

On the basis of the above evidence, it seems clear that colicin molecules have some form of elongated structure in solution and the implications of this for their possible mode of action will be briefly considered. Firstly, with major axes of 140 Å to over 300 Å for colicins I and E1, respectively, a single colicin molecule, lying vertical to the plane of the cell surface, should in principle be capable of spanning the 200 Å distance through the cell envelope in order to interact with the inner membrane, provided that the outer layer of the envelope can somehow be breached. Whether the less elongated colicin E2 and E3 molecules can also do this is more problematical (see Fig. 6, p. 129). Nevertheless, it seems very likely that the inherent asymmetry of all colicins, including colicins E2 and E3, greatly facilitates their penetration of the cell surface, providing the molecules are aligned with their major axis perpendicular to the plane of the outer membrane. How interaction with the inner membrane is finally achieved is unknown, but one may speculate that edgeways insertion into the inner membrane of a narrow, presumably hydrophilic, cage could provide a highly specific drain for cellular cations, which, as discussed below, could explain many of the properties of colicins of the E1 type.

Before concluding this section, attention should be drawn to the remarkable multiplicity of functions which appear to be built into a single colicin molecule. First of all, the molecules possess receptor recognition sites unique for most colicins, although identical for colicins E2 and E3. Secondly, each colicin promotes, probably by direct action, a specific biochemical change at the target site. Thirdly, genetic studies demonstrate that even colicins related either in structure or function respond differently to the genetic blocks present in colicin-tolerant mutants in which some step in the second stage of colicin action is inactivated. The possibility of colicin recognition of a specific bacterial "uptake" system, which precedes interaction with the target, cannot therefore be ruled out. Colicin E3 also appears to interact with a specific immunity protein, and this seems likely to be a general phenomenon affecting other colicins. Although not all of these properties need necessarily be determined by separate and specific regions of the colicin molecule, elucidation of the functional organization of these curious polypeptides promises to be extremely interesting.

Finally, since a most puzzling aspect of colicins is their ability to negotiate impermeable lipid-protein barriers, the possibility remains that these molecules carry a built-in phospholipase activity which is used to effect penetration. Such activities should be sought, as discussed in Section X (p. 107).

IV. Colicin Receptors

Much of the early work on colicin action was concerned with the idea that colicins, like bacteriophages, require the mediation of specific surface receptors in order to act upon sensitive bacteria. The classification of different colicins, largely worked out by Fredericq, depended upon the isolation of a whole range of mutants, each resistant to, and presumably unable to adsorb, a specific class of colicins. This work, whose basis has now been fully confirmed by the biochemical studies of colicin binding, discussed below, has been extensively reviewed by Fredericq (1957), and also by Nomura (1967) and Reeves (1965b, 1973), and will not be repeated here.

A. Localization

Little was known concerning the nature of colicin receptors until Maeda and Nomura (1966) made the first quantitative measurements of colicin adsorption using radioactive colicin E2. Sensitive bacteria were shown to bind between 2 and 3000 molecules of colicin per cell, more than 80% of which was recovered in the envelope fraction after subsequent disintegration of the cell. This study also demonstrated that colicins E2 and E3 competed for binding sites indicating that the colicin E2 and E3 receptors are probably identical (see also Šmarda and Adler, 1971). This view was supported by the later findings that the loss of the colicin E2 receptor, by mutation, was invariably accompanied by loss of the colicin E3 receptor, and *vice versa* (Hill and Holland, 1967).

Two techniques have been recently described which have greatly encouraged attempts to localize colicin receptors in specific layers of the cell envelope. The most effective technique, developed by Osborn *et al.* (1972a), involves the separation, by equilibrium centrifugation in sucrose, of the outer membrane (which, by virtue of its lipopolysaccharide content, has a density of 1·22 g/cc) from the inner membrane (which has a density of 1·14–1·16 g/cc). A rather less specific, but more convenient, technique developed by Schnaitman (1971b) involves the treatment of envelopes with Triton X-100 which solubilizes the inner membrane and leaves the outer membrane-wall complex relatively intact. Using this latter technique, Sabet and Schnaitman (1971, 1973) have shown that an envelope component, which neutralizes colicins E2 and E3, and which is present in sensitive but not in resistant strains, is localized in the outer membrane. The presence of specific receptors, primarily in the outer membrane rather than in the cytoplasmic membrane, has also been reported for colicins Ia and Ib (Konisky *et al.*, 1973) for colicin K (Weltzien and Jesaitis, 1971) for colicin E1 (Di Masi *et al.*, 1973), and for colicin M (Braun and Wolff, 1973). Some earlier studies, based upon the colicin sensitivity of bacterial L-forms (Šmarda and Taubeneck, 1968), or disrupted sphaeroplasts (Bhattacharyya *et al.*, 1970), suggested that colicin receptors might, on the contrary, reside in the inner membrane. It appears possible, however, that significant amounts of residual outer membrane, containing colicin receptor sites, were overlooked in these studies. Moreover, actual adsorption to the inner membrane, or even to the L-forms or vesicles, was not clearly tested in these studies. As discussed below, the "sensitivity" of a membrane vesicle to the action of colicins of the E1 type need not have an obligatory requirement for colicin receptors. The weight of evidence therefore appears to indicate that in the majority of cases, at least, colicin receptors are localized in the outer membrane of the cell envelope.

Since sphaeroplasts, which lack the majority of the peptidoglycan layer, still bind normal amounts of colicins E2, E3 and K (Nomura and Maeda, 1965), and since the colicin K receptor can be solubilized from sphaeroplasts (Weltzien and Jesaitis, 1971), it may be safely concluded that the peptidoglycan layer does not constitute a major part of colicin receptors. At least one report (Chang and Hager, 1970) has suggested that bacterial lipopolysaccharide can neutralize colicin E2 activity, but this effect appears to be non-specific. The surface receptor for colicin K (Weltzien and Jesaitis, 1971) is in fact a poor antigen, suggesting that lipopolysaccharide is not a major constituent of this receptor. Finally, neither colicin K (Weltzien and Jesaitis, 1971) nor colicin I (J. Konisky, personal communication) is capable of binding to isolated lipopolysaccharide. In fact, as discussed in the following section, all available evidence suggests that protein is a major constituent of the functional receptors for colicins E, K and I, although the participation of additional non-protein components cannot be excluded.

Although the presence of several types of colicin receptors in the outer membrane now appears firmly established, the lateral and vertical distribution of receptors within this layer has received little attention so far. In contrast to solubilized receptors, those *in situ* are not inactivated by trypsin and, presumably, therefore they are buried to some extent within the outer membrane. Unfortunately, this kind of experiment is not very informative since the degree to which trypsin can penetrate the various layers of the intact bacterial cell envelope is not known. Examination of the binding of fluorescent conjugates of colicin E2 to the surface of sensitive cells has indicated that colicin E2 receptors are uniformly dispersed throughout the surface, but the low resolution of the technique precluded detection of their precise distribution (Samson, 1970). The use of ferritin-labelled (anti-E2) γ-globulin conjugates should provide the necessary resolution, and would settle the important question of whether colicin receptors, like many phage receptors, are clustered around the tubular structures connecting the inner and outer membranes in the cell envelope (Bayer, 1968a, b).

B. Isolation and Properties

With the localization of colicin receptors in the outer membrane, attempts are now being made to purify the receptor and to study quantitatively its interaction with the colicin molecule. The colicin E receptor is of particular interest in this connection since, as discussed in later sections, genetic and other studies with whole cells indicate that, whilst sharing at least one component in common, the complete receptor for colicin E1 is quite distinct from the colicin E2 and E3 receptor, and from the receptor for phage BF23. Sabet and Schnaitman (1973) have carried out an extensive purification of a colicin E3-binding complex which was extracted from isolated outer membrane fractions with Triton X-100 and EDTA. The final product obtained consisted predominantly of a 60,000 dalton hydrophobic protein, and control experiments demonstrated that the anti-colicin activity resided in this major component and not in other minor constituents. As anticipated, the complex was equally effective in neutralizing colicin E2 but had little activity against colicin E1 or the unrelated colicin K. Further experiments showed that this protein was apparently absent from resistant mutants which fail to adsorb colicin E. The receptor complex also contained small amounts of carbohydrate, and functional activity was destroyed by periodate. The active complex may therefore be a glycoprotein. Finally, calculations showed that the receptor fraction was quite a minor component of the intact outer membrane, being present in approximately 200–300 copies. This figure is considerably less than the 2000–3000 colicin E2 binding sites detected by Maeda and Nomura (1966) in whole cells, but the significance of this discrepancy cannot be assessed until the stoicheiometry of colicin binding to the isolated receptor has been established by *in vitro* binding studies.

Since the fixation of colicin molecules by sensitive bacteria is normally a lethal event, the presence of colicin receptors in Nature has long been a puzzle. A possible explanation of this seemingly paradoxical situation has recently emerged from the findings of Di Masi *et al.* (1973) which show that at least one portion of the colicin E receptor complex also constitutes a vitamin B_{12} binding site, a site which is essential for the first stage of uptake of this molecule. Bacterial strains unable to synthesize vitamin B_{12} grow normally when supplied with exogenous vitamin, whilst mutants which also lack the vitamin B_{12} receptor grow poorly. Presumably, therefore, cells producing the vitamin B_{12} receptor have a selective advantage over strains without this receptor, and this could explain the maintenance of the colicin E binding site in Nature if the two receptors are indeed identical, or overlapping. In fact, Di Masi *et al.* (1973) provide considerable evidence that vitamin B_{12} acts as a competitive inhibitor of both colicin E1 and colicin E3 for binding sites in the cell envelope. Furthermore, Kadner and Liggins (1973) have shown that mutants selected for colicin E-resistance simultaneously lose their capacity to bind vitamin B_{12}, and *vice versa*, and that the locus *btu*B (determining vitamin B_{12} binding) maps at the same position on the *E. coli* K12 chromosome as does the *bfe* locus which determines the presence or absence of both the colicin E and phage BF23 binding sites (Buxton, 1971; Jasper *et al.*, 1972). The simplest interpretation of all these results is that the 60,000 dalton protein complex isolated by Sabet and Schnaitman (1973) may be a major constituent of both the vitamin B_{12} binding site (although this has not been directly established) and of all colicin E and phage BF23 receptors. This protein may even be the sole constituent of the colicin E2–E3 binding site, whilst the colicin E1 receptor constitutes a distinct complex apparently containing additional components. The phage BF23 site may well constitute a third variant with its own unique constituents in addition to the 60,000 dalton protein. The isolation of specific vitamin B_{12} and colicin E and phage BF23 receptors, and careful binding studies *in vitro* with all of the homologous and heterologous binder molecules, will be necessary to clarify the precise relationship between all of these receptor complexes.

Colicins Ia and Ib, although distinguishable by the immunity of the producing strain to the homologous colicin (Stocker, 1966), are very similar chemically and they appear to compete for identical binding sites on the surface of sensitive bacteria (Konisky and Cowell, 1972). Using a purified colicin Ia and Ib, radioactively labelled *in vitro* with [^{125}I]-iodine, Konisky *et al.* (1973) have shown that these colicins also bind efficiently to isolated outer membrane fractions, but only poorly to isolated cytoplasmic membranes. This binding is specific since outer membranes isolated from strains resistant to colicin Ia or Ib do not bind either colicin. J. Konisky and C.-T. Liu (personal communication) have achieved a partial purification of the colicin I receptor by extraction of cell envelopes with Triton X-100 and EDTA. The binding activity of the isolated complex, which appears quite hydrophobic, was destroyed by trypsin but was unaffected by periodate or phospholipase, indicating that the major constituent of the colicin I receptor is a protein. Konisky and Liu have also obtained some preliminary data suggesting that the receptor-colicin complex has a highly asymmetric, oblate or prolate structure. Further studies with highly purified receptor complexes will be of great interest.

Konisky has also carried out an extensive quantitative study of the binding of [^{125}I]-iodine to *E. coli* cells (Konisky and Cowell, 1972). At saturation, sensitive bacteria bind an average of 5000 molecules of colicin Ia per cell. An analysis of the efficiency of binding to a constant amount of cells, over a wide range of colicin concentrations, has shown that binding to about half of the receptors was extremely efficient, with an association constant (K) of approximately 1×10^{10} M^{-1} at 37°C. Similar results were obtained which were independent of the degree of iodinization of the colicin. Binding to the remaining receptors appeared very inefficient with an association constant (K) of about 1×10^{9} M^{-1}, and this closely resembled the pattern of low-level binding of colicin I to resistant bacteria which lack the normal colicin I receptor. Finally, since binding to the efficient receptors was associated with the killing of 96% of the bacteria, it was concluded that each cell in the population probably carried both types of receptor, rather than 50% of the bacteria having one class or the other.

V. Formation of Colicin-Target Complexes

We have previously proposed (Holland and Holland, 1970) that the interaction of colicins with sensitive bacteria is essentially a two-stage process. In the first step, Complex I is formed as colicin binds to specific surface receptors. Complex I formation in some cases may require the presence of magnesium ions (Reynolds and Reeves, 1969) and is largely, but not completely, irreversible (Reynolds and Reeves, 1969; Shannon and Hedges, 1973). In the form of Complex I, however, colicin does not promote any detectable biochemical changes, and the bacteria can be "rescued" by treatment with trypsin which degrades the surface-bound colicin (see Fig. 2, p. 75).

The second stage in colicin action, which is dependent upon several factors, results in the expression of the specific biochemical effects of the colicin concerned. These effects may be detected at both the surface and at the intracellular level. When this stage is reached, the capacity for trypsin rescue decays rapidly (at least for most colicins) even though the bulk of the colicin may still be present in the cell surface and accessible to trypsin (Maeda and Nomura, 1966). Complex II therefore represents the form of a relative minority of colicin molecules which actually interact with the target site. The precise mechanism which leads to completion of this second step is unclear, and several alternative possibilities can still be envisaged. Some of these will now be discussed, and in a following section various factors known to affect Complex II formation by different colicins will be examined.

A. Possible Mechanisms

Since the "transmission model" has stimulated so much thought about colicin action in particular, and about the properties of biological membranes in general, it will be examined first. According to this model, colicin molecules actively promoting intracellular changes remain in the surface layers of the cell and the effects of their presence are presumed to be transmitted along the membrane to the appropriate target site. In this case, the action of colicin is seen to be indirect and dependent upon a specific transmission mechanism. This theory has been most carefully expounded by Nomura (Nomura, 1964; Nomura and Maeda, 1965) and Luria (1964), primarily on the basis of the trypsin reversibility of colicin K action for long periods of time after the complete suppression of macromolecule synthesis (Nomura and Nakamura, 1962). Furthermore, in order to reconcile the single-hit killing of sensitive bacteria by colicin with the concomitant inactivation of large numbers of target sites (e.g., ribosomes in the case of colicin E3 or sites of energy metabolism in the case of colicin E1), amplification of the effects of a localized interaction of colicin with the membrane, by some mechanism, was also postulated to occur. More specifically, Changeux and Thiery (1967) proposed that localized interaction of colicin with the membrane might result in a long-range molecular re-arrangement of protein subunits, culminating in the altered activity of specific membrane proteins at many sites, or, alternatively, in a change in the distribution of a specific cellular component between membrane and cytoplasm. However, since the first formulation of the transmission model, knowledge (not only of the structure of the *E. coli* envelope and the molecular basis of colicin E3 action, but also of the fundamental nature of the cell membrane) has increased enormously. Thus, the recognition that Gram-negative bacteria possess a well defined outer membrane has focused attention upon this obstacle to colicin interaction with cellular targets. Moreover the concept of a fluid lipid-protein membrane (Singer and Nicolson, 1972), which envisages little or no long-range order in most membranes and, at the same time, provides the possibility for relatively rapid lateral diffusion of colicin (or a colicin-activated membrane component), is not compatible with the Changeux and Thiery model. Finally, the realization that colicin E3 is an enzyme, or an enzyme cofactor, provides a simpler explanation for the presumed amplification mechanism, namely that a catalytic inactivation of numerous target sites is triggered directly by the colicin molecule or by enzyme molecules activated by colicin.

In the light of all of these fundamental conceptual changes, and the accumulation of new facts, it now seems preferable simply to restate the possible sequence of events which occurs upon adsorption of colicin molecules rather than to present the transmission model, as such, as an alternative to other mechanisms. Inevitably some features of the model will remain, but no attempt will be made here to emphasize these specifically.

Binding of colicin molecules to outer membrane receptors is an essential first step which eventually permits some molecules to penetrate this layer. The subsequent interaction with the inner membrane, which is assumed to occur with all colicins, is likely to involve components of this membrane that are specific for each colicin. Any subsequent steps are then likely to depend upon the location of the target of the colicin concerned. For colicins of the E1 type, partial penetration of the inner membrane might be achieved, given the elongated form of the molecule, without necessitating its release from the receptor (see Fig. 6, p. 129). This may be followed by localized disruption of the permeability barrier which, if not sufficient by itself to suppress cellular energy metabolism, may activate enzymes which then diffuse within the plane of the membrane to accomplish this. For colicin E3, penetration of the outer membrane must be followed either by partial penetration of the inner membrane, so as to reach the cytoplasm, or by complete translocation of colicin E3 molecules to the interior of the cell. In either event, since colicin E3 induces the same specific cleavage of ribosomal RNA *in vitro* as it does *in vivo* (Section VII, p. 83), a colicin E3 molecule must act directly upon its ribosomal target. The precise target site for colicin E2 has not yet been identified. Nevertheless it seems likely that this colicin also must either partially penetrate the inner membrane or be fully translocated to the interior of the cell in order to promote DNA degradation, perhaps by direct interaction with the chromosome. To test some of these basic alternatives, it will be necessary to detect unequivocally the presence of colicin molecules within the cytoplasm, or to demonstrate the presence of colicin-membrane-target complexes which can still be inactivated by extracellular trypsin, for example, even though rescue of viable cells by this treatment might not be possible.

Possible mechanisms for the penetration of the cell envelope by colicin will now be considered in a little more detail. The outer membrane must first be breached, and it seems likely that the receptor complex plays a crucial role in this process, coupled with the orienting of colicin molecules necessary for their interaction with the cytoplasmic membrane. Similarly, irrespective of whether the colicin molecule merely bridges the surface envelope between inner and outer membranes, or is translocated into the cell, it appears most likely that specific components (probably proteins) of the inner membrane are essential in the final stages of Complex II formation. The role of such components presumably would be to complete the correct orientation of the colicin molecule, relative to the inner membrane, in order to form part of a specific transport system, or in the case of the E1 type colicin, to constitute part of the final disruptive complex itself.

The mechanism of colicin penetration of the outer membrane is unknown but the properties of this layer do not rule out the possibility that sufficiently large pores are present to allow, in the presence of the appropriate receptor, the passage of molecules of 10–20 Å in diameter. The interchain spacing (10–13 Å; see Braun *et al.*, 1973) between chains of the underlying peptidoglycan layer should also allow penetration of needle- or plate-shaped molecules. However, translocation of at least part of the colicin molecule from the primary binding site to the cell interior, which now appears certain for colicin E3, is still extremely difficult to envisage. Specific transport mechanisms may be involved in order to provide passage for a charged molecule through the highly hydrophobic lipid-protein membrane. Further examination of the structure and function of the "Bayer" tubules will be extremely important, both in relation to general protein export mechanisms and to the specific problem of colicin translocation. Although the precise mechanism of penetration of the cell envelope is still obscure, undoubtedly the key lies in an understanding of the inhibitory effects of energy uncouplers and cyanide on Complex II formation. Although, as will be discussed subsequently, ATP itself may not be necessary for the penetration or translocation step, a suitable membrane potential does appear to be required. The actual penetration step may of course include in some cases the specific cleavage of the surface-bound bacteriocin. The entry of an active fragment of diphtheria toxin into eukaryote cells appears to be promoted by such a mechanism (Gill *et al.*, 1973). Examination of the ability of proteolytic enzymes to generate specific colicin E3 fragments with increased specific activities in the *in vitro* assay (see Section VII, p. 83), and the screening of colicin-tolerant mutants for diminished proteolytic enzyme activities, could be carried out to test this hypothesis.

Finally, a general point concerning the action of colicins of the E2 and

E3 type, in particular, should be considered. The complete or partial transport of polypeptides of large molecular weight through the inner membrane may in itself be sufficient to disrupt the functioning of that membrane in some way. Thus, in addition to the specific destruction of an intracellular target, colicin action might be accompanied in some cases by adverse physiological changes resulting from membrane damage. This might occur, for example, if penetration of the inner membrane was facilitated by localized phospholipase activity, stimulated by colicin or inherent to the molecule itself. In fact, without the intervention of such a mechanism, it is extremely difficult to see how highly polar molecules could penetrate a highly non-polar membrane, unless alternative routes in the form of hydrophilic channels were available. Whatever the mechanism, any membrane damage incurred during transit of colicin could explain the apparent independence of the inhibition of cell division, and the degradation of DNA, induced by colicin E2, and the reported promotion of potassium efflux by Cloacin DF3, in addition to its colicin E3-like action on ribosomes (De Graaf, 1973).

Although several alternative mechanisms may still be invoked to explain Complex II formation, and its lethal consequences, one mechanism at least may be firmly ruled out. Inhibition of protein or nucleic acid synthesis, immediately prior to and during colicin treatment, has no affect upon Complex II formation for colicins E1 and K (Fields and Luria, 1969b), colicin E2 (Nomura and Maeda, 1965; Holland and Holland, 1970) or colicin E3 (Senior *et al.*, 1970, and unpublished data). Induced enzyme synthesis is therefore not involved in any way in facilitating colicin penetration of the surface, or in the actual process of target disruption.

B. Factors Affecting Complex II Formation

Adsorption of colicin to sensitive bacteria, at 37°C, is essentially complete in 3–5 minutes under most conditions (Marotel-Schirmann *et al.*, 1970; Cavard *et al.*, 1971; I. B. Holland, unpublished data) and occurs almost as rapidly at 4°C (Maeda and Nomura, 1966; Wendt, 1970). Magnesium ions are required for optimal adsorption of colicin E2 (Reynolds and Reeves, 1969) but no cofactors have been reported for other colicins. The majority of colicin molecules appear to bind irreversibly (Maeda and Nomura, 1966; Reynolds and Reeves, 1969; Shannon and Hedges, 1973) but many factors, including the physiological state of the bacteria, come into play before a lethal hit is sustained by a cell (Fig. 3). Thus, although colicin killing is a single-hit process (for detailed discussion see Reeves, 1973) the probability that one adsorbed molecule of colicin will promote a lethal hit varies enormously. As shown in Figure

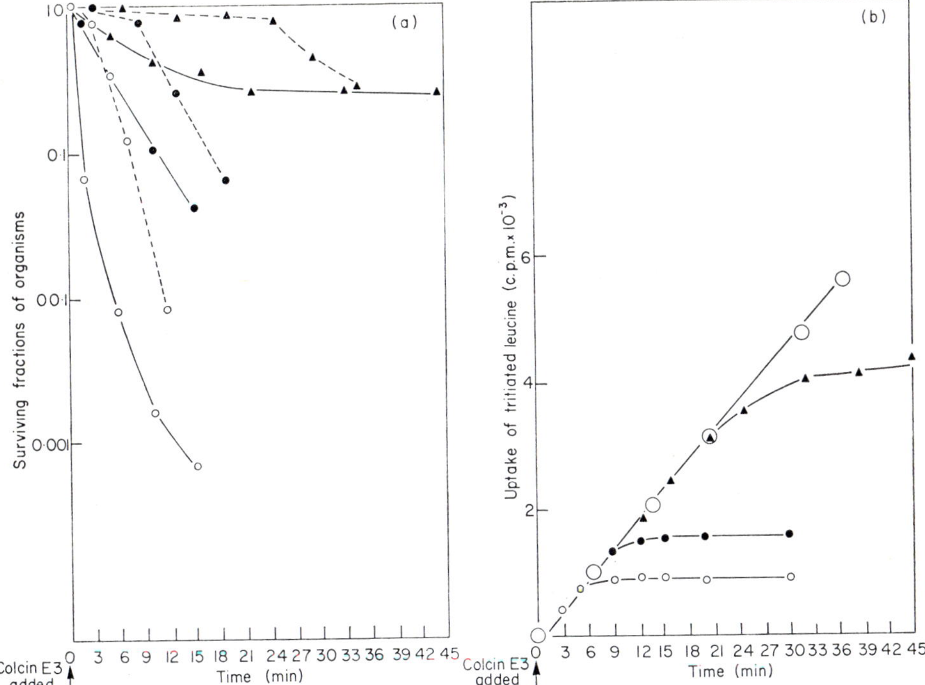

FIG. 2. Kinetics of Complex II formation in colicin-treated cells of *Escherichia coli*. (a) Exponential-phase cultures of *Escherichia coli* K12 were treated with different amounts of colicin E3, and cell survival determined at intervals. The survival of bacteria treated with 50 units of colicin E3 per ml is indicated by (○), that of bacteria treated with 10 units per ml by (●), and that treated with 2 units per ml by (▲). Continuous lines indicate survival in the absence, and dashed lines in the presence, of trypsin. (b) Tritiated-leucine was added to the cultures at zero time, and the subsequent incorporation of label into protein, in both untreated and colicin E3-treated cultures, was determined at intervals. The symbols represent the same dose levels of colicin E3; (◯) is the untreated control culture.

1a, 50–800 molecules of colicin E2 may be required in order to achieve a single hit, depending on whether the tested bacteria are grown in liquid or harvested from nutrient-agar plates. Only part of this effect may be ascribed to differences in numbers of receptors (Mayr-Harting and Shimeld, 1965) and, as shown in Figure 1b, in the case of mutants of the Cet type, which adsorb amounts of colicin comparable to that by wild-type strains, the probability of an adsorbed colicin molecule achieving a lethal hit varies from about 0·01%–1% depending on the cultural conditions. In the former case, since a lethal unit of colicin now corresponds to about 10^4 molecules, this exceeds the maximum number of colicin E2

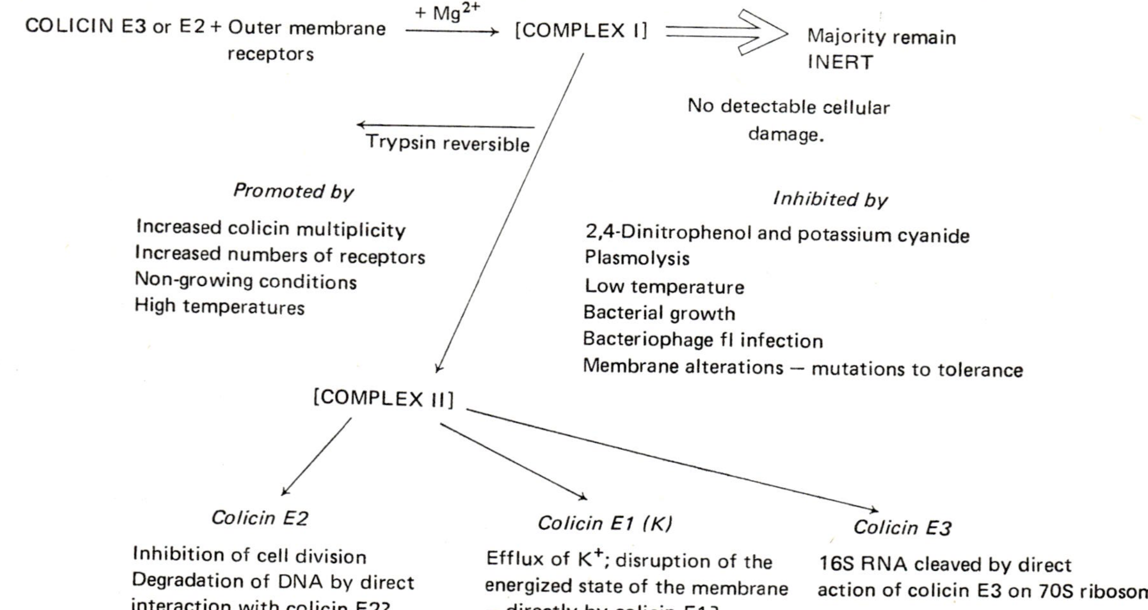

FIG. 3. Factors affecting Complex II formation in colicin-treated *Escherichia coli*. The formation of Complex II probably results from penetration of colicin molecules into the inner membrane and perhaps, in the case of colicins E2 and E3, into the cell interior. Under most conditions, Complex II formation results ultimately in a lethal change, and trypsin rescue of viable cells is not obtainable. The probable colicin targets are also indicated as are the major consequences resulting from their disruption.

receptors present in the cell surface and the bulk of the bacteria in the population survive.

A major feature of the action of all colicins is the ability, first demonstrated by Nomura and Nakamura (1962), to rescue cells which have adsorbed a potentially lethal dose of colicin by treatment with trypsin. Nevertheless, as discussed below, trypsin rescue of viable cells in most cases is limited to those bacteria which have not yet undergone an irreversible change. The length of the rescue period therefore defines an important first step in colicin action, the transition from an inert Complex I to an active Complex II. As discussed in the following section (p. 82), Complex II formation is also signalled by the onset of primary biochemical changes at the presumed target site.

1. *Colicin Multiplicity and Complex II Formation*

In cultures growing exponentially at 37°C, adsorption of either colicin E2 or E3 is followed by a lag period of at least three minutes before DNA degradation or inhibition of protein synthesis, respectively, takes place. During this lag period the great majority of colicin E2-treated cells can be rescued by trypsin (Ringrose, 1970; Obinata and Mizuno, 1970). Similar results can be obtained with colicin E3, as shown in Figure 2. Moreover, by progressively lowering the multiplicity of colicin E3, this lag period can be extended for at least 20 minutes, after which time the specific intracellular effect is abruptly triggered. This effect of multiplicity is characteristic of both colicins E2 and E3 (Nomura and Maeda, 1965; Holland and Holland, 1970; Senior *et al.*, 1970). Furthermore, as also shown in Fig. 2, the timing of the onset of inhibition of protein synthesis precisely parallels the period of maximum trypsin rescue for each colicin E3 concentration used. Previous studies (see Section VII, p. 83) have shown that the characteristic effects of colicin E3 (polyribosomal instability and 16S RNA cleavage) cannot be detected during the lag period. This indicates that the final step in Complex II formation is not the culmination of a progressive deterioration of the protein synthesizing machinery, but results from an abrupt change of state of the adsorbed colicin molecules.

Evidence for a similar Complex I to Complex II transition has been obtained recently for both colicins E1 and K. Plate and Luria (1972; see also Dandeu *et al.*, 1969) have shown that the kinetics of two primary effects of these colicins (efflux of K^+ and thiomethylgalactoside) precisely parallel the decay in the capacity of the cells to be rescued by trypsin. However, in contrast to colicins E2 and E3, Complex II formation with colicins E1 and K can be initiated at 37°C, without a detectable lag period even at low multiplicities (Wendt, 1970; Cavard *et al.*, 1971,

Phillips and Cramer, 1973; Plate and Luria, 1972; Plate, 1973). Nevertheless, at lower temperatures, Complex II formation is seen to be progressively delayed to an extent which is dependent upon the colicin multiplicity (Wendt, 1970). Therefore, any differences in this respect between colicin E2 and E3, and colicins of the E1 type, are likely to be quantitative rather than qualitative, indicating the requirement for a larger activation energy for Complex II formation in the case of colicins E2 and E3. Although the basis of these multiplicity effects is not understood, the effects are consistent with the temperature-dependent diffusion of a mobile colicin-receptor complex within the membrane, either to specific sites or to form more complex aggregates which finally trigger colicin penetration of the outer membrane.

2. *Trypsin Reversibility*

Before considering in detail other factors affecting Complex II formation, a point of possible confusion concerning the trypsin reversibility of colicin K action should be clarified. As indicated above, Plate and Luria (1972) found that the ability to rescue colicin K-treated cells decays rapidly after a lethal hit has been incurred. In contrast, Nomura and Nakamura (1962) demonstrated that the lethal effect of colicin K was actually *reversed* by trypsin treatment, with consequent resumption of growth in cultures in which protein synthesis and biomass increase had been completely suppressed. This result has recently been confirmed by M. Nomura and C. M. Bowman (personal communication), and also by Foulds and Shemin (1969), for bacteriocin bc. The reversibility of the lethal effects of colicin K is therefore formally analogous to the reversibility of uncoupling agents and ionophores like valinomycin (see Section X, p. 101) which act on the membrane to block energy metabolism without provoking irreversible damage to the cell. Consequently, the conflicting results obtained for reversibility of colicin K action can be reconciled if it is assumed that, under certain conditions, Complex II formation is accompanied by secondary effects causing membrane damage which cannot subsequently be repaired. Nomura (1967) has, in fact, pointed out that the efficiency of trypsin reversibility of colicin K action is dependent upon the growth medium and is best expressed when resting cells are used. In any event, the fact that a positive result was obtained by Nomura and Nakamura (1962) is sufficient cause to seek explanations for the irreversibility of colicin K action in other studies. The implications of these findings are, of course, extremely important since, as originally suggested by Nomura and Nakamura, the results clearly indicate that Complex II leaves active colicin K molecules in the cell-surface layers still accessible to trypsin.

Colicin E2 promotes rapid and extensive degradation of DNA (see p. 90) and the rescue of viable cells after completion of Complex II formation is not to be expected, and has not been reported. Nevertheless, the possibility remains that colicin E2 molecules located largely external to the cytoplasmic membrane can directly promote DNA degradation, and in order to test this hypothesis we have examined the effect of addition of trypsin to cells already showing colicin E2-induced DNA degradation (Holland and Holland, 1970). Trypsin effectively restricts the rate of further DNA breakdown when added to cells at time intervals up to 30 min after the addition of the colicin; after 30 min the rate of DNA breakdown is maximal. Since, under these conditions, the majority of cells can no longer be rescued by trypsin treatment after 10 min exposure to colicin E2 (I. B. Holland, unpublished results), it is concluded that those colicin E2 molecules actively engaged in promoting DNA breakdown are still accessible to trypsin. Furthermore, Ringrose (1970) reported that the early addition of trypsin to colicin E2-treated cells halted further fragmentation of DNA, and molecules already containing single-strand nicks were then repaired. Although suggestive, these experiments are nevertheless inconclusive since similar results would be obtained if maximal rates of DNA degradation were dependent upon the continued uptake of colicin E2 into the cytoplasm over a long period.

3. *Effect of Bacterial Growth*

The delayed action effect, particularly at low multiplicities of colicins E2 and E3, in exponentially growing cultures suggests that growth of the bacteria may in some way contribute directly to the triggering of Complex II formation (see also Reynolds and Reeves, 1969). However, the delay in the onset of DNA degradation is not observed when resting cells in buffer are treated with colicin E2 (Holland and Holland, 1970). Under these conditions, the rate of adsorption of colicin E2 is increased, and the length of the trypsin-rescue period is greatly decreased (Reynolds and Reeves, 1969). Consequently these results suggest that, in growing bacteria, Complex II formation is actually inhibited rather than the reverse. In an attempt to establish the nature of this inhibitory effect on colicin E3 action, Complex II formation in cells in which RNA, DNA, protein or peptidoglycan synthesis was inhibited were examined, but none of the treatments had any significant effect (B. W. Senior and I. B. Holland, unpublished results). The effect of inhibition of phospholipid metabolism on colicin action should now also be examined, since the insertion of colicin molecules into the expanding cell membrane of growing bacteria could delay a crucial interaction of a mobile colicin-receptor complex with a minor surface component.

4. *Energy Requirement*

Complex II formation for all colicins studied so far appears to be an energy-dependent process. Bacteria treated with colicin E2 (Nomura and Maeda, 1965; Reynolds and Reeves, 1969), colicin E3 (B. W. Senior and I. B. Holland, unpublished data) and colicin K (Nomura, 1963), in the presence of 2-4-dinitrophenol or potassium cyanide, remain in the trypsin-rescuable state indefinitely. Dinitrophenol also blocks the colicin E1- or K-induced efflux of magnesium ions (Lusk and Nelson, 1972) and the initiation of DNA degradation by colicin E2 (Nomura and Maeda, 1965; Holland and Holland, 1970). The significance of the latter result is, however, difficult to assess since, as discussed elsewhere, dinitrophenol also promptly blocks ongoing DNA degradation, even when added at late times. In all of the above studies with the uncoupling agent dinitrophenol, experiments were carried out either in nutrient broth or in minimal-salts plus glucose cultures. Considerable amounts of ATP generated by glycolysis should therefore be available in the treated cells, and the effectiveness of dinitrophenol under these conditions suggests that some form of active transport, or some system driven by a suitable membrane potential (see Section X, p. 101), is involved in Complex II formation. In the case of colicin E3, and possibly with colicin E2, this may indicate the presence of a specific uptake system for partial penetration of the inner membrane by the colicin, or even for its complete translocation to the cell interior. For colicins of the E1-type it is still extremely difficult to envisage any mechanism which requires energy that could be necessary to ensure interaction with the cytoplasmic membrane.

5. *Effect of Temperature*

If adsorption of colicin E1 is first completed at 4°C, and the cells then kept at this temperature, they remain fully trypsin-rescuable until the temperature is raised, whereupon the ability to rescue the cells immediately begins to decline (Plate, 1973). Furthermore, K^+ efflux, or the inhibition of thiomethylgalactoside uptake in colicin E1-treated cells (Wendt, 1970), or in colicin K-treated cells (Plate, 1973), is progressively delayed as the incubation temperature is lowered below 26°C. The induced fluorescence increase in the membrane probe analinonaphthalene 8-sulphonate (ANS) (see Section X, p. 106) is both delayed, and the initial rate of increase slowed, when cells are treated with colicin El at low temperatures (Cramer and Phillips, 1970). Finally, the lag period before the onset of inhibition of protein synthesis in cells treated with a high multiplicity of colicin E3, increases about threefold for a 10°C drop in temperature (B. W. Senior and I. B. Holland, unpublished data). The effect of temperature on the rate of Complex II formation, once initiated within a bacterial population (a parameter which can be measured independently of the lag period which precedes it), has also been examined. In experiments described by Plate (1973) involving colicins E1 and K, the rate constant for the Complex I to Complex II transition was obtained by measuring the rate of decay of trypsin-rescue at different temperatures. The results showed that the value of the rate constant undergoes an abrupt transition at about 20°C. Similar biphasic Arrhenius plots have been obtained for several membrane functions, including transport (Machtiger and Fox, 1973) and respiration (Overath *et al.*, 1970). This phenomenon has been interpreted as order–disorder phase changes involving the hydrocarbon chains of membrane fatty acids (Chapman and Wallach, 1968; Linden *et al.*, 1974). Moreover, the "melting" point is a function of the fatty-acid composition of the membrane and, utilizing this fact, Plate (1973) has also measured the effect of temperature on colicin killing in fatty-acid auxotrophs grown in the presence of different fatty acid supplements. The results obtained demonstrated that Arrhenius plots for the Complex I–Complex II transition, under these conditions, reflected the altered phospholipid composition of the bacteria in a way identical to that seen previously for well characterized membrane functions. Essentially similar results for colicin E1 have been obtained by Cramer *et al.* (1973), who measured the initial rate of increase of membrane-bound ANS, as a function of temperature, in cells with different phospholipid compositions. These findings demonstrate that the relative fluidity of the membrane–lipid phase markedly affects the rate of Complex II formation for both colicins E1 and K. Unfortunately these studies do not pinpoint the step in Complex II formation involved. Thus, physical penetration of the outer or inner membranes, activation of inner or outer membrane enzymes by colicin, or simply the efficiency of the respiratory system in providing the necessary energy for Complex II formation, or a combination of these three factors, could all be dependent upon membrane fluidity (see also the discussion by Plate, 1973).

6. *Osmolarity of the Medium*

Complex II formation for colicin E2 is also blocked in plasmolysed cells. Beppu and Arima (1967) reported that the trypsin-rescue period for colicin E2 is extended, and DNA degradation decreased, in the presence of 0·8 M-sucrose, although adsorption of the colicin was not affected. Complex II formation for colicin K (Beppu and Arima, 1967) and colicin E3 (B. W. Senior and I. B. Holland, unpublished data) is similarly inhibited by sucrose, indicating the importance of the close apposition of outer and inner membranes in some early step in colicin action subsequent to adsorption. Colicin E2-induced DNA breakdown, in buffer suspension of cells, is delayed or completely suppressed by increasing salt concentrations. This effect may also be ascribed to plasmolysis but, curiously, exposure of cells to high salt concentrations for periods of time up to 30 minutes after addition of colicin E2 also greatly decreases ongoing DNA degradation (Holland and Holland, 1972). These findings again provide circumstantial evidence for the existence of some kind of colicin E2-membrane complex which is active in promoting DNA degradation and which is disrupted under conditions that cause separation of the inner and outer layers of the cell surface.

7. *Other Factors*

The initiation of Complex II formation for colicin E2, in resting cells in Tris-HCl buffer, has been shown to be stimulated by addition of Na/K-phosphate (Holland Holland, 1970), but the basis of this stimulatory effect has not been further investigated. An interesting report by Zinder (1973; see also Smilowitz, 1974) has shown that the efficiency of Complex II formation of colicins E3, K and E2 is also greatly decreased in cells infected with the filamentous DNA phage f1. Preliminary evidence was presented specifically implicating a minor protein component of the virion, normally involved in phage penetration, in the induced "tolerance" to the colicins. The basis of the observed tolerance is obscure but it seems likely to involve some step in the penetration of the bacterial surface by colicin, and further study should be profitable. Finally, as found by Mayr-Harting and Schimeld (1965), and also indicated in Fig. 1 (p. 63), the frequency of surface receptors is important in determining Complex II formation. With few receptors the chances of a colicin molecule achieving a lethal hit may be very small and, in growing cultures, in particular, many cells will survive by producing colicin-free progeny before a lethal hit is sustained.

VI. Biochemical Changes and Single-Hit Killing

Since the adsorption of a one killing unit of colicin can promote cell death, any biochemical change related to the lethal event should also be triggered by one killing unit which, under certain conditions, may be equal to a single molecule. This is difficult to measure directly, but some data are available, particularly in relation to colicins E1 and K. Wendt (1970) measured the efflux of labelled potassium from cells treated at 37°C with varying multiplicities of colicin K. With multiplicities of up to three killing units per cell, the total amount of potassium lost from the cell population correlated very closely with the fraction of cells actually killed, suggesting that potassium efflux is an all or none event and that the adsorption of one killing unit virtually empties a cell of its potassium content. Furthermore, these studies showed that the initial rate of potassium efflux is a linear function of the colicin concentration, confirming that the promotion of potassium loss is a non-co-operative process. Studies by Plate and Luria (1972), who compared the extent of thiomethylgalactoside (TMG) efflux with the fraction of cells killed at various times by colicins E1 or K, showed that the adsorption of a single killing unit was apparently sufficient to promote maximal release of TMG from individual cells. In a different approach, Cramer and Phillips (1970) demonstrated that the amplitude of the increase in fluorescence (induced by colicin E1) of a hydrophilic probe bound to the cell surface also was relatively independent of colicin multiplicity. In contrast, the *rate* of change in fluorescence increased linearly with colicin concentration. All these results suggest that the observed biochemical changes associated with early stages of colicin E1 and K action are initiated by an all or none response, and that the major effect of increasing colicin multiplicity is to accomplish in a shorter time that which one active molecule would do over a longer period. This conclusion is therefore quite consistent with the observed one-hit killing action of colicin E1 and other colicins. Comparable data are not available for colicins E2 and E3 although, as indicated previously (p. 77), DNA degradation and inhibition of protein synthesis promoted by colicins E2 and E3, respectively, can be detected at extremely low molecular multiplicities, and the main effect of increasing this multiplicity is to accelerate the final triggering of these effects rather than to affect their final magnitude. Since colicin E3 appears to interact directly with its ribosomal target (presumably also the lethal target), this requires that, in order to kill a cell, a single molecule must be capable of acting catalytically upon large numbers of ribosomes. The colicin E2 biochemical and/or lethal target has not yet been clearly identified, but if this proves to be DNA then the conclusion appears inescapable that this colicin also must act catalytically to produce multiple lesions in the chromosome which ultimately prove lethal. For colicins of the E1 and K types, the precise target is even less clear and, whilst speculation is probably not warranted at this stage, Wendt (1970) has posed the possibility that a single localized lesion within the cell membrane might be sufficient to promote the collapse of the energized state throughout the whole cell membrane.

VII. Molecular Basis of Colicin E3 Action

The mode of action of colicin E3 is the best understood of all the bacteriocins and, in recent years, a virtually complete account of the molecular basis of the lethal effects of this colicin has been obtained. Nomura and his coworkers (Nomura, 1963; Nomura and Maeda, 1965) first demonstrated that protein synthesis was rapidly halted in colicin E3-treated cells, whilst RNA and DNA synthesis was not immediately affected. Respiration and potassium transport continued normally, indicating the absence of any general cellular damage. Colicin E3 was also shown to block T4 phage-directed protein synthesis (Nomura, 1963), suggesting that a basic component of the protein-synthesizing machinery was affected rather than, for example, host-specific transcription. The action of colicin E3, like that of some other colicins, shows a strong multiplicity effect even at 37°C. With high doses (up to $2{\cdot}5 \times 10^3$ molecules/cell), inhibition of protein synthesis is triggered within three to four minutes in exponentially growing cultures (Nomura and Maeda, 1965; Senior *et al.*, 1970). At low doses (100 molecules/cell), protein synthesis continues normally for up to 20 minutes and then inhibition commences abruptly. As shown in Figure 2 (p. 75), the variation in the length of the lag period before colicin E3 action is triggered precisely reflects, for each colicin E3 dose, the period of maximum trypsin rescue. The delay in the intracellular changes promoted by colicin E3 therefore reflects the kinetics of Complex II formation and the presumed translocation of the colicin to its target site.

A. Ribosome Modification *in vivo*

Ribosomes were first identified as the probable target site for colicin E3 by Konisky and Nomura (1967) who showed that 30S ribosomal subunits, but not 50S particles or supernatant enzyme fractions from colicin E3-treated cells, were largely inactive in reconstituted protein-synthesizing systems *in vitro*. Colicins K and E2 did not produce these effects and colicin E3 failed to induce such changes in a colicin-resistant strain lacking the colicin E receptor. Specific inactivation of ribosomes, which is accompanied *in vivo* by marked instability of 30–50S complexes, was also demonstrated by Senior *et al.* (1970), although in this case 70S ribosomes from fully inhibited bacteria still retained 30–40% *in vitro* activity when tested in a poly-U-directed system. This suggested that much of the functional integrity of the ribosome was retained. Inactivation of ribosomes by colicin E3 treatment proceeds normally in amino acid-starved cells or in the presence of actinomycin D (Senior *et al.*, 1970). Potentiation of colicin action by enzymes synthesized *de novo*, after colicin treatment, can therefore be ruled out. These results also demonstrate that ribosomes need not be actively engaged in protein synthesis in order to be inactivated.

In 1971 Nomura and his coworkers (Bowman *et al.*, 1971a) and Senior and Holland (1971) independently demonstrated that 16S RNA from either 30S or 70S ribosomes, isolated from colicin E3-treated cells, contained a single fracture which yielded a large fragment with a sedimentation coefficient of approximately 15·5S and a small fragment consisting of about 50 nucleotides. Finger-print analysis of large and small fragments clearly demonstrated that the small fragment was homogeneous, and consisted of the 3′-terminal region of the 16S RNA molecule. Further studies of the large fragment failed to reveal any additional structural or chemical modifications of the molecule and, in addition, no changes in 30S ribosomal proteins were detected by acrylamide gel electrophoresis. These results indicated that the functional defect in colicin E3-treated 30S ribosomes therefore lay in the RNA moiety rather than in the protein constituents, and this was confirmed by Bowman *et al.* (1971a) who demonstrated that ribosomes reconstituted from the 30S ribosomal RNA of colicin E3-treated cells and control 30S subunit proteins were inactive *in vitro*, whilst control RNA plus colicin E3-treated 30S proteins formed fully active particles. The kinetics of appearance of the small fragment in 30S particles, isolated from treated cultures, have been examined and shown to parallel exactly the kinetics of inhibition of protein synthesis at both high and low colicin E3 multiplicities (Samson *et al.*, 1972). Thus, fragmentation of 16S RNA is, in all probability, the specific and primary event which blocks protein synthesis in colicin E3-treated bacteria. Moreover, since in at least some studies (Samson *et al.*, 1972; Boon, 1972, but in contrast see Bowman *et al.*, 1971a) the 3′-terminal fragment can be recovered quantitatively from purified 30S particles, without loss or further degradation, simple cleavage of the 16S RNA molecule appears sufficient to block protein synthesis *in vivo*. As discussed later, this suggests strongly that the 3′-terminal end of the molecule, which probably lies very close to the ribosomal surface (Santer and Santer, 1973), may be directly involved in polypeptide chain biosynthesis.

Size estimates for the small RNA fragment from colicin E3-treated cells, by either oligonucleotide analysis (Bowman *et al.*, 1971a) or by mobility in acrylamide gels (Samson *et al.*, 1972), agree closely and suggest a molecular weight of $1{\cdot}74 \times 10^4$ daltons, equivalent to 50–52 nucleotides. The small fragment, which appears to have a 5′-hydroxyl terminus, contains three of the twelve methylated bases of the whole 16S RNA molecule (Bowman *et al.*, 1971a). One methylated base lies close to the position of fracture, but the presence of the methyl group does not appear essential for the cleavage reaction (Dahlberg *et al.*, 1973).

B. Ribosome Modification *in vitro*

The recent demonstration that purified colicin E3, when added to 70S ribosomes *in vitro*, can promote both the scission and functional inactivation of 16S RNA exactly as *in vivo* (Boon, 1971; Bowman *et al.*, 1971b) has necessitated a complete re-examination of the way in which colicins are thought to interact with sensitive cells. For this colicin, at least, direct interaction of the colicin molecule (or an active fragment of it) with the intracellular target must take place *in vivo*.

The *in vitro* studies have, in addition, brought to light the remarkable substrate specificity of the RNA-cleavage reaction and the probable basis of immunity of Col $E3^+$ bacteria to exogenous colicin E3. Both Boon (1971) and Bowman *et al.* (1971b) showed that, whilst purified 70S ribosomes from wild-type bacteria and from Col $E3^+$ bacteria were equally sensitive to colicin E3, a crude extract from the Col^+ bacteria blocked the *in vitro* action of the colicin. Using this *in vitro* assay system, Nomura and his coworkers (Bowman *et al.*, 1973, and personal communication) have purified an apparently acidic protein from mitomycin C-induced lysates of Col^+ bacteria which specifically blocks colicin E3 action *in vitro*. This protein, which is not present in Col^- bacteria, may be the substance proposed by Nomura (1963) to explain the immunity of Col^+ bacteria. Failure to remove the inhibitor from partially purified preparations of colicin E3 blocks the *in vitro* action, but not the *in vivo* action, of the colicin (Konisky and Nomura, 1967; Bowman *et al.*, 1973). The inhibitor does not apparently bind to ribosomes *in vitro*; it does not irreversibly inactivate colicin E3 but nevertheless it most probably acts by binding to the free colicin (M. Nomura, personal communication).

The studies of Boon (1971, 1972) and Bowman (1972) have revealed that the cleavage reaction is highly specific. When colicin E3 is added to 70S ribosomes, or to mixtures of 50S and 30S subunits, *in vitro*, strand cutting of 16S RNA occurs. In contrast, neither 16S RNA alone nor 30S subunits alone undergoes RNA cleavage upon addition of colicin E3, thus attesting to the unique structural specificity of the reaction (Table 1). Perturbation of ribosome structure, induced by binding of several antibiotics at apparently quite distinct sites on the 30S subunit, has been shown to block the cleavage reaction both *in vitro* and *in vivo* (Dahlberg *et al.*, 1973). These results all suggest that cleavage of the small subunit RNA depends upon a highly specific structural organization of ribosomal components. Evidence which suggests that structural organization, rather than specific RNA base sequence, may be of primary importance in cleavage of the RNA has also been obtained by Sidikaro and Nomura (1973). They showed that ribosomes from *Bacillus stearothermophilus* (an organism which lacks colicin E3 receptors and has a 16S RNA base composition and nucleotide sequence quite different from that of *E. coli*; Nomura *et al.*, 1968), also undergo fragmentation of 16S RNA in the expected position close to the 3′-terminus when treated with colicin E3. An intriguing report by Turnowsky *et al.* (1973) that colicin E3, *in vitro*, inactivates ribosomes from Krebs Ascites cells suggests that

TABLE 1. Specificity of colicin E3 Action *in vitro* and *in vivo*

Ribosomes (70S) extracted from sensitive, immune, or resistant cells (including species unrelated to *Escherichia coli*) are all inactivated by purified colicin E3 *in vitro*, and the 16S RNA is cleaved close to the 3′-terminus. The *in vitro* reaction is blocked by an inhibitor present in immune Col E3$^+$ bacteria, but not in Col$^-$ bacteria, and cleavage of 16S RNA is not obtained unless both 30S and 50S ribosomal subunits are present.

	Action *in vivo*	*In vitro* cleavage of 16S RNA			
		E3 + 70S ribosomes	E3 + 50S + 30S ribosomes	E3 + 30S ribosomes	E3 + 16S RNA
Escherichia coli (Col$^-$)	Sensitive, 16S RNA cleaved	+	+	−	−
Escherichia coli (Col E3$^+$)	Immune	+	/	/	/
Escherichia coli (*bfe*$^-$)	Resistant	+	/	/	/
Bacillus stearothermophilus	Resistant	+	/	/	/

colicin E3 may interact with a specific ribosomal region highly conserved in both eukaryotes and prokaryotes. Confirmation that colicin E3 actually promotes 18S RNA fragmentation close to the 3′-terminus in this system is therefore urgently needed.

Before completing this section, a final comment is necessary concerning the apparent efficiency of colicin E3 *in vitro*. It has been claimed (Boon, 1971; Bowman *et al.*, 1971b) that highly purified colicin E3, at multiplicities of less than 0·1 molecules/ribosome can promote 16S RNA cleavage, *in vitro*, suggesting that the colicin is acting catalytically as presumably it does *in vivo*. Nevertheless it appears from an examination of the published data (although these are rarely quantitative) that the rate of *in vitro* cleavage is very slow compared with that observed *in vivo*. Moreover, in all studies reported so far, maximum cleavage is only observed at colicin E3 multiplicities per ribosome of about 100 to 1000-fold higher than those likely to be encountered *in vivo*. Additional cellular constituents, including the colicin E3 surface receptor, may therefore be required before optimal *in vitro* conditions are achieved. Actual cleavage of the colicin E3 molecule itself, in order to release an active fragment, may also be necessary *in vitro* if a similar event occurs *in vivo* during Complex II formation. Further intensive study of colicin E3 action *in vitro* should resolve all these questions, including final identification of the molecule(s) which promotes the hydrolytic cleavage of the 16S RNA component.

C. Possible Involvement of Cellular Nucleases

Despite the great advances in the understanding of colicin E3 action at the ribosomal level, the identity of the "ribonuclease" actually involved remains an enigma. Since colicin E3 has no degradative action on isolated 16S RNA, one must consider the possibility that colicin E3 activates a ribosomal ribonuclease or that colicin E3, in combination with one or more integral ribosomal proteins, constitutes the hydrolytic complex. To test the former possibility Meyhack *et al.* (1973) examined the *in vitro* action of colicin E3 in the absence of ribonuclease II, ribonuclease III and polynucleotide phosphorylase. In all these cases, and also in previous studies using mutant strains lacking ribonuclease I (Senior and Holland, 1971; Bowman *et al.*, 1971a) normal cleavage of 16S RNA was observed. Therefore Meyhack *et al.* (1973) concluded that colicin E3 alone, or in combination with a ribosomal protein, must constitute a specific endoribonuclease. In this connection, it is interesting to note that small fragments of RNA, equal in size to the colicin E3 fragment, have been recovered from 30S particles of untreated cells (Samson *et al.*, 1972), and significant levels of a 15·5S RNA fraction have also been observed under similar conditions (Bowman *et al.*, 1971a). This indicates that "spontaneous" cleavage of 16S RNA, directed by endogenous ribonucleases, may therefore take place in normal cells under certain conditions, or, alternatively, that 16S RNA is particularly prone to fracture at this point in the molecule during isolation of ribosomal subunits. The former proposition may be supported by the fact that at a late stage in the maturation of bacterial ribosomes the precursor 16S RNA (pRNA), already incorporated into the partially assembled particles, is tailored to optimum size (Pace, 1973) by nucleases which to my knowledge have yet to be identified. Tailoring of pRNA to the appropriate size, by ribosomal proteins transiently activated by specific ribosomal conformations, could thus be an essential part of the normal maturation process. If binding of the colicin E3 molecule induces a ribosomal alteration which mimics an immature conformation, this could induce a further and now lethal reduction in size of the 16S RNA.

D. Functional Defects in Ribosomes

Why are colicin E3-modified 30S particles inactive in protein synthesis? This is a particularly intriguing question since only the RNA moiety appears modified and the change involved is such a relatively minor one. Ribosomes isolated from treated cells, completely blocked in protein synthesis, nevertheless show quite high levels of activity *in vitro* with a polyuridylic acid-directed system (Konisky and Nomura, 1967; Senior *et al.*, 1970). However, when natural messenger RNA is used, 30S particles from colicin E3-treated cells are virtually inactive *in vitro* (Konisky and Nomura, 1967). This suggests that specific polypeptide chain initiation or termination may be defective in colicin E3-derived ribosomes. Measurements of specific binding capacities of 70S particles from colicin E3-treated cells, by Konisky and Nomura (1967), revealed that polyuridylic acid and phage MS2 mRNA binding was only slightly less than that of the controls, but polyuridylic acid directed binding of phenylalanine tRNA was greatly decreased. This clearly indicates some alteration in one or other of the two tRNA binding sites. Pulse labelling of colicin E3-treated cells, during the period when the inhibition of protein synthesis is not yet complete, demonstrated that nascent polypeptides become progressively more resistant to chasing from the polysome fraction (Senior *et al.*, 1970). This result is compatible with the hypothesis that modification of a tRNA site by colicin E3 blocks polypeptide extension or its normal termination, but does not appear to be consistent with any mechanism which specifically involves the initiation reaction. Dalgarno and Shine (1973) have recently reported that the 3′-terminal octanucleotide sequence of 18S rRNA is completely conserved in three quite distinct eukaryotic species, and contains triplets complementary to the nonsense codons UGA and UAA. These authors suggest that the known composition of the 3′-terminus of *E. coli* 16S RNA (see Ehresmann *et al.*, 1971) is also compatible with a similar content of nonsense anticodons and that this region of the molecule may therefore be similarly concerned in specific termination of polypeptide chains in bacteria. This suggestion is clearly compatible with the mode of action of colicin E3 discussed above, which indicates a direct role for the 3′ end of 16S RNA in protein synthesis. An alternative possibility, which is not mutually exclusive with this interpretation, is suggested by some preliminary data reported by Bowman *et al.* (1971a). In reconstituting 30S ribosomes with the large 15·5S RNA fragment from 30S particles of colicin E3-treated cells, at least one protein (P15) was not incorporated into the re-assembled ribosome. Since protein P15 appears to be located on the ribosome surface (Mizushima and Nomura, 1970) the possibility arises that both protein P15 and the 3′-terminus of the RNA are closely associated with the region of the amino-acyl tRNA site. If so, cleavage of the RNA moiety could in any case lead to ribosomal inactivation, by disturbing the functional organization of proteins located in this region, without requiring the direct involvement of 16S RNA in protein biosynthesis. Studies on the action of colicin E3-derived ribosomes *in vitro*, of the kind discussed above, should resolve these problems as well as revealing more of the organization of ribosomal components.

VIII. Primary Effects of Colicin E2

Nomura was the first to show that low multiplicities of colicin E2 caused DNA solubilization in sensitive cells, but not in immune (Col E2$^+$) bacteria (Nomura, 1963; Nomura and Maeda, 1965). Respiration, protein synthesis and RNA synthesis were not initially inhibited, and growth of bacteriophage T4 proceeded normally in treated cells, all indicating that the major biosynthetic capacity of the cells was unimpaired. Subsequent studies by Swift and Wiberg (1971, 1973a, b) have demonstrated that the colicin E2-induced breakdown of the host DNA is greatly inhibited by T4 phage infection, and hence the absence of phage DNA breakdown, originally noted by Nomura (1963), can be largely ascribed to this inhibitory effect without having to postulate a limited substrate specificity for the colicin E2 nuclease system. Earlier studies demonstrated that low doses of colicin E2 induced the development of bacteriophage λ in lysogenic bacteria (Nomura, 1963; Endo *et al.*, 1963), a finding consistent with the idea that the primary effect of colicin E2 was upon DNA metabolism. The first detectable biochemical change in treated cells still appears to be degradation of DNA, although the time of onset of DNA breakdown is very much dependent upon the multiplicity of colicin E2 used. Thus, especially at low doses (50–100 molecules/cell), breakdown may be delayed for up to 40 min. (i.e. one generation) after the completion of adsorption (Holland and Holland, 1970; Nomura and Maeda, 1965). Consequently, the possibility might be raised that DNA degradation is a secondary, non-specific, effect of colicin E2 action. However, several studies have now shown that, in the case of both colicins E2 and E3, the delay in appearance of specific intracellular biochemical changes can be fully accounted for by the relatively slow, and multiplicity-dependent, second stage in colicin action (Section V, Fig. 2, p. 75). This second step, leading to the formation of Complex II, probably represents the "vertical" translocation of an active colicin molecule to the target site. In fact in buffer suspensions, where colicin adsorption and Complex II formation (Reynolds and Reeves, 1969; Holland and Holland, 1970) appear to proceed more rapidly than in growing cultures, DNA solubilization can be detected at 37°C as early as one minute after addition of quite low multiplicities of colicin E2 (Holland and Holland, 1970, 1972).

A. DNA Degradation

Despite the large amount of data concerning induction of DNA degradation in colicin E2-treated cells, the precise nature of the process, and particularly the identity of the enzymes involved, remains unclear. Some important questions that require clarification are as follows: (i) is there substrate specificity at the chromosomal level, i.e. bacterial, phage or plasmid DNA? (ii) what is the specificity of nucleotide sequences or nucleoprotein tertiary structures at the target site? (iii) if cellular nucleases are involved, what is their normal functional role in the cell? We cannot claim that any of these questions can as yet receive clear answers, but some relevant data are available and these will now be considered.

1. Mechanism of DNA Breakdown

The degradation process, in fact, involves both solubilization and fragmentation of DNA. This probably means that both endo- and exonucleases are active, although the relative role of each, and the precise sequence of events, is still controversial. Ringrose (1970) has demonstrated, by sucrose-gradient analysis, that an early stage of colicin E2 action *in vivo* involves fragmentation of DNA molecules. On the basis of his results, Ringrose proposed three consecutive stages in the degradative process; single strand *nicking*, followed by double-strand cleavage or *cutting*, and finally the rapid solubilization of the fragments by a presumed exonuclease. Obinata and Mizuno (1970), in a more limited analysis, failed to detect nicked molecules as intermediates and concluded that colicin E2 action primarily involved double strand cleavage of DNA, followed by rapid solubilization. Subsequently, in an extensive analysis of the kinetics of both solubilization and fragmentation of DNA, measured simultaneously in the same culture, Holland and Holland (1972) confirmed the sequence of nicking and then cutting observed by Ringrose. However, due to the greater sensitivity of the technique used in this study, DNA solubilization was found to commence simultaneously with fragmentation. These experiments also showed that since DNA from a colicin E2 resistant receptor-minus mutant, present in the same lysate, was not degraded, the observed fragmentation did not result from non-specific, post-lysis cleavage, but truly reflected events *in vivo*. In addition, the demonstration that DNA solubilization occurs early in E2-treated cells may indicate that endo- and exo-nucleolytic activities may be triggered independently. There is some support for this since it is found that maximal rates of DNA solubilization are relatively independent of colicin multiplicity (Holland and Holland, 1970; J. M. Silver and P. L. Kuempel, personal communication), whilst the rate of endonucleolytic cleavage is proportional to the concentration of colicin E2 (J. M. Silver and P. L. Kuempel, personal communication). Moreover, DNA cutting and nicking at high colicin E2 doses is frequently completed in 30–40 min (culminating in the accumulation of 10^6–10^7-dalton fragments, a considerable fraction of which are not further degraded; Obinato and Mizuno, 1970; Ringrose, 1970; Swift and Wiberg, 1973a, b) whereas solubilization may continue for 80–90 min until 70–80% of DNA has been degraded. We cannot therefore rule out the possibility that the exonuclease-like activity in colicin E2-treated cells generates its own nicks, or proceeds from pre-existing breaks in DNA quite independently, at least initially, of the major endonuclease activity. However, other results suggest that endo- and exonuclease are closely linked in some way. Both activities are promptly and severely diminished by addition of dinitrophenol, even at late times after the onset of DNA breakdown, and both activities are enhanced in $RecA^-$ bacteria (Holland and Holland, 1970, 1972). In addition, the infection of treated cells with phage T4 mutants, which unlike wild-type T4 are unable to carry out an early stage of phage directed nicking of host DNA, inhibits both colicin E2-induced solubilization and chain cutting of DNA. On the other hand, T4 infection of treated cells in the presence of chloramphenicol only appears to block colicin E2-induced DNA solubilization (Swift and Wiberg, 1971, 1973a, b). Under these latter conditions, double strand pieces (of 10^6 daltons) produced by colicin E2 action accumulate in the infected cells. This finding strongly suggests that both endo- and exonucleases are active in colicin E2-treated cells, rather than a single enzyme responsible for both activities, as suggested by some workers (see Almendinger and Hager, 1972).

Studies by Hull and Reeves (1971), and by J. M. Silver and P. L.

Kuempel (personal communication), have shown that replicating DNA of phage λ and covalently closed F′lac DNA, in colicin E2-treated cells, are also cleaved by an endonuclease activity similar to that seen to degrade chromosomal DNA. In the case of phage λ-DNA, degradation is frequently less extensive than that of chromosomal DNA in the same culture. This may be due to the smaller target for nucleolytic attack presented by the plasmid DNA, or to there being relatively few specific sites for the initiation of plasmid DNA breakdown compared to the number of such sites on the chromosome. Alternatively, the possible compartmentalism of plasmid DNA, restricting it to the close vicinity of the cell membrane, may have a protective effect against the colicin E2 nuclease. Ringrose (1973) has presented evidence that membrane-bound regions of chromosomal DNA, including the replication fork, are more resistant to colicin E2-induced DNA solubilization than the bulk of the chromosome. Finally, some studies (Nose and Mizuno, 1971) have indicated that colicin E2-induced DNA breakdown is considerably restricted in phage λ lysogenic bacteria. The basis of this phenomenon is not clear but it suggests the production of a colicin E2 inhibitor determined by the prophage. It seems unlikely, however, that such an inhibitor, if present in cells infected with phage λ, would protect the phage DNA more efficiently than it would the chromosomal DNA.

2. *Possible Involvement of Cellular Nucleases*

Since double stranded DNA fragments with a molecular weight of 10^6–10^7 daltons appear to be the end product of the fragmentation activity in colicin E2-treated cultures, Ringrose (1970) has suggested that the site of the endonucleolytic attack might be at the cytosine-rich clusters which appear in DNA at these spacings (Szybalski *et al.*, 1966). In *E. coli*, an enzyme which appears specific for such sites is endonuclease II (Kutter and Wiberg, 1968). Unfortunately *E. coli* mutants lacking this enzyme are not yet available, and therefore a direct test of the involvement of endonuclease II in colicin E2-induced DNA degradation is not possible. Many other kinds of mutants lacking specific nucleases have, however, been tested with colicin E2 including UvrA⁻, UvrC⁻, Hss⁻, Hsp⁻ (restriction-modification genes) RecB⁻, RecC⁻, PolA⁻, and several mutants defective in DNA replication at high temperature. In all of these cases DNA solubilization was largely unaffected (Holland and Holland, 1970; Hull and Reeves, 1971). Thus, major nucleases involved in DNA repair, recombination, or in restriction-modification mechanisms are not specifically "activated" by colicin E2 action, contrary to an earlier suggestion (Holland, 1968). Similarly, DNA-ligase also appears not to be involved in the ultimate triggering of DNA degradation (R. L. Swift and J. S. Wiberg, personal communication). Endonucleolytic cleavage of

DNA promoted by colicin E2 also appears to be independent of a functional RecA gene (Holland and Holland, 1972); as indicated earlier, both solubilization and endonucleolytic cleavage actually proceed faster in RecA$^-$ mutants compared with wild-types.

The role of the major nuclease in *E. coli* (endonuclease I) in colicin E2-induced DNA breakdown has been examined by several groups, but with contradictory results. Obinata and Mizuno (1970), using a mutant strain *E. coli* 1100 deficient in endonuclease I (Dürwald and Hoffmann-Berling, 1968), found no difference in the rate and extent of either solubilization or cutting of DNA compared with wild-type strains. In contrast, Almendinger and Hager (1972), using the same mutant strain, showed that log-phase cultures were two- to threefold more resistant to colicin E2 (on the basis of viable counts) than the wild-type parent. However, using late log-phase and stationary-phase cultures, no differences in colicin E2 sensitivity between mutant and wild-type strains were obtained. Under these latter conditions it was claimed that residual endonuclease levels in the mutant were higher than in exponential-phase cultures. Further circumstantial evidence for the participation of endonuclease I in colicin E2 action was provided by the finding that wild-type bacteria failed to show colicin E2-induced DNA breakdown after being submitted to osmotic shock which released the bulk of endonuclease I from the periplasm. On the basis of these results, the authors proposed an ingenious explanation for colicin E2 action, namely that the translocation or redistribution of active endonuclease I molecules, from the periplasm into the cell interior, was potentiated by the colicin. The data in support of this theory are, however, open to criticism on several grounds. The parent strain used in this study is unfortunately atypical in being ususually resistant to several colicins, including colicin E2, making difficult the measurement of any increase in resistance resulting from loss of endonuclease I activity (Almendinger and Hager, 1972). In addition, much of the data concerning the osmotic shock treatment of wild-type cells is indirect, and is open to the objection that many proteins or other cofactors (see Holland and Holland, 1970) essential for some step in colicin E2 action may be released at the same time as endonuclease I. Finally, although the authors data indicate that endonuclease I activity in the mutant strain 1100 was as high as 2-3% of the wild-type level under certain growth conditions, the variation in residual enzyme level was only twofold over the range where maximum resistance to colicin E2 of strain 1100 was claimed to be expressed. In a more recent attempt to correlate mutationally induced changes in endonuclease I levels with altered DNA degradation patterns, the effect of colicin E2 on several mutant strains, including amber mutants containing $<0{\cdot}1\%$ of wild-type levels of endonuclease I, have been examined (Buxton and Holland,

1974a). The experiments were conducted using a range of colicin E2 concentrations, and in two mutants a slight lowering in the rate of DNA solubilization was observed; however in two other mutants the rate of DNA solubilization actually increased. A similar stimulation of DNA degradation has been observed by me to occur in the endonuclease I mutant, *E. coli* 1100 (unpublished results). Although it is difficult to rule out the possibility that a few residual molecules of endonuclease I per cell may suffice to facilitate colicin E2 action, these results, together with the absence of any direct evidence implicating endonuclease I, suggest that this enzyme does not play a major role in promoting DNA solubilization by colicin E2. However, a role for the enzyme in the fragmentation process is not ruled out, and the examination of this aspect of the breakdown process in strains with little or no endonuclease I activity is clearly needed.

In conclusion, the process of DNA degradation in colicin E2 treated cells may be summarized as follows. Breakdown appears to involve at least two major enzymes in the early stages—an endo- and an exonuclease, which do not necessarily act sequentially. Additional enzymes may participate at later stages when the secondary and tertiary structure of the chromosome is destroyed. *Escherichia coli*, F′*lac* and phage λ DNA are all attacked via a similar mechanism. Bacteriophage T4 DNA is not degraded, but this is due at least in part to phage directed inhibition of general cellular deoxyribonuclease activity. Solubilization of DNA does not appear to be initiated in membrane-associated regions of DNA, including the replication fork. The site of action of the colicin E2-induced endonuclease may be specifically localized in the cytosine-rich clusters of the DNA molecule. However, attempts so far to identify either of the two postulated nucleases with a specific cellular enzyme have been unsuccessful, although some evidence points to a possible role for endonuclease I in the fragmentation process (see also Almendinger and Hager, 1973). As will be discussed in the following section, we cannot rule out the possibility that colicin E2, or an active fragment of the molecule, penetrates sensitive cells and directly promotes DNA degradation. Further studies may reveal that colicin E2 alone, or in conjunction with some cellular protein, may initiate enzymatic degradation of the *E. coli* chromosome provided that, as in the case of colicin E3, highly specific substrate requirements are met. Examination of colicin E2 action, both *in vitro* and *in vivo*, upon highly folded largely "native" chromosomes (Stonington and Pettijohn, 1971; Worcel and Burgi, 1972) should be extremely instructive.

3. *Inhibition of Colicin E2-induced DNA Degradation*

In addition to preventing the initiation of DNA degradation and the conversion of colicin E2-treated cells to a state no longer susceptible to trypsin rescue, 2,4-dinitrophenol also immediately blocks ongoing DNA solubilization when added at any time after the colicin. The effect is immediately reversible if the inhibitor is diluted out, and to a considerable extent potassium cyanide or colicin K act similarly (Holland and Holland, 1970). Moreover, DNA fragmentation in colicin E2-treated cells is also blocked upon addition of dinitrophenol (Holland and Holland, 1972). Similar effects have been obtained with mitomycin C-induced DNA breakdown (Buttin and Wright, 1968), and the basis of the apparent energy requirement was ascribed to the activity of the ATP-dependent Rec, B, C enzyme (Wright *et al.*, 1971). In contrast to that provoked by mitomycin C, colicin E2-induced DNA solubilization proceeds normally in Rec B, C mutants (Holland and Holland, 1970). A very similar pattern of DNA solubilization to that promoted by colicin E2 is obtained with the peptide antibiotic phleomycin (Farrell and Reiter, 1973) and this suggests that energy-dependent nucleases other than the Rec B, C enzymes are involved in both cases. Alternatively, the dissipation of the energized membrane state which occurs upon the addition of uncoupling agents (see Section X, p. 101), may lead to the efflux of previously accumulated magnesium ions in sufficient amounts to limit severely further nuclease activity. Since dinitrophenol inhibits colicin E2-induced DNA degradation very effectively in nutrient broth cultures, where high levels of cellular ATP are maintained independently of oxidative phosphorylation, the latter alternative appears to have considerable merit.

Ongoing DNA solubilization, induced by colicin E2 in buffer suspensions of cells, is also immediately and severely restricted by the addition of 0·5 M-NaCl or KCl (Holland and Holland, 1972). This inhibitory effect can be obtained by addition of salt at times up to at least 45 min after the onset of DNA breakdown. This result, although not unequivocal, is consistent with the view that a colicin-membrane complex, which promotes DNA degradation, is either directly dissociated by the high external salt concentration or indirectly dissociated due to the plasmolysis of cells, with the consequent separation of the inner and outer membranes which may occur under these conditions. Thus far, a definitive experiment which would discriminate clearly between intracellular colicin E2 (which promotes DNA degradation) and an active colicin-membrane complex (in which at least part of the colicin molecule is external to the cytoplasmic membrane) has not been reported.

B. Inhibition of Cell Division

Colicin E2 treatment does not produce any immediate effect upon RNA synthesis or protein synthesis (Nomura, 1963). RNA degradation has been reported for certain strains, but this always occurs after the onset of DNA breakdown (Nose and Mizuno, 1968). Somewhat unexpectedly for an agent which promotes DNA degradation, colicin E2 does not apparently cause an immediate inhibition of DNA synthesis (Nomura, 1963; Holland, 1968). Kinetic experiments have shown that DNA solubilization precedes any reduction in the rate of thymine incorporation by several minutes (Holland and Holland, 1970), and transfer replication in Hfr × F^- crosses appears equally resistant (É. M. Holland, unpublished observations). In contrast to its delayed effect upon DNA synthesis, colicin E2 treatment leads to an early inhibition of cell division. This is best seen at low multiplicities of colicin E2, when growth of the cultures continues and inhibition of cell division is detected as the cells become filamentous (Holland, 1968; Holland and Holland, 1970; Beppu and Arima, 1971). Kinetic studies have shown that this effect takes place a few minutes after the onset of DNA solubilization, but several minutes before the inhibition of thymine uptake (Holland and Holland, 1970). Inhibition of cell division is also multiplicity-dependent and may therefore be a specific effect of colicin E2, quite independent of DNA breakdown induced, for example, during translocation of the colicin molecule to or through the cytoplasmic membrane. Alternatively, inhibition of division in colicin E2-treated cells may be dependent upon the degradation of a specific region of the DNA, for example near the chromosomal terminus (see Jones and Donachie, 1973). Studies of mutants tolerant to colicin E2 have, however, provided some evidence for a specific and independent effect of this colicin upon cell division. Thus when treated with colicin E2, at a high temperature, Cet mutants (see Section XI, p. 122) show very little colicin E2-induced DNA degradation, but division is fully blocked (Holland, 1968; Buxton, 1973). Furthermore Beppu *et al.* (1972) have recently isolated a mutant (DB312) tolerant to colicin E2 which, if grown at 30°C and treated with colicin E2, shows no detectable DNA solubilization or fragmentation, and prophage λ is not induced. Nevertheless, division is fully blocked and viable cells, equivalent to approximately one per filament, can be rescued if the cells are treated with trypsin at intervals up to 90 min after the addition of colicin. This is a quite remarkable result, demonstrating, on the one hand, the apparent separation of the effect of colicin E2 upon cell division from its effect on the chromosome, and, on the other hand, that, in this mutant at least, the active colicin molecule remains at the cell surface and accessible to trypsin for long periods. The observation by Ringrose (1970), which has been largely ignored, that the very early stages of colicin E2-induced nicking of DNA strands can also be halted by trypsin treatment, and the gaps or nicks then resealed, should now be re-examined in view of the properties of the mutant strain DB312.

C. Cell Surface Changes Accompanying Colicin E2 Action

In contrast to colicins E1 and K (Cavard *et al.*, 1968), colicin E2 action is not accompanied by any significant changes in the phospholipid composition of surface membranes (Cavard and Barbu, 1969; Nose *et al.*, 1970). Similarly, colicin E2 does not stimulate fluorescence changes of lipophilic probes previously bound to the cell membrane (Cramer and Phillips, 1970). However, bacteria treated with colicin E2 are frequently more difficult to lyse than untreated cells. In an attempt to quantify this phenomenon Nose *et al.* (1970) have shown that sphaeroplasts prepared from colicin E2-treated cultures are more resistant, both to spontaneous lysis and to EDTA-induced lysis, than are sphaeroplasts prepared from untreated cells or colicin E2-resistant cells treated with colicin E2. The development of this stabilizing effect is blocked by dinitrophenol but is otherwise complete in about five minutes. This period, which presumably reflects quite extensive surface changes, co-incides with the period of maximum trypsin reversal, and therefore with the formation of Complex II.

Two groups have investigated the possibility that colicin E2 displaces the whole, or a part, of the bacterial chromosome from membrane binding sites, thereby rendering it susceptible to attack by nucleases. Unfortunately the results obtained have been conflicting, presumably due to the general lack of rigid criteria for the identification of different forms of DNA-membrane binding site either *in vitro* or *in vivo*. Ringrose (1973) failed to demonstrate any specific dissociation of either uniformly-labelled or pulse-labelled DNA from "fast sedimenting" complexes in lysates from colicin E2-treated cells. On the contrary, newly synthesized DNA appeared less liable to solubilization than bulk DNA. In a different approach, Beppu and Arima (1972) demonstrated that purified colicin E2, but not colicins E3 or K, induced a rapid dissociation of DNA from a particulate complex *in vitro*. This effect, which was strikingly dependent upon added ATP, or other nucleoside triphosphates, was not accompanied by any detectable endonucleolytic cleavage of DNA. The complex was in fact dissociated above 30°C by either colicin E2 or ATP acting alone but in combination the releasing activity, which is strongly temperature dependent, was already maximal at 20°C. This phenomenon is similar to a finding reported recently by Worcel and Burgi (1974) that "folded chromosomes" can be melted out of isolated DNA-membrane complexes by a slight raising of the temperature. Curiously, dissociation of the DNA-envelope complex, studied by Beppu and Arima (1972), was also achieved with preparations obtained from both colicin E2-resistant and colicin E2-tolerant mutants, suggesting some direct effect of colicin E2 on the membrane which was independent of the receptor or of the normal translocation step. Unfortunately, in the absence of any precise knowledge of the nature of the complex being studied in these partly *in vitro* experiments, it is difficult to assess the significance of the results obtained. The properties of the complex, which appeared to contain 50% of the total cell protein in addition to 90% of the DNA, are reminiscent of the sieve-like ghost particles obtained after Brij lysis of sphaeroplasts (Godson and Sinsheimer, 1967). A relatively non-specific disruption of the ghost particles by the colicin preparation, leading to the release of trapped DNA, cannot therefore be ruled out. Nevertheless, this approach remains an important means of establishing the possible nature of colicin E2 interaction with the cell membrane, if not with the DNA itself. Careful characterization of the complexes used is, however, essential if the results are to be related to the possible *in vivo* action of the bacteriocin.

IX. Action of Colicin E2 *in vitro*

Maeda and Nomura (1966) found that after treatment of cells with radioactively-labelled colicin E2 only a small percentage of the labelled material was subsequently found in the cytoplasmic fraction, whereas over 80% was located in the envelope fraction. Nevertheless, since only 1–2% of bound colicin molecules appeared capable of promoting a lethal hit, the authors pointed out that these "active" colicin E2 molecules could be represented exclusively by the labelled material detected in the cytoplasmic fraction after breaking open the bacteria. However, to demonstrate unequivocally that a small minority of colicin molecules specifically enter the cell and interact directly with DNA is extremely difficult to accomplish. As an alternative approach to test the possibility that colicin E2 can act directly on the intracellular target, some *in vitro* experiments with purified colicin E2 and DNA have been carried out. Ringrose (1972) could not detect either fragmentation or solubilization of *E. coli* DNA with highly purified colicin E2, and several other attempts to measure a nuclease activity in colicin E2 preparations have also been unsuccessful (Nomura, 1964; Almendinger and Hager, 1972). Experience with colicin E3 tells us, however, that latent nuclease activity may go undetected in the absence of an appropriately structured substrate or essential cofactors. In fact, a recent study by L. Saxe and S. E. Luria (personal communication) indicates that colicin E2 preparations may indeed cause limited single-strand nicking in supercoiled phage λ DNA molecules *in vitro*. Full details of this experiment, and in particular whether this effect is limited to supercoiled forms of phage λ DNA are awaited with great interest.

Other studies by Ringrose (1972) have revealed some kind of interaction between colicin E2 and *E. coli* DNA *in vitro*. Thus, in low ionic-strength buffer, colicin E2 appears to bind to double-stranded DNA at many sites along the molecule. The significance of this is not clear, however, since colicin I, which has no apparent interactions with the chromosome *in vivo*, also binds to DNA (J. Konisky and Cuo-Tung Liu, personal communication). More interestingly, colicin E2, but not colicin E3, was found to lower the melting temperature of *E. coli* DNA to an extent proportional to the amount of colicin E2 added. This result indicates the ability of colicin E2 to destabilize helical regions of DNA, although under the test conditions used this could only be achieved when traces of phenol remained in the DNA preparation (Ringrose, 1972; see also Ringrose, 1971). This casts some doubt on the specificity of the effect and, in any case, indicates that the colicin acts only on DNA molecules which are already partially destabilized. The minimum concentration of purified colicin E2 required in these studies to produce detectable changes in the melting profile is equivalent to about 100 molecules per 10^7 daltons of DNA. This may be compared with the much smaller intracellular concentration of colicin E2 which is probably about 5–50 molecules per cell, if we assume that only 1% of colicin E2 molecules penetrate the cell surface (see Maeda and Nomura, 1966; Holland and Holland, 1970). The intracellular concentration is therefore likely to be 1000-fold less than the *in vitro* levels in these experiments. If colicin E2 acts in this way *in vivo*, it must be strictly limited to highly localized regions of DNA, and whether one single molecule could be capable of promoting extensive damage to DNA by this mechanism seems somewhat doubtful. However, as Ringrose concludes, these results are consistent with the idea that colicin E2 can induce localized destabilization of the DNA helix which may then facilitate its degradation by endogenous nucleases. Unfortunately, it is impossible to determine directly whether or not this mechanism operates *in vivo*, particularly if the nucleases "activated" are numerous and relatively non-specific.

A report by Seto *et al.* (1973) indicates that purified colicin E2 also interacts with single-stranded DNA of bacteriophage ϕX174. Thus, incubation of phage ϕX174 DNA with colicin E2, in low ionic-strength buffer, rapidly destroys the ability of the DNA to infect bacterial sphaeroplasts. This effect of colicin E2 is not accompanied by any detectable fragmentation of the phage DNA and, in fact, the DNA can be re-activated by centrifuging through alkaline sucrose gradients, suggesting that the loss of biological activity was due to the binding of colicin molecules. However, the significance of these results is difficult to assess since control experiments using other colicins were not carried out, and the step in the infective process blocked by colicin E2 treatment was not established.

These various studies provide some evidence that, *in vitro*, colicin E2 can act directly upon DNA, a molecule which appears to be the primary target *in vivo*. The results are nevertheless inconclusive, suffering primarily from the lack of a specific assay of *in vitro* activity which could be directly compared with specific changes promoted by colicin E2 *in vivo*. The establishment of the precise substrate (including its secondary and tertiary structure) for the primary nucleolytic cleavage *in vivo* may be necessary before the possible *in vitro* action of colicin E2 can be examined realistically.

X. Action of Colicins of the E1 Type

The great majority of colicins, and indeed of all bacteriocins studied so far, appear to act in a similar way by the suppression of energy supply derived from respiration. In consequence, growth of treated bacteria ceases and many small molecules, normally accumulated by active transport, are lost from the cells. Colicins E1, K, A and I are typical of this group, and the mode of action of colicins E1 and K, which have been extensively studied (particularly by Luria and his coworkers), will be examined in some detail. Adsorption of these colicins at 37°C appears to be extremely rapid (Wendt, 1970; Phillips and Cramer, 1973), and the specific biochemical changes which follow may appear within one or two min after the addition of colicin to either growing cultures or to cells suspended in buffer. The effect of multiplicity upon the kinetics of colicin E1 or K action has not received much attention, but several reports have indicated that, as with colicins E2 and E3, the effects promoted by colicins E1 and K also occur faster at higher multiplicities (Fields and Luria, 1969a). Conversely, in one study Wendt (1970) has shown that the rapid efflux of potassium provoked by colicin K becomes progressively more and more delayed as the incubation temperature is lowered, but this delay can be overcome by increasing the multiplicity of the colicin. In addition, Plate and Luria (1972) and Phillips and Cramer (1973) have demonstrated that cells treated with colicins E1 and K can be rescued by trypsin treatment at intervals up to several minutes subsequent to adsorption of colicin. Thus, colicins of the E1 type also appear to act via a two-step process involving an inactive Complex I and an active Complex II. In the case of colicin K, however, Complex II is not always lethal since, under certain conditions, Complex II can be reversed by trypsin treatment (Nomura and Nakamura, 1962). The implications of this effect will be considered later.

A. Effects of Colicins E1 and K on Metabolism

Jacob *et al.* (1952) first showed that colicin E1 immediately blocked growth and protein synthesis of treated bacteria without having any major effect upon respiration and without causing cell lysis. Nomura (1963) reported similar effects for colicin K. Levinthal and Levinthal (cited by Luria, 1964) later confirmed these results for colicin E1 but, in addition, made the important observation that bacteria growing strictly anaerobically are "insensitive" to colicin. Levinthal and Levinthal concluded that colicin E1 specifically inhibits the supply of energy from oxidative phosphorylation and that the site of action is therefore located in the cell membrane. In support of this proposition, Luria (1964) described the inhibition by both colicins E1 and K of energy-dependent permeases for uptake of isoleucine, lactose and K^+. Nomura and Maeda (1965) confirmed that K^+ efflux is an early effect of colicin E1 which is not observed with either colicin E2 or colicin E3. This effect, which can virtually empty treated bateria of potassium within 5 min (Wendt, 1970), is probably sufficient by itself to explain the inhibition of protein synthesis, a process which is dependent upon intracellular potassium (see Lubin, 1964; Harold and Baarda, 1968). The reported prompt inhibition of nucleic acid synthesis by colicins E1 and K cannot be explained on this basis. However, in view of the findings by Fields and Luria (1969a) that colicins E1 and K block a wide range of energy-dependent transport systems, measurement of nucleic acid or protein synthesis in colicin-treated cultures, by methods which depend upon the uptake of radioactive precursors, must be considered highly suspect. In fact Takagaki *et al.* (1973) observed considerable ^{32}P-orthophosphate incorporation into both RNA and DNA after colicin K treatment, and similar results are obtained when the synthesis of DNA and RNA is measured chemically in colicin I treated cells (I. B. Holland, unpublished data). In consequence, and on the basis of published data, it seems likely that inhibition of nucleic acid synthesis is not an early effect, but a secondary effect of colicin E1 or colicin K action.

B. Disruption of Energy Metabolism

1. *Nature of Possible Colicin E1 Targets*

The mechanism of energy production and its coupling to active transport is extremely complex and is far from being completely understood. In order to identify possible targets involved in cellular energy metabolism, which might be inactivated by colicins of the E1 type, it will therefore be necessary at this point to examine, at least in outline, the nature of energy transduction in bacteria as it is now generally conceived. The simple scheme presented in Fig. 4, the explanation of which draws heavily upon the excellent reviews of Harold (1970, 1972), shows that energy derived from substrate oxidation by the electron-transport chain, during aerobic growth, is trapped in the form of a high-energy intermediate designated ("~") which can both be harnessed directly to drive active transport or quite independently to drive forward the synthesis of ATP by oxidative phosphorylation (see also Klein and Boyer, 1972). However, in cells grown on glucose, ATP is also generated by glycolysis, whilst oxidative phosphorylation is repressed. Operating by the reverse pathway, ATPase functions under anaerobic conditions to synthesize ("~") from ATP supplied by glycolysis in the absence of an active electron transport chain. The exact nature of the high-energy intermediate ("~") is

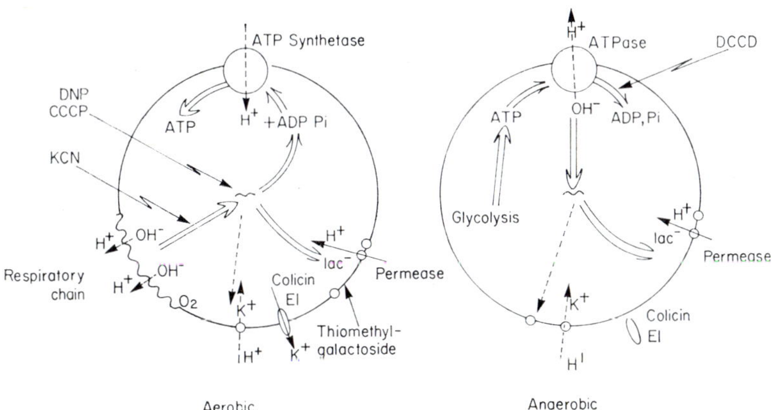

FIG. 4. Scheme for the major energy-generating process and its coupling to active transport in *Escherichia coli*. As discussed in the text, the high-energy intermediate ("~") is considered to be the energized state of the membrane which develops through the extrusion of hydrogen ions from the cell exterior during respiration. The sites of action of the uncouplers carbonyl cyanide *m*-chlorophenylhydrazone and 2,4-dinitrophenol (CCCP and DNP), the respiratory inhibitor (potassium cyanide), and the ATPase inhibitor N,N′-dicyclohexylcarbodiimide (DCCD), are as shown. The exact nature of the coupling of potassium-ion transport to the membrane potential is not yet established, and is indicated here by the broken line. Colicin E1 is seen here as intercalating into the cytoplasmic membrane to promote efflux of potassium ions under aerobic, but not under anaerobic, conditions. It is not yet clear how this relates to the dissipation of the energized state of the membrane by colicin E1.

still controversial, but many data concerning colicin action can be most easily presented and understood against a background which assumes, on the basis of the chemiosmotic theory of energy transduction (see Mitchell, 1972), that ("~") constitutes a membrane potential generated by the separation of hydroxyl ions from protons which are extruded to the cell exterior (for further discussion see Harold, 1972). On this model, uncoupling agents such as 2,4-dinitrophenol or carbonyl cyanide *m*-chlorophenylhydrazone dissolve in the membrane and act as proton conductors, thereby inducing collapse of the gradient and therefore the dissipation of the energized state. Major disruption of energy metabolism during aerobic growth can conceivably arise in two additional ways, namely direct inhibition of the electron-transport chain (e.g. by potassium cyanide), or the inhibition of ATP synthetase or any alternative means of ATP production (e.g. glycolysis). In the latter case, macromolecular synthesis would be affected as well as transport processes like the phosphoenolpyruvate-dependent phosphotransferase system which is essential for uptake of some sugars (including glucose) and non-metabolizable compounds like α-methylglucoside (Roseman, 1972). On the other hand, uncoupling agents, or inhibitors of electron transport, block active transport of many small molecules such as amino acids, β-galactosides, potassium ions and magnesium ions.

2. *Colicin E1 Action on Respiration and ATP Synthesis*

As already indicated above, colicins E1 and K specifically abolish active uptake and retention of amino acids, β-galactosides like thiomethyl-β-D-galactoside and K^+. In addition, these colicins also cause the suppression of cellular motility (Fields and Luria, 1969b). Furthermore, treated cells become permeable to magnesium ions and cobalt ions, although these effects appear somewhat more slowly (Lusk and Nelson, 1972). On the other hand, treated bacteria still accumulate α-methyl-glucoside and continue to respire (Jacob *et al.*, 1952; Fields and Luria, 1969b; Nomura, 1963), and haemin-deficient mutants, which lack cytochromes when grown partly anaerobically on glucose, remain fully sensitive to colicin under these conditions (Fields and Luria, 1969b). The functioning of the electron transport chain therefore seems unaffected by colicin. Treated cells also continue to carry out glycolysis (Fields and Luria, 1969b), to synthesize ATP (Feingold, 1970), and to incorporate ^{32}P-orthophosphate into both nucleotide and non-nucleotide compounds (Nomura and Maeda, 1965; Takagaki *et al.*, 1973). However, Fields and Luria (1969a) made the important observation that cellular ATP levels fall by at least 50% in the first 5–10 min after addition of colicin E1 or colicin K. In cells grown on glycerol, in which ATP is derived almost exclusively from oxidative phosphorylation, the drop in ATP levels is much more dramatic (Hirata *et al.*, 1969). Feingold (1970) confirmed the effects of colicin E1 upon cellular ATP levels, but his further investigations indicated that the decrease in ATP could be ascribed to the activation of ATPase by colicin, rather than to the inhibition of ATP synthesis. The addition of the ATPase inhibitor, N,N′-dicyclohexylcarbodiimide to colicin E1 treated cells was shown to prevent the disappearance of cellular ATP, and ATP levels were then observed to increase slightly. Nevertheless the induction of potassium efflux by colicin E1 was still observed. Moreover, *uncA* mutants, which lack [Ca^{2+} + Mg^{2+}]-dependent ATPase are still killed by colicin E1, although the biochemical basis of sensitivity in these mutants has not yet been analysed (B. Rolfe, personal communication). Thus, the decrease in ATP levels in colicin E1- and colicin K-treated cells is a secondary effect and is not a cause of the inhibition of energy-requiring processes which takes place. Contrary to the suggestion by Levinthal and Levinthal (see Luria, 1964), these results also indicate that oxidative phosphorylation is not specifically affected by colicins of the E1 type.

If respiration, and respiration-linked ATP synthesis, are not directly affected by colicin, we have the paradox that cells grown under strict anaerobiosis, and therefore lacking these processes, are in fact insensitive to both colicin E1 and colicin K. However, careful examination of this question by Fields and Luria (1969b) has indicated that the presence or absence of oxygen is perhaps more crucial in this connection than the actual pattern of ATP synthesis which is operating. Thus, although cultures growing semi-anaerobically generate ATP predominantly by a fermentative rather than by an oxidative pathway, the cells remain colicin sensitive. On the other hand, strict anaerobiosis renders the bacteria insensitive to colicin but sensitivity is expressed immediately upon admission of oxygen, suggesting that oxygen is required directly at the target site to promote colicin E1 or colicin K action.

3. *Colicin E1 as an Uncoupling Agent*

As discussed above, the colicin E1 target does not appear to involve specifically the electron-transport chain or sites of ATP synthesis. Consequently, we are left with the only viable alternative, that is that colicin acts directly on the membrane to dissipate the energized state ("~"). Unfortunately, colicin does not act as a simple uncoupling agent since analysis of colicin action reveals at least two important characteristics not found with classical uncouplers. Firstly, although dinitrophenol or carbonyl cyanide M-chlorophenly-hydrazone (CCCP) block K^+ uptake, efflux of this cation is not observed unless a K^+ conductor, like valinomycin, is also added. (see Feingold, 1970). Secondly, according to the studies of Feingold (1970), colicin E1, unlike CCCP, does not act as a proton conductor across the membrane, but selectively renders the cell permeable to potassium ions. This raises two extremely important points. Firstly, if colicin E1 does act directly to dissipate the membrane potential, this does not appear sufficient *per se* to induce K^+ efflux, but indicates that colicin may in addition specifically promote K^+ conduction (as found with the alkali–metal ionophores, like nigericin or the less specific gramicidins; Harold, 1970). Secondly, if colicin E1 does not promote H^+ conduction, how then is the presumed discharge of membrane potential achieved since clearly promotion of K^+ permeability alone cannot explain the effects of colicin action on energized transport reactions (see also discussion by Feingold, 1970). One may plead perhaps that proton conduction by colicin-treated membranes is small, and therefore undetectable, although still sufficient to dissipate effectively the energized state. There is again some precedence for this in the mode of action of gramicidins which, although primarily conducting K^+, also to some extent conduct protons in exchange for potassium, and the inhibition of several energy-dependent uptake systems is observed (Chappell and Crofts, 1966). A possible way out of this impasse is indicated by the suggestion (P. J. F. Henderson, personal communication) that the ion which is taken up in colicin-treated cells in exchange for K^+, if not H^+, may be NH_4^+, or a related ion. The transport of ammonium ions into the cell interior can lead to discharge of both membrane potential and pH gradients in the absence of any detectable net movement of protons (for detailed mechanism see Henderson, 1971). Finally, it may be noted that ionophores of the gramicidin type have been shown to catalyse very effectively translocation of NH_4^+ ions (Chappell and Crofts, 1966). The examination of NH_4^+ translocation in colicin-treated cultures, although still not revealing the whole story of colicin action as will be evident from further discussion below, may prove quite rewarding.

C. Effects of Colicins E1 and K on the Cell Surface

Bacteria treated with colicins E1 or K do not become permeable to exogenous substrates such as *o*-nitrophenyl-β-D-galactoside, and neither ATP nor most of the phosphorylated compounds or cellular enzymes appear to leak out (Fields and Luria, 1969a; Nomura and Maeda, 1965). Therefore, generalized breakdown of the membrane does not take place, a conclusion which is strengthened by the fact that colicin K-treated cells can, under certain conditions, be fully rescued by trypsin for long periods after protein synthesis has been completely inhibited (Nomura and Nakamura, 1962; Nomura, 1963). Nevertheless, widespread changes do appear to take place in one or both of the surface membranes in the early stages of colicin action, as indicated by the following evidence. Cramer and Phillips (1970) and Phillips and Cramer (1973) have demonstrated that the addition of colicin E1 to sensitive cells results in a large increase in fluorescence of the membrane probes, ANS (analinonaphthalene 8-sulphonate) and NPN (N-phenyl 1-naphthylamine), previously bound to the cells. This effect is not obtained with colicins E2 or E3 and is also not observed when colicin E1 is added to immune (Col E1$^+$) bacteria. This effect of colicin E1 on ANS fluorescence is in marked contrast to the decrease in fluorescence which accompanies the setting up of an energized state in membrane vesicles upon addition of D-lactate (Kaback, 1972). Fluorescence changes in dyes of this kind have been postulated to reflect structural or conformational changes in hydrophobic regions of proteins (see Edelman and McClure, 1968; Brand and Gohlke, 1972), which could presumably be brought about in colicin-treated cells by overall changes in membrane potential. Other explanations are possible, however, particularly since Cavard *et al.* (1968) have reported activation of a phospholipase in the cell membrane of bacteria treated with either colicin E1 or colicin K. Such effects, which have also been observed with other bacteriocins (see p. 111), could, even at low colicin multiplicities, spread within the plane of the membrane to produce large changes in fluorescence of lipophilic probes as observed in the studies of Cramer and Phillips (1970). Unfortunately, although Cavard *et al.* (1968) reported that alterations in membrane phospholipid composition occur rapidly and prior to the release of phosphorylated compounds from colicin-treated cells, the kinetics of these changes, relative to other major effects of colicin action, were not established. Nevertheless, as discussed below, serious consideration must be given to the possibility that a primary effect of many colicins is the modification of membrane phospholipids. Phillips and Cramer (1973) also compared the kinetics of several changes induced by colicin E1, and were able to conclude that under most conditions the increased fluorescence of the lipophilic probe NPN, the decay in ATP levels and efflux of K^+ all quite closely parallel the loss of the capacity of cells for trypsin rescue. In the presence of the ATPase inhibitor dicyclohexylcarbodiimide, ATP levels rose rather than fell in colicin E1-treated cells, as found by Feingold (1970), but the increased fluorescence of the dye was not affected. These results indicate therefore that potassium efflux, activation of ATPase and presumably the loss of membrane potential, all occur more or less simultaneously with extensive structural changes in the cell surface, as indicated by the altered behaviour of the fluorescence probe. One must emphasize, however, that the significance of these structural changes, and the altered phospholipid composition of colicin-treated cells observed by Cavard *et al.* (1968), cannot be fully assessed until it has been established whether either or both of the *E. coli* surface membranes is involved.

D. Colicins and Phospholipid Metabolism

From the foregoing it is clear that specific changes in membrane phospholipids do accompany the action of colicins of the E1 type. It appears equally clear that the functional integrity of membrane transport systems, and their coupling to potential gradients generated by the electron-transport chain, will be dependent upon the composition of the associated phospholipids (see Gale, 1971; Rothfield and Romeo, 1971). Consequently several questions may now be posed in relation to colicin action. Are the observed changes in phospholipid metabolism associated with the penetration of colicin through the outer membrane, or with its interaction with sites in the inner membrane? If the latter, are such changes due to the activation of endogenous enzymes or to an inherent enzymic activity of the colicin molecule? Finally, are the changes in phospholipid metabolism associated with the causes or the effects of inhibition of energy metabolism? These are mainly questions for future research, but since an example of a bacteriocin with a phospholipase activity has already been well documented, the implications of this for the possible mode of action of colicins should be considered here. Purified megacin A, a highly potent bacteriocin produced by *Bacillus megaterium* (Holland, 1967b), carries a phospholipase A2 activity which can be measured *in vitro* (Ozaki *et al.*, 1966). Cells or protoplasts treated with megacin A gradually lyse (Ivánovics *et al.*, 1959), and it appears very likely that cell death is associated with the enzymic activity of the bound molecule (Holland, 1962). Unfortunately, measurements of metabolic changes in megacin treated cultures have not been carried out and consequently any concomitant effects upon active transport, prior to cell lysis, would have gone undetected. Megacins are seemingly only active against strains of *Bacillus megaterium* (Nagy *et al.*, 1959), and this strict specificity, which is maintained at the protoplast level, indicates that the phospholipase activity is extremely discriminatory and may only be active on biological membranes containing certain very specific structural configurations. In relation to colicin action and phospholipid metabolism, the major change apparent in *E. coli* membranes after treatment with either colicin E1 or colicin K is the conversion of phosphatidylethanolamine to the lyso-derivative, a reaction which is indeed catalysed by phospholipase A2 (Cavard *et al.*, 1968). Phosphatidylethanolamine is particularly abundant in the outer membrane of *E. coli* where it constitutes a major part of the lipid A component (see Rothfield and Romeo, 1971); however, phosphatidylethanolamine also occurs in the inner membrane. Unfortunately, kinetic experiments using whole cells are unlikely to establish whether changes in phospholipids are primary or secondary effects of colicin action and, in any case, such experiments will not discriminate between changes induced in the outer rather than in the inner membrane. Thus, an examination of the action of colicins of the E1 type upon isolated membranes will be required. Such studies should include a search for phospholipase activities arising from activation of endogenous enzyme, or executed directly by colicin, when presented with various membrane preparations such as those derived from mutants tolerant to colicins of the E1 type. In the light of previous studies with colicin E3, one might anticipate that highly purified colicins, free of immunity factors and used with relatively intact membrane structures, will be required if enzyme capabilities are to be revealed. Enzyme action which leads to disruption of essential membrane functions need not, of course, be restricted to phospholipases, and the search may have to be broadened to include other enzymes. As indicated in a following section, the major changes in phospholipid composition which accompany colicin A and bacteriocin 1580 action appear to involve conversion of phosphatidylglycerol to diphosphatidylglycerol, which may proceed by the condensation of CDP-diacylglycerol and a molecule of 3-phosphatidylglycerol, and is presumably mediated by a membrane-bound enzyme. Finally, the possible involvement of oxygen in colicin E1 action must be taken into account and, as suggested by Feingold (1970), this may be indicative of an oxidative attack upon unsaturated fatty acids mediated, for example, by hydrogen peroxide resulting from localized activation of NADH oxidase by colicin.

E. Subcellular Systems and the Action of Colicins E1 and K

Clearly, the examination of colicin action upon sphaeroplasts, membrane vesicles or other membrane fractions, could reveal important information concerning the direct effects of colicin upon the membrane. Some preliminary studies of this kind with colicins of the E1 type have, in fact, been quite illuminating with regard to the role of the outer membrane receptor. Thus, sphaeroplasts of colicin-sensitive strains have been shown to be resistant to colicins E2 and E3, but sensitive to colicins K (Nomura and Maeda, 1965) and E1 (Obdržalek *et al.*, 1969). Sphaeroplasts adsorb normal amounts of colicins E2 and E3 but the disposition of the residual surface layers of these "organisms", which leaves the inner membrane well separated from remnants of the outer membrane (Birdsell and Cota-Robles, 1967), appears to exclude any interaction between the inner membrane and colicin molecules bound to receptors. In the case of colicins E2 and E3, this is probably sufficient to render the sphaeroplasts insensitive. In the case of colicins of the E1 type, however, the sensitivity of sphaeroplasts may indicate that colicin can act directly on target sites present on the exposed surface of the cytoplasmic membrane, without the intervention of the receptor. In support of this, Bhattacharyya *et al.* (1970) have shown that membrane vesicles prepared from colicin E-resistant (*bfe*$^-$) cells are just as sensitive to colicin E1 inhibition of proline uptake as are vesicles prepared from sensitive bacteria. One may question the specificity of such an effect in a subcellular system, but the characteristics of proline uptake under these conditions appear to be very similar to those of whole cells (Kaback, 1972). Moreover TolC mutants, specifically tolerant to colicin E1, produce equally tolerant membrane vesicles. Essentially similar results for the action of colicin K upon vesicles derived from resistant, sensitive or tolerant bacteria have been obtained by Takagaki *et al.* (1973). Unfortunately, the ability of vesicles derived from various mutants actually to adsorb colicin has not been measured, and in some cases it has merely been assumed that colicin receptors must reside in the cytoplasmic membrane and not in the outer membrane (Bhattacharyya *et al.*, 1970). In view of the evidence now available for the localization of colicin-binding sites in the outer membrane, it seems reasonable to conclude with Takagaki *et al.* (1973) that the outer membrane in *E. coli* actually constitutes a barrier to colicin K which in sensitive cells is somehow circumvented by colicin K binding to the receptor sites. Consequently, in the absence of the outer membrane, this colicin can act directly upon target sites in the inner membrane. Should this action prove to be enzymic, or at least catalytic in nature, irreversible binding to the target is unlikely to occur. Interestingly, in support of their hypothesis, Takagaki *et al.* (1973) describe the properties of a mutant, isolated from a parent bacterial strain lacking the colicin K receptor, which had regained colicin K sensitivity. This revertant, which has an altered lipopolysaccharide component of the outer membrane and which is hypersensitive to novobiocin, still does not adsorb measurable amounts of colicin K. Thus, although the surface receptor may prove to play a more positive role in the mode of action of colicins E2 and E3, its function in the case of colicin K appears to be dispensable if the target site is immediately accessible. This situation is perhaps similar to the case of amino-acid and sugar-binding proteins present in the periplasm of Gram-negative organisms which act as efficient scavengers of essential nutrients but do not participate in the actual uptake process when this is measured in vesicle preparations (Kaback, 1972).

Isolated membrane fragments or vesicles have not so far been extensively used as systems for the direct examination of colicin E1 or K action upon membrane components. However, the energy-dependent uptake of proline, as discussed above (p. 109), has been shown to be blocked by colicins E1 and K in vesicle preparations. In addition, Takagaki *et al.* (1973) have described inhibition of the energy-dependent uptake of ^{32}P-orthophosphate into disrupted sphaeroplasts by colicins E1 and K. In the latter studies, however, colicin E2 appeared just as effective an inhibitor, which suggests caution in accepting the specificity of these effects. Some mention should be made here of the extensive studies on the effects of partially purified colicin A, and a highly purified *Staphylococcus* bacteriocin (1580), on vesicle preparations which have been carried out by Jetten and Vogels (1973). Colicin A and bacteriocin 1580 are very similar to colicin E1 in their action upon both intact cells and vesicles. Active transport of proline and glutamate is blocked, and increased fluorescence of a lipophilic probe (ANS) is observed, upon addition of colicin A or bacteriocin 1580 to membrane vesicles. Moreover, in the case of colicin A, vesicles derived from tolerant mutants show no increased ANS fluorescence whilst, as might now be anticipated, vesicles derived from a colicin A-resistant mutant do show increased fluorescence with the probe. Interestingly, the species specificity of the two bacteriocins for whole cells was preserved at the membrane vesicle level, attesting to the specificity of the effects observed in these subcellular systems. This result strongly suggests that the interaction of these bacteriocins with the target site will prove to be highly specific. This feature, which is reminiscent of the substrate specificity of megacin A discussed above (p. 108), stands in contrast to that of the colicin E3 target which appears to be conserved in the structure of ribosomes of several bacterial species. (See Section VII, p. 85.)

F. Action of Other Bacteriocins of the E1 Type

Colicin A has few binding sites on sensitive bacteria, and irreversible binding takes place more slowly than with colicins E1 or K (Cavard and Barbu, 1970). Nevertheless, Nagel de Zwaig and Vitelli-Flores (1973) have shown that isoleucine uptake is blocked within a few minutes of addition of colicin A. Colicin A also blocks the energized uptake of other amino acids and sugars, and halts cell motility, but has no effect upon the enzyme-dependent phosphoenolpyruvate-mediated uptake of α-methylglucoside (Nagel de Zwaig, 1969; Jetten and Vogels, 1973). Colicin A, like colicins E1 and K, does not promote H^+ conduction across the membrane, although the membrane becomes abruptly permeable to protons upon subsequent addition of carbonyl cyanide-*m*-chlorophenylhydrazone, as observed with colicin E1 (Jetten and Vogels, 1973). This suggests that colicin A too causes selective permeability of the membrane to potassium, but this has not yet been determined. Finally, increased levels of diphosphatidylglycerol and decreased amounts of phosphatidylglycerol occur in colicin A-treated cells, but these changes are less dramatic than those obtained with colicins E1 or K (Cavard *et al.*, 1968).

Colicins Ia and Ib have been studied by Levisohn *et al.* (1967). These colicins compete for an identical receptor site present in both colicinogenic and non-colicinogenic bacteria and can only be distinguished by the immunity of the producing strains to the homologous I colicin. Both colicins immediately block incorporation of radioactive leucine into protein and, at high multiplicities, ^{32}P-orthophosphate incorporation into RNA and DNA is also markedly inhibited. As found with colicins E1 and K, colicin I has little immediate effect upon respiration and, in cells grown on glucose, ATP synthesis continues at nearly half the maximal rate. Colicins Ia and Ib have been shown to inhibit active transport of thiomethylgalactoside and proline (J. Konisky, personal communication). Therefore, these colicins probably block energy supply derived from respiration by a similar mechanism to that promoted by colicins E1 and K. Colicin I killing is also subject to rescue by trypsin treatment, but the kinetics of this process have not been established. Levisohn *et al.* (1967) were able to show that immunity to colicin I is not complete; by increasing the dose of homologous colicin I at least ten-fold above that used for sensitive strains, "immunity breakdown" of the colicinogenic strains was achieved. Nevertheless, under these conditions the effects induced by colicin I appeared identical to those observed with sensitive non-colicinogenic cultures. These results are analogous to those obtained with bacteriophage λ, when high multiplicities of phage are seen to titrate out some endogenous repressor or immunity factor.

Three other bacteriocins with properties similar to colicin E1 have been described. Colicin S8 (Nagel de Zwaig and Vitelli-Flores, 1973) inhibits uptake into the cells of protein and nucleic-acid precursors and methyl-thio-β-D-galactoside, but accumulation of α-methylglucoside is not blocked. However, these biochemical changes appear relatively slowly when compared with the kinetics of colicin E1 or colicin K action, suggesting that complex II formation in this case is quite inefficient. The action of colicin S8 is greatly diminished at 20°C or below, and fatty-acid auxotrophs of *E. coli* grown on elaidate are less sensitive to the action of colicin S8 than oleate grown cells. These results indicate the importance of the phospholipid composition of the cell surface for complex II formation in the case of this bacteriocin. Foulds (1971) has described the properties of a bacteriocin JF246 produced by *Serratia marcescens* which blocks active transport and induces a large reduction in ATP levels, whilst uptake of α-methylglucoside is unaffected. Extensive studies have also been carried out on a bacteriocin (1580) produced by *Staphylococcus epidermis* which is active against several species of Gram-positive bacteria but has no effect upon *E. coli* (Jetten and Vogels, 1972a, b, 1973; Jetten, 1973). Bacteriocin 1580 appears to have identical effects to colicin E1 upon active transport and upon the ability of treated bacteria to retain accumulated small molecules, including rubidium ions. Levels of ATP are also drastically decreased and motility is blocked. Generalized breakdown of the membrane is not observed and respiration, although decreased, is much less affected than energized transport. Bacteriocin 1580 does not cause H^+ conduction across the membrane, although the proton gradient collapses immediately upon subsequent addition of carbonyl cyanide-*m*-chlorophenylhydrazone. This effect is independent of the external potassium concentration, and appears to be identical to that promoted by colicins A and E1. Consequently, it seems most likely that bacteriocin 1580 also causes a selective permeability of the cell membrane to a major cation other than H^+, and the most likely candidate is potassium. Jetten and Vogels (1973) have examined the composition of phospholipids from bacteriocin 1580-treated cells. Under conditions where the total phospholipid content changed very little, they observed a marked decrease of phosphatidylglycerol and a virtually equivalent increase of diphosphatidylglycerol, suggesting direct conversion of the former to the latter, within the membrane. These changes took place over the same time period as the other biochemical effects of bacteriocin 1580, but it is impossible to deduce whether they are primary or secondary effects of its action.

XI. Insensitivity to Colicins

The majority of *E. coli* strains isolated from natural sources are sensitive to a wide variety of colicins. Colicin *insensitivity* can, however, arise in three different ways: (i) colicinogenic strains are *immune* to the colicin they produce; (ii) loss or inactivation of surface-binding sites leads to *resistance* through failure to adsorb specific colicins; and (iii) colicin *tolerant* mutants adsorb normal amounts of colicin but some subsequent step in colicin action is not completed and the bacteria survive. The basis of these three forms of colicin insensitivity will be considered in turn.

A. Immunity

Colicins of the E group, although originally classified together because of their identical activity spectra against sensitive bacteria, were subsequently subdivided into groups E1, E2 and E3 with the finding that the respective producing strains were immune to the homologous colicin but sensitive to the other members of the group (Fredericq, 1957, 1958). Colicins of group I were similarly subdivided, in this case into groups Ia and Ib on the basis of the immunity of the producing strains to the homologous colicin (Stocker, 1966). Colicinogenic strains still adsorb homologous colicin (Maeda and Nomura, 1966; Levisohn *et al.*, 1967; Konisky and Cowell, 1972); the surface receptors are therefore present and apparently intact. Moreover, since colicins E2 and E3 and colicins Ia and Ib, respectively, each adsorb to identical binding sites, immunity specificity is not expressed at the level of the receptor. Bowman *et al.* (1971b) have shown that purified ribosomes obtained from bacteria carrying the Col E3 factor are just as sensitive to added colicin E3 *in vitro* as are the ribosomes from sensitive strains. Therefore, immunity, in this case at least, does not involve modification of the colicin target either. As discussed in Section VII (p. 85), a protein which appears to neutralize colicin E3 action *in vitro* can be isolated from lysates of induced Col E3 cultures, but not from similarly treated Col⁻ cultures. Consequently, the inactivity of crude preparations of colicin E3 in the *in vitro* assay (Bowman *et al.*, 1971b) can be ascribed to the presence of this inhibitor. Since unpurified colicin E3 is nevertheless active on intact cells, separation from the inhibitor must take place as colicin penetrates the cell surface. Purification of colicin E3 similarly leads to removal of the inhibitor without loss of colicin E3 activity, demonstrating that the inhibitor does not irreversibly inactivate the colicin. Bowman *et al.* (1971b, 1973) have proposed that this inhibitor is the factor responsible for the immunity of Col E3 bacteria. Although persuasive, this evidence is not unequivocal and the possibility has not been ruled out that Col E3 bacteria produce an additional factor responsible for immunity which, for example, is normally expressed at the cell membrane level. The isolation of Col factor mutants, deficient in immunity to colicin E3, which also produce an altered inhibitor inactive against E3 *in vitro*, may be necessary to establish the identity of the *in vivo* immunity substance. Whatever the normal role of the inhibitor protein in Col E3 cultures, it appears unlikely that this molecule plays a crucial part in transcriptional control, analogous to that of the phage λ repressor (Ptashne, 1972). Thus, high levels of the inhibitor occur in induced cells, whereas only low levels are apparently present in non-induced cultures (Bowman *et al.*, 1971b), when colicin synthesis at least is repressed.

An inhibitor of megacin A, present in cell lysates from induced cultures of the producing strain, has also been reported (Ochi *et al.*, 1971). However, information has not so far been obtained concerning the possible basis of immunity to any other colicins. In the case of colicins of the E1 type which act on the membrane, it will be particularly interesting to determine whether immunity involves direct interaction between colicin and the immunity factor within the membrane.

B. Resistance

Mutants strains which have lost a functional surface receptor and fail to adsorb colicin can very easily be isolated by cross-streaking sensitive bacteria against a streak of the appropriate colicin. A few isolated colonies of resistant bacteria are almost invariably observed which, upon purification, are shown to be non-adsorbers. Mutants specifically resistant to one of a variety of different colicins have been obtained in this way and a number of genetic loci on the *E. coli* chromosomal map, each determining colicin resistance, have been identified (see Taylor, 1970). The only detailed information which is available concerns the basis of resistance to the E group colicins, and this will now be examined. The great majority of mutants selected for resistance to any one of the E colicins, or to phage BF23, prove to be simultaneously resistant to colicins E1, E2, E3 and BF23 (Fredericq, 1949; Reeves, 1965b; Hill and Holland, 1967). These mutants fail to adsorb any of the E colicins and a locus, *bfe*, determining this character has been mapped between *arg* (HBCE) and *thi* at 79 minutes on the *E. coli* chromosome (Buxton, 1971; Jasper *et al.*, 1972). Mutations at this locus are recessive to the wild-type allele, and 14 independent mutations isolated by Buxton (1971) all failed to complement when tested in merodiploids, suggesting that the *bfe* locus constitutes a single complementation group. As indicated in Section IV (p. 66), strains mutant at the *bfe* locus lack a minor envelope constituent, a 60,000 dalton polypeptide, which, when isolated from wild-type strains, binds colicin E2 and colicin E3 *in vitro*. Whilst this evidence suggests that the 60,000 dalton polypeptide is the product of the *bfe* locus, other evidence indicates that the colicin E receptor is nevertheless quite complex. The isolated receptor has little neutralizing activity for colicin E1 (Sabet and Schnaitman, 1971) which suggests that binding of colicin E1 to the receptor requires components in addition to the *bfe* product. Furthermore, competition binding experiments (Maeda and Nomura, 1966) have shown that colicins E2 and E3 bind to identical receptors in the intact cell, whilst colicin E1 binds to different receptors. Finally, mutants occur at low frequency which fail to bind colicins E2 and E3 but still bind colicin E1 (Hill and Holland, 1967). Unfortunately, genetic analysis has not been carried out on the latter mutants and so it is not possible to know whether they represent a specific class of Bfe mutants or whether they represent a class of mutations affecting another constituent of the colicin E receptor. On the basis of all the data, it appears that the receptor for colicins E1, E2, E3 and BF23 all contain the *bfe* product, but that specific modifications of this polypeptide, or the presence of additional components, are necessary in order to specifically bind colicin E1 on the one hand, or colicins E2 and E3 on the other. Finally, of the mutants resistant to colicins E2 and E3, but still adsorbing colicin E1, some are sensitive and some resistant to phage BF23, indicating that the phage receptors, although containing the *bfe* product, are also complex and distinct from the colicin binding sites (Hill and Holland, 1967).

C. Tolerance

Mutants insensitive to colicin, which nevertheless retain the surface receptor and bind normal amounts of colicin, are now called "tolerant". Earlier synonyms, such as "refractory" (Hill and Holland, 1967) or "mutationally immune" (Clowes, 1965), are no longer in general use. Attempts to isolate colicin-tolerant mutants were initiated in order to reveal the nature of the steps, subsequent to fixation by the surface receptors, which lead to the killing of sensitive bacteria by colicin. The isolation and study of tolerant mutants in several laboratories have indeed confirmed that prior to disruption of the target site there are intermediate steps in colicin action which appear to involve the cell surface. The properties of these mutants have, however, so far failed to uncover any precise details of these intermediate steps. For example, the presence of specific surface proteins which recognize bound colicin molecules and subsequently promote their translocation to the cell interior, or signal their presence to the cytoplasmic membrane by the initiation of specific conformational changes, has not been revealed. In addition, efforts to identify tolerant mutants actually altered in the target sites have so far been unsuccessful. This latter failure is perhaps less unexpected since the frequency of mutations affecting, for example, ribosomal proteins which constitute part of the colicin E3 target, is likely to be extremely low. Despite these disappointments, and the unexpected complexity of some of the mutants, the isolation and study of tolerant strains is being actively pursued since it is clearly a convenient method for probing into the wider problems of the organization and assembly of bacterial membranes.

Colicins of the E type have been most frequently used in the selection of tolerant mutants since potential mutants can be readily identified as strains insensitive to at least one E-type colicin, but still sensitive to other E-type colicins or to phage BF23. Screening large numbers of insensitive clones is therefore quite simple and the mutants obtained may be purified and their ability to adsorb the appropriate colicin confirmed, as far as possible, by mixing them with colicin and measuring the disappearance of the latter from the medium. These preliminary studies sometimes reveal mutant strains which excrete colicin inhibitors into the medium. A series of such mutants have been isolated by Guterman (Guterman and Luria, 1969; Guterman, 1971) and shown to hyperproduce the iron chelator enterochelin which appears to neutralize colicins I, B and V. This type of insensitive mutant, however, is relatively rare and has not so far been recognized amongst the colicin E and K-tolerant mutants, which will now be considered in detail.

When selection is made against any one of the E type colicins, 70–80% of the insensitive mutants obtained are of the *bfe*$^-$, receptor negative, type. The rest constitute a heterogeneous collection of tolerant mutants of which about 80% are usually *tol*A,B or C mutants. The great majority of tolerant mutants do, in fact, show cross tolerance to a variety of different colicins, including some with quite different receptors and modes of action from the colicin used in the selection. An exhaustive survey involving 18 different colicins, by J. K. Davies and P. Reeves (personal communication, and in preparation) has shown, however, that cross tolerance patterns are restricted to two major groups—the E, K, group and the I, B group. Cross-tolerance is the rule within each group, but is not observed between the groups. The basis for this separation of multi-tolerant types is not clear. In contrast, a few classes, namely TolI, TolC and Cet mutants are mainly tolerant to only a single colicin. These mutants are most likely to be specifically defective in Complex II formation or even to have an altered colicin target. The characteristics of the major ten groups of tolerant mutant which have been isolated so far are shown in Table 2, and some of their physiological and genetical properties will be considered in turn.

1. TolP,A and B Mutants

A large number of single point mutants tolerant to colicins E2, E3, K and A (TolP,A), and another group (TolB) which, in addition, are tolerant to colicin E1, have been mapped near *gal* on the *E. coli* K12 chromosome. Extensive studies by Rolfe and coworkers (Bernstein *et al.*, 1972, 1973) have now confirmed by complementation analysis that three closely linked cistrons are involved and that the map order is *suc*AB, p, *tol*P, *tol*A, p, *tol*B, *aro*G, *gal*. The presence of the promotor regions (designated p) was inferred from the properties of polar mutations obtained by integration of bacteriophage μ into various sites of the *tol*P.A.B. region (Bernstein, 1973). The results indicated that the *tol*P.A cistrons constituted an operon transcribed independently of the *tol*B cistron. These studies were taken an important stage further by the isolation of specialized λ-transducing phages carrying varying extents of the *tol*P.A.B. cluster (Bernstein *et al.*, 1972). Using these phages, identification and characterization of the specific gene products of *tol* genes, and their mode of transcription and translation, should now be possible both *in vivo* and *in vitro*. Bernstein (1972 and personal communication), in some preliminary experiments, has obtained evidence for the synthesis of three specific polypeptides in cells infected with various λd*tol* transducing phages. Control experiments indicated that these polypeptides were not determined by phage λ cistrons and they could therefore be the products of the TolP.A.B. genes. Amber mutations affecting both *tol*A and *tol*B have been isolated (Schwarz, 1972) and a temperature-sensitive *tol*A mutant has been described whose properties suggest that synthesis of some protein is defective at high temperatures. Other evidence, although indirect, suggests that the *tol* gene products are protein constituents of the cell envelope or, as suggested by Nagel de Zwaig and Luria (1969), enzymes involved in envelope synthesis. Thus TolA,B mutants grow slowly, particularly at high

TABLE 2. Properties of colicin-tolerant (Tol) Mutants of *Escherichia coli*

Initial symbol	Locus	Map position (min)	Tolerance to colicin	Major additional characters	Known surface alteration	Dominant/ recessive	Major references
TolII, Ref. VI	*tol*A	16·5	E1, E2, E3, K, A	Doc^s	—	R	a, b, c, d
TolIII, Ref. Va	*tol*B	16·5	E2, E3, K, A	Doc^s	—	R	a, b, c, d
TolVIII, Ref. I	*tol*C	58	E1	mb^s	lacks envelope polypeptide*	—	a, c, e
	*tol*D	20	E2, E3	Doc^s, amp^r	—	D?	f
	*tol*E	20	E2, E3	Doc^s, amp^r	defective lipopolysacharide	—	f
TolIa?	*tol*F	21–23	A, K	Doc^s	—	—	b, g
	*tol*G	21–23	bc	$Novobiocin^s$	lacks a major outer-membrane polypeptide	—	g
TolII	*tol*P	16·5	partial tolerance to E1, E2, E3, K, A	Doc^s	—	R	a, b, c, d
Ref II, TolVIIt	*cet*B	0	E2	None	Increased levels of inner-membrane polypeptide	D	a, b
TolIV t		64	E3, E2	Col E3, E2, $tol^{42°}$	—	—	b
TolIV 1, strain 483		about 64	E3, E2	Col E32, $tol^{42°}$; no growth at 42°	lacks an envelope polypeptide	—	b

Several additional loci unlinked to *gal* at 16·5 min and probably located in left half of chromosome have also been isolated namely $TolIb^b$; $TolIVb^b$; $TolV^b$; $TolVI^b$; Ref III^a; Ref IV^a; Ref Vb^a and Ref VIII^a.

References: a, Hill and Holland (1967); b, Nomura and Witten (1967); c, Nagel de Zwaig and Luria (1967); d, Bernstein *et al.* (1972); e, Whitney (1971); f, Eriksson-Grennberg and Nordström (1973); g, Foulds (1974).

*This organism is deletion mutant of TolC.

temperatures unless a medium of high osmolarity is employed. The cell envelope is quite fragile and β-galactosidase can be readily detected in the growth medium. Finally, the mutants, at least when first isolated, are hypersensitive to deoxycholate and EDTA (Nagel de Zwaig and Luria, 1967) and to various antibiotics (Bernstein *et al.*, 1972). These properties indicate that TolP.A.B. mutations, like those of most colicin tolerant classes, are pleiotropic. Reversion studies and cotransductional analyses have confirmed that all the newly acquired characteristics of these mutants are indeed due to a single mutational event (Nomura and Witten, 1967; Bernstein *et al.*, 1972). Moreover, in partial diploids, both deoxycholate sensitivity and colicin tolerance are recessive to the wild-type allele, and *tol*A mutants, which are tolerant to colicin only at high temperatures, are also sensitive to deoxycholate at high but not at low temperatures.

The altered permeability of *tol*P,A,B mutants to a wide range of antibiotics and other compounds suggests some fairly non-specific changes in the outer membrane of the cells. Nevertheless, preliminary examination of the gross lipopolysaccharide content of the surface of tol^{+} and tol^{-} cells has not revealed any significant differences (Nagel de Zwaig and Luria, 1967). Studies concerning the phospholipid composition of the mutants have not so far been reported. However, studies by Bernstein *et al.* (1972) have shown that, in addition to their other properties, *tol* A,B mutants suppress the defect in lysis defective bacteriophage λ.S mutants. Consequently, since the S-gene product is thought to act upon the cytoplasmic membrane (Reader and Siminovitch, 1971), Bernstein *et al.* (1972) suggest that the *tol*A,B gene products may also be located in that region. Unfortunately, an analysis of the membrane polypeptides of these mutants has not yet been reported, and so one may only speculate on the precise localization of the defect in the envelope which renders the cells tolerant to colicins E, A and K.

Many Tol mutants, including those of the *tol*P,A,B type, are frequently found to be unstable, reverting fully to wild-type or, in some cases, regaining resistance to detergents whilst retaining the original tolerance pattern (Schwarz, 1972; I. B. Holland, unpublished observations). Care should therefore be taken to preserve mutant stocks as deep frozen or lyophilized preparations in order to avoid out-growth of revertants.

2. *TolC Mutants*

TolC mutants are tolerant to colicin E1 and slightly tolerant to colicin A (Nagel de Zwaig and Luria, 1967; Davies and Reeves, 1974) and therefore present a much more specific class of tolerant mutants than do the *tol*P,A,B group. The *tol*C locus has been placed at min 58 on the *E. coli* map, 5% cotransducible with *met*C (Whitney, 1971). These mutations appear to be pleiotropic since TolC mutants are hypersensitive to deoxycholate and to certain dyes including methylene blue (Clowes, 1965; Nagel de Zwaig and Luria, 1967). Revertants and wild-type recombinants, from crosses involving two *tol*C mutations, can be selected for by plating on deoxycholate-containing plates, which confirms that tolerance and detergent sensitivity are due to a single mutational change. Using this technique, Whitney (1971) has produced a fine-structure map of the *tol*C gene, including the positioning of several large deletions. Temperature-sensitive (C. Hill and I. B. Holland, unpublished data; Schwarz, 1972) and *tol*C amber mutants (Nagel de Zwaig and Luria, 1969) have been isolated indicating that the *tol*C gene product is a protein. The properties of these mutants, as described above, indicate that this protein is concerned with the cell surface and an analysis of envelope proteins of a *tol*C deletion mutant by polyacrylamide-gel electrophoresis, in the presence of sodium dodecylsulphate, has indeed shown that a specific polypeptide band is absent from gel profiles when these are compared with the wild-type parent (Rolfe and Ondera, 1971). But the localization of this polypeptide within the bacterial envelope, and its precise relationship to the tolerance of the mutant, remain to be established.

The nature of the dye sensitivity of TolC mutants has been examined in some detail since this may be an important indicator of the envelope defect in these mutants. On the basis of his results, Clowes (1965) concluded that a defect in the electron-transport chain prevented normal reduction of the dye methylene blue in TolC mutants. Since colicin E1 blocks utilization of energy derived from electron transport, this conclusion suggested that *tol*C mutants might be altered in the colicin E1 target. However, further studies by Nagel de Zwaig and Luria (1967) showed clearly that *tol*C mutants took up much greater amounts of both methylene blue and acridine orange than did colicin E1-sensitive strains. Therefore it was concluded that the mutant strains were simply more permeable to these dyes. Nevertheless, the extensive changes in cellular permeability seen with some TolA,B mutants are not observed with TolC strains, and the possibility remains that these mutants do possess an important functional defect in the cytoplasmic membrane which promotes increased uptake of the dye. Working on this hypothesis, B. Rolfe and his coworkers (personal communication) have initiated an important series of experiments, examining the detergent sensitivity of various respiratory enzymes in membrane fragments obtained from TolC mutants. The outcome of these experiments are awaited with great interest.

TolC mutations are unstable and the strains are frequently slow growing, particularly under anaerobic conditions, which may indicate a functional defect in the cell membrane affecting, for example, ATPase activity. Mutant strains kept at room temperature in agar stabs are especially unstable, and are rapidly lost or revert to wild-type (I. B. Holland, unpublished results). The sensitivity of the mutants to acriflavin also frequently reverts, independent of colicin E1 tolerance (Whitney, 1971), pointing to strong selective pressures for compensatory changes in the cell surface to relieve some of the more lethal consequences of the Tol mutations.

3. TolD and E Mutants

Originally isolated as mutants showing enhanced levels of ampicillin resistance, these strains were shown to be also tolerant to colicins E2 and E3 but sensitive to colicin E1 (Burman and Nordström, 1971; Eriksson-Grennberg and Nordström, 1973). Genetic studies placed both loci close to minute 20 on the *E. coli* map, between *pur* B and *gal*. Since the mutants were distinguishable phenotypically, two distinct loci (*tol*D and *tol*E) were designated. If TolE mutants are grown on glucose, the bacterial lipopolysaccharide component is significantly depleted in rhamnose, glucose and galactose and, under these conditions, the cells are E2-tolerant and hypersensitive to deoxycholate (Eriksson-Grennberg and Nordström, 1973). In contrast, when grown on galactose the lipopolysaccharide composition of the mutants is normal and wild-type colicin E1 and deoxycholate sensitivities are regained. TolE mutants therefore, appear to carry a single pleiotropic mutation somehow involved in lipopolysaccharide assembly and galactose metabolism. In consequence, colicin tolerance most likely derives from defective orientation of the colicin E2–E3 receptor in the outer membrane.

The tolerance towards colicin E2 of TolD mutants is not suppressed by growth on galactose, and no changes in lipopolysaccharide content have been detected (Burman and Nordström, 1971). TolD mutants are also sensitive to deoxycholate and this characteristic, together with ampicillin resistance, does not segregate from colicin tolerance in genetic crosses. Therefore this mutation, too, appears to be pleiotropic. Finally, some evidence indicates that the *tol*D mutation may be dominant over the wild-type allele in partial diploids, but this still requires to be confirmed. TolD strains also show some degree of instability in their mutant characteristics (J. Foulds, personal communication) and spontaneous revertants frequently arise which revert to deoxycholate resistance, independent of tolerance to colicin E2 (Burman and Nordström, 1971).

4. TolF and TolG Mutants

Foulds and Barrett (1973) and Foulds (1974) have recently identified two further *tol* loci, close to *tol*D,E, determining tolerance to either bacteriocin bc (produced by *Serratia marcescens*) or to colicin K and colicin

A. TolF and TolG mutants are also hypersensitive to dyes and several antibiotics. Examination of envelope proteins of several TolG mutants, and their transductants, by SDS-polyacrylamide gel electrophoresis, has revealed that the isolated outer membrane fraction (see Osborn *et al.*, 1972a) invariably lacks a single major polypeptide (apparent molecular weight of 38,000 daltons) which normally constitutes about 15% of the outer membrane protein (J. Foulds, personal communication and in preparation). The simplest interpretation of all these results is that *tol*G is indeed the structural gene for this polypeptide and that its presence in the outer membrane is specifically required for the formation of complex II by bacteriocin bc. However, the precise correlation of gene and gene-product, when the latter may be a membrane polypeptide of no known biological activity, is extremely difficult to establish unequivocally, this being a general problem in membrane studies. Thus, the exact relationship between the missing protein and the *tol*G gene, although very suggestive, is not yet established.

As indicated in Table 2, *tol*D,E,F and G are extremely closely linked and, moreover, a class of TolI mutants, which have some properties in common with TolF strains has been reported to be linked to *gal* (Nomura and Witten, 1967). Careful complementation studies will therefore be required to establish the number of distinct cistrons, and possible operons, present in this region.

5. *Cet Mutants*

The most intensively studied class of tolerant strains are Cet mutants, specifically tolerant to colicin E2 at low temperature but predominantly sensitive at high temperature. The mutants are therefore cold-sensitive for colicin E2 tolerance, and colicin sensitivity is immediately expressed upon shifting the mutants to a high temperature (Holland, 1968) even in the absence of protein synthesis (Nomura and Witten, 1967). This, and other results discussed below, indicate that these mutants have an altered cell envelope which, at low temperature, fails to respond to the presence of bound colicin; at high temperature Complex II is formed, albeit with decreased efficiency.

Genetic analyses, involving several independently isolated Cet mutants, have now mapped a *cet*B locus between *ser*B and *thr* at minute 0 on the *E. coli* K12 map (Buxton and Holland, 1973). This result supersedes earlier data which placed *cet* to the left of *ser*B (Threlfall and Holland, 1970). Although Cet mutants constitute about 10% of all colicin E-tolerant isolates obtained from certain *E. coli* strains, they occur extremely rarely in other strains (Hill and Holland, 1967; Nomura and Witten, 1967). Similarly, when some strains are transduced for *cet*B, colicin E2-tolerance is poorly expressed (Buxton, 1973). Some preliminary studies have tentatively identified a second gene, designated *cet*A, located between *thr* and *leu*, which appears to enhance the expression of colicin E2-tolerance (Threlfall and Holland, 1970).

An extensive analysis of partial diploid strains carrying a *rec*A mutation to prevent recombination, has clearly demonstrated that *cet*B mutations, whether present on the episome or on the chromosome, are dominant over the wild-type allele (Buxton and Holland, 1973). This finding, together with the failure so far to obtain amber mutants of the Cet type (R. S. Buxton, personal communication), may indicate that the *cet* locus normally has a regulatory function, perhaps active only in *cis*. However, other alternative explanations are still tenable and further genetic analysis is needed. Nevertheless, dominance of Cet mutations lends support to our earlier suggestion for a regulatory role for the *cet* locus since envelopes of Cet mutants contain a specific polypeptide, apparent molecular weight 44,000 daltons, which is increased at least five-fold over any similar polypeptide in wild-type envelopes (Samson and Holland, 1970; Holland and Tuckett, 1972). We have now shown (Holland and Darby, 1973 and unpublished data) that this protein is localized predominantly, if not exclusively, in cytoplasmic membrane fragments isolated from tolerant strains by the method of Osborn *et al.* (1972a). Remarkably, this "Cet" protein appears as the major polypeptide of the inner membrane, constituting about 12% of the total protein (Fig. 5). This latter result appears to conflict with our earlier finding that trypsin treatment of whole cells of a Cet mutant resulted in the absence of the "Cet" protein in the subsequently isolated envelope (Holland and Tuckett, 1972). Such a result is difficult to reconcile with an inner membrane localization for the "Cet" protein, and therefore this effect of trypsin is being reinvestigated.

For the reasons discussed above in the case of TolG mutants, establishing a precise relationship between the "Cet" protein and the *cet*B gene is not possible on the basis of existing data. The situation is particularly complicated since these Tol mutants contain enhanced levels of the "Cet" protein, whether grown at low or at high temperature, even though the bacteria are largely sensitive to colicin E2 under the latter conditions. Thus, although the properties of the mutant suggest that an altered inner membrane constitutes the basis of its colicin E2 tolerance, direct interaction between colicin E2 and the Cet protein may not necessarily occur.

Unlike most Tol strains, CetB mutants have not so far been reported to have any unusual growth characteristics, and the strains are quite stable. However, several colicin E2-tolerant mutants, designated CetC, which also produce enhanced levels of the "Cet" protein (Samson and Holland, 1970), were initially found to have several other features indicative of an altered cell envelope, namely slow growth, deoxycholate- and ultraviolet-sensitivity, poor growth of bacteriophage λ, and filament formation (Holland, 1967a; Holland *et al.*, 1970). Genetic analysis placed the *cet*C locus close to *cet*B (Threlfall and Holland, 1970; see also Buxton

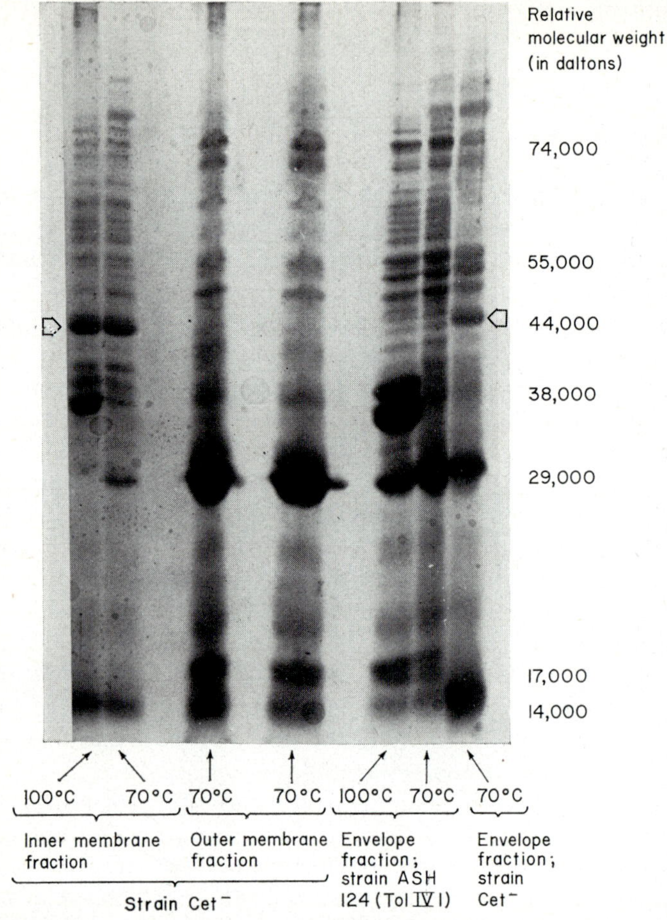

FIG. 5. Altered inner membranes of *Escherichia coli* Cet mutants. Envelopes were isolated from a CetB mutant and separated into inner and outer membrane fractions by sucrose density-gradient centrifugation. After dispersing in sodium dodecyl sulphate, samples were separated electrophoretically on a 9% polyacrylamide slab gel with a 5% stacking gel (not shown) and finally stained with Coomassie Blue. Prior to electrophoresis, samples were heated to 70°C or 100°C in sodium dodecyl sulphate as indicated. Relative molecular weights of some major bands are indicated, including the 44,000 dalton polypeptide (arrowed) which appears only in the inner membrane of Cet mutants. For comparison, the pattern obtained with the total envelope of ASH 124, a conditional lethal TolIV mutant, is also included. This strain, like the wild-type strain, does not overproduce the 44,000 daltons band, but has enhanced levels of another polypeptide with apparent molecular weight equal to 74,000 daltons.

and Holland, 1973) and also demonstrated that virtually all of the altered properties of the *cet*C mutants could be ascribed to a single mutational change. Some *cet*C mutants also appeared to be recombination deficient when acting as recipients in genetic crosses with Hfr male strains, but this feature is quite often encountered amongst Tol mutants, some of which are fully sensitive to colicin E2 (Nomura and Witten, 1967; Schwarz, 1972). This property may therefore reflect the increased sensitivity of recipients with defective membranes to "lethal zygosis" in conjugation experiments (Alföldi *et al.*, 1957; Skurray and Reeves, 1973), rather than to defects in recombination enzymes. The properties of CetC mutants are nevertheless extremely interesting since they indicate the alteration of an important functional element of the cell surface. CetC mutations have, however, proved to be extremely unstable, spontaneously "reverting" at a high frequency to strains with a CetB phenotype, although in a few cases such "revertants" are still ultraviolet-sensitive as well as colicin E2 tolerant. A re-investigation of three "revertants" of this type, obtained from independently isolated CetC mutants, has shown that the tolerant phenotype and the map position of colicin E2 tolerance appears identical to that of typical CetB strains. Somewhat surprisingly, the ultraviolet sensitivity of these strains was found to be due to the presence of an unlinked *lon*⁻ locus in two of the strains, and to a locus near *rec*A in a third "revertant" (Buxton, 1973; Buxton and Holland, 1974b). The nature of the changes which result in the formation of "revertants" of this type is not understood and it is not possible to establish retrospectively when the Lon mutation actually appeared in these strains. These results do, however, suggest that in certain Cet mutants strong selective forces can be set up, favouring in particular the survival of strains which carry *lon*⁻ in addition to *cet*⁻. The *lon* locus has previously been implicated in cell-surface synthesis and cell division (Walker and Pardee, 1967), and one may speculate that direct or indirect products of the *lon* and *cet* genes interact in some manner in the cytoplasmic membrane. Finally, these studies provide a cautionary tale for those engaged in the study of envelope mutants; the properties of such mutants are frequently complex and great care is required in unravelling effects due to real pleiotropic changes, on the one hand, and to additional, unlinked mutations on the other, which may also occur as a result of unknown selective forces.

6. *Other Tolerant Mutants*

In addition to the mutants discussed above, several other distinct classes of colicin E-tolerant mutants have been recognized, but their properties have not yet been studied (see Table 2, p. 118). A strain of *E. coli* tolerant to colicin E2 has, however, been recently isolated with rather unusual properties. *Escherichia coli* strain DB312 was selected by Beppu *et al.* (1972) for the ability to remain rescuable by trypsin after long periods of exposure to colicin E2. The mutant is fully sensitive to colicin E2 if not treated with trypsin, but cannot be rescued from colicin E3 action by this method. Examination of the properties of the mutant in liquid culture has shown that cell division is effectively blocked by even low doses of colicin E2, but neither solubilization nor fragmentation of DNA can be detected. Subsequent treatment of the filamented cells by trypsin leads to the rescue of at least one colony-forming cell per filament, suggesting that colicin is able to block cell division whilst remaining in the cell-surface layers and still accessible to trypsin. The properties of this mutant therefore provide strong support for the notion that inhibition of cell division is an independent and, perhaps, a primary effect of colicin E2 on sensitive bacteria, and that this effect is promoted by the surface-bound form of the colicin.

Mutants tolerant to colicin A, and to a bacteriocin formed by *Klebsiella pneumoniae,* have also been isolated in *Citrobacter freundii* (De Graaf *et al.*, 1973). The mutations appear to be pleiotropic and the dye sensitivity or filament formation of the mutants indicate some modification of the cell envelope. Several differences between mutant and wild-type cytoplasmic membrane proteins were observed in SDS-poly-acrylamide gel profiles, but the significance of these changes cannot be assessed in the absence of a more detailed genetic analysis. The analysis of the major phospholipids of mutant and wild-type envelopes, however, revealed no differences.

Several classes of mutants tolerant to an aeruginocin have also been isolated in *Pseudomonas aeruginosa* (Holloway *et al.*, 1974). These mutations too appear to be pleiotropic, and the mutants have defective cell surfaces.

7. *Conditional Lethal Mutants and Colicin Tolerance*

On the assumption that mutations to tolerance might in some cases lead to loss of an essential cellular function, Nomura and Witten (1967) set out to isolate conditional lethal mutants in which tolerance was expressed at the non-permissive temperature. Several mutants, designated TolIV1, were obtained which were unable to grow at 42°C (possibly through the inability to make RNA) and tolerant to colicins E2 and E3 at that temperature. Preliminary data from genetic crosses and reversion studies indicated that tolerance and the inability to grow at high temperature were indeed due to a single mutational change. Rolfe *et al.* (1973) extended these studies with a similar TolIV1 mutant obtained from M. Nomura (Table 2). Mutant strain 483 can grow at high temperatures in the presence of high concentrations of salt under which conditions it is hypersensitive to detergents and to several antibiotics. More significantly, mutant 483 shows a strong bias towards the lysogenic response when infected with wild-type phage λ. This effect is obtained despite the fact that the mutant has lower levels of cyclic-AMP (cAMP), a situation which is contrary to the proposal (Grodzicker *et al.*, 1972) that elevated levels of cAMP are required to promote lysogenization. In contrast, phage λ mutants, lacking a functional repressor, plate normally on mutant 483. On the basis of these results, Rolfe *et al.* (1973) concluded that the enhanced lysogenization of the Tol mutant, by phage λ, results from a preference for transcription from repressor and related promotors, compared to promotors linked to genes directly concerned with phage multiplication. This in turn led to the conclusion that strain 483 carries, perhaps in addition to other mutations, an altered RNA polymerase. Unfortunately, detailed genetic data has not been reported for this mutant, but when forthcoming this should provide a convenient test of this most intriguing hypothesis, and also reveal the precise relationship of colicin tolerance to abnormal growth of phage λ in this strain.

Another conditional mutant of the TolIV1 type (colicin E2 and colicin E3 tolerant) has been isolated recently in my laboratory. In contrast to strain 483 this TolIV mutant is tolerant at low temperature but becomes largely sensitive to colicin when shifted to a high temperature. However, at high temperatures, division of the mutant is immediately blocked, whilst macromolecule synthesis continues almost undisturbed; consequently the cells form filaments. Examination of the cell envelope of this mutant, grown at either high or low temperatures, has revealed that the outer membrane has an unusually low density, similar to that reported for some mutants deficient in lipopolysaccharide (Osborn *et al.*, 1972b; Kulpa and Leive, 1972). The mutant envelope also appears to contain increased levels of at least two polypeptides with molecular weights of about 74,000 and 63,000 daltons, respectively (I. B. Holland and V. Darby, unpublished data). Genetical and physiological studies of this mutant, and other cell division mutants which are frequently obtained by selection for colicin E2 or colicin E3 tolerance, are being actively continued in the hope that basic information concerning the division process may thereby be obtained.

XII. Summary and Prospects

A. Colicin Action is a Stepwise Process

The action of different colicins upon sensitive bacteria can be divided into three major phases: (1) binding to specific receptors in the outer membrane; (2) penetration or translocation of a whole or a part of the colicin molecule to sites within the cytoplasmic membrane, followed in some cases by penetration of at least a part of the molecule into the cytoplasm; (3) upon completion of the second phase, a biochemical, and ultimately a lethal, change in the cellular target takes place which is specific for each colicin. The first stage constitutes the formation of Complex I, an apparently innocuous state which is nevertheless largely irreversible unless the exposed colicin is digested with trypsin. The final stage, involving only a minority of colicin molecules, culminates in the interaction of colicin and target to form Complex II, a state of uncertain structure which is, however, no longer reversible by trypsin under most conditions. Complex II formation is energy dependent and is also dependent upon several factors which affect the physical state of the cell surface. For colicins E1 and K, Complex II formation probably coincides with the insertion of colicin into the inner membrane, whilst for colicin E3, interaction with ribosomes constitutes the final step. Whether this involves the complete entry of intact or cleaved fragments of colicin E3 molecules into the cell interior, or whether the active form is membrane bound (with the colicin protruding into the cytoplasm), is open to speculation at the present time. These questions could perhaps be approached by examination of the effect of proteolytic enzymes and surface components, including the colicin E receptor, upon the action of colicin E3 *in vitro*. Although DNA is most probably the target for colicin E2, inhibition of cell division may also be a primary and specific effect, and mutants have been isolated which do not degrade DNA after addition of colicin E2, although division is still blocked. Under these conditions, inhibition of division can be relieved by treatment of the cells with trypsin, indicating that Complex II corresponds to a form of colicin which incompletely penetrates the inner membrane. Some evidence, albeit weak, suggests that a similar complex, with colicin E2 piercing the membrane sufficiently to interact with the DNA, may occur in wild type strains. Unfortunately, this possibility is extremely difficult to test but definite evidence that colicin E2, or an active fragment of it, is present within the cytoplasm of treated cells will be necessary before this possibility can be rejected.

1. Penetration of the Cell Envelope

Although dispensable for the action of colicin E1 on vesicle preparations, specific receptors in the outer membrane are essential for penetration of the intact cell surface by all colicins. Receptors are complex structures but appear to contain a major protein constituent which is specific for each colicin. Binding of colicin to the receptor, which is a rapid and largely irreversible process, is eventually followed by the breaching of the outer membrane. Since colicins are elongated molecules, this step, and the subsequent interaction with inner membrane, may be accomplished without the necessity to disrupt the receptor-colicin complex. As indicated in Fig. 6, this may be especially true for colicins of the E1 type which have the most elongated form and whose target probably lies within the inner membrane itself. The actual method of penetration of either of the surface membranes by colicin molecules remains a mystery, although some possible mechanisms can be envisaged. Entry may be effected through surface channels of the type described by Bayer (1968a, b) or by localized degradation of the membranes. In the latter case phospholipase activity, inherent to the colicin molecule, cannot yet be ruled out. Alternatively if, as proposed by Capaldi and Green (1972), cell membranes contain polypeptide clusters intercalated into the phospholipid matrix, a rather different mechanism can be postulated. Thus, insertion of a colicin molecule into the centre of such an aggregate may facilitate

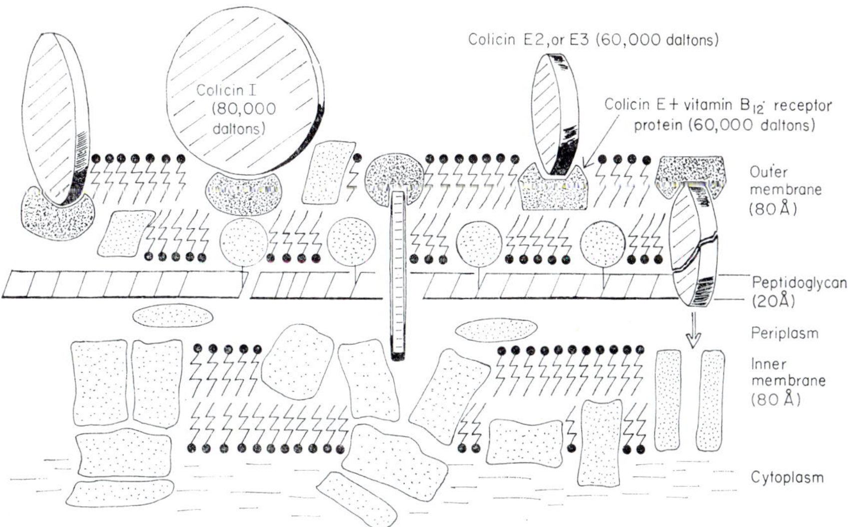

FIG. 6. Schematic representation of the envelope of *Escherichia coli* drawn approximately to scale. Close association of outer membrane and the peptidoglycan layer, through covalent linkage of many lipoprotein subunits to cross links in the latter are shown; lipopolysaccharide units in the outer membrane, and Bayer tubules, are omitted for clarity. The approximate size and shape of colicin molecules are presented on the basis of available data. Colicin I, as an example of the E1 type of colicin, is shown acting directly upon the inner membrane in the course of partial or complete penetration of that layer. The mechanisms which may convert inactive colicin I receptor complex (Complex I) to the active Complex II are discussed in the text, and no attempt is made to portray them. Although direct interaction of a receptor-bound form of colicin E2 with the inner membrane is not ruled out, this colicin probably penetrates the membrane to act upon DNA. The mechanism of entry suggested here involves insertion of a whole or a cleaved fragment of colicin E2, into a specific protein cluster which facilitates uptake. For alternative mechanisms of entry see the text.

its subsequent penetration by the lateral displacement of components of the cluster, thereby opening up a relatively hydrophilic channel for passage of the colicin molecule. Such a mechanism may require energy, which could explain the observation that Complex II formation is apparently dependent upon a suitably energized state of the membrane.

The transition from Complex I to Complex II, which need not occur immediately, takes place with a certain probability which varies with the cultural conditions, but is constant for each bound molecule of colicin. Indeed, quite complex events, subsequent to fixation, appear to precede the final triggering of colicin penetration of the surface envelope. To understand these events, several observations, which at first sight seem contradictory, must be accommodated. These are: (i) that colicin killing is a one-hit phenomenon; (ii) that frequently up to 50 molecules must be adsorbed before killing can occur; (iii) that the transition from Complex I to Complex II is time dependent, particularly at low temperatures; and (iv) that the exact length of time which elapses before this transition is triggered is dependent upon the colicin multiplicity.

In order to reconcile all of these observations, the following sequence of events is suggested to occur. Colicin binding to the receptor is followed by the lateral diffusion of this complex within the outer membrane until, with a certain probability reminiscent of a biological clock mechanism, a second interaction takes place involving other specific surface components. Upon formation of this aggregate, colicin penetration is promoted and Complex II formation takes place. The successful outcome of this process should depend upon the fluidity of the membrane, the frequency of receptors (and perhaps other membrane components), and the number of colicin molecules in the cell surface. This multiplicity effect does not imply a co-operative action between colicin molecules since one molecule is sufficient to kill the cell; rather that increasing the multiplicity increases the probability of the early formation of the penetration complex. One final point in relation to the penetration mechanism, and/or target disruption, concerns the possible specificity of the interaction of the invading colicin molecule with the inner membrane. It seems quite likely that, for different colicins, the partial or complete penetration of this membrane will take place only at preferred sites, and that the frequency of these will also affect the probability of Complex II formation. No evidence for the existence of such sites has yet been obtained, but it is hoped that the examination of the properties of Tol mutants with altered inner membranes will reveal the presence of these sites, if they exist.

2. *Targets*

The colicin E3 biochemical target has been clearly identified as 16S ribosomal RNA which, in the presence of colicin E3, is cleaved close to the 3′ terminus both *in vivo* and *in vitro*. The precise substrate for this reaction appears to involve a unique structural configuration of the 16S RNA which is expressed only in the presence of both ribosomal subunits. This structure is conserved in ribosomes of several bacterial species, some quite unrelated to *E. coli*. Specific collaboration between colicin E3 and a ribosomal protein may be needed to promote the hydrolytic event, but known *E. coli* ribonucleases do not appear to be involved. The available data suggests that scission of the RNA molecule is sufficient, directly or indirectly, to block the elongation and/or termination of polypeptide chains. Cleavage of the RNA is therefore probably a lethal event. This effect of colicin E3 on ribosomal function suggests that the 16S RNA moiety may normally play a direct role in the synthesis of polypeptide chains.

For colicins of the E1 type, the primary effect on sensitive cells is the dissipation of a high-energy state (presumably in the form of a membrane potential), with the consequent disruption of energized transport and oxidative phosphorylation. Unlike classical uncouplers, however, colicin E1 does not promote H^+ permeability of the cell membrane, whilst, in contrast, rapid efflux of K^+ is facilitated. Induction of K^+ efflux may, in fact, be the primary effect of this and similar colicins, although how this should effect the collapse of the membrane potential is uncertain. The role of molecular oxygen, which appears to be essential for the action of colicins E1 and K on sensitive cells, appears equally unclear. Colicins E1 and K probably attack target sites within the cytoplasmic membrane, and the total loss of cellular K^+, which occurs in treated cells, may result directly from the insertion of even a single molecule of colicin into the membrane. It is not clear whether the action of the E1 type of colicin is specifically accompanied by enzyme action on the membrane but, if so, the ensuing damage must be repairable since, under some conditions at least, the inhibition of energy metabolism by colicin K is reversible by trypsin treatment. Enzyme activity has not so far been reported for purified preparations of colicin E1-like colicins, but the presence of phospholipase activity, with complex structural substrate requirements, cannot yet be ruled out. Colicins E1 and K promote widespread changes in surface membranes, as indicated by the enhanced fluorescence exhibited by hydrophobic probes present in the membrane. This effect may be due to the collapse of the energized state throughout the inner membrane, or to other consequences of colicin action (for example, the widespread activation of membrane-bound phospholipases). Thus, there is no compelling reason at the moment to attribute specific and primary *causes* of colicin E1 action to the structural changes which underlie the observed fluorescence increase of bound probes.

As already indicated, colicin E2 may block cell division and promote

DNA degradation by independent mechanisms which are activated during the passage of the molecule through the cell surface. In consequence, colicin E2 may prove to be a useful tool for exploring some aspects of the division process. Although the nature of DNA breakdown which occurs in colicin E2-treated cells has been extensively studied, the mechanism which specifically initiates this process remains elusive. However, the ability of a single "active" colicin E2 molecule to promote extensive DNA degradation, and the possibility that the active form is a bound molecule which transfixes the inner membrane, strongly suggest that any interaction between colicin E2 and DNA will involve an enzyme-like action rather than localized changes in DNA secondary structure. Furthermore, some evidence indicates that colicin E2, *in vitro*, does promote single-strand breaks in specific DNA tertiary structures. Nevertheless, in order to account for the extensive endo- and exonucleolytic degradation of DNA observed *in vivo*, specific cellular deoxyribonucleases must also participate. The identity of these enzymes remains unknown, although the involvement of several major deoxyribonucleases has been ruled out.

B. Tolerant Mutants

Colicin tolerant mutants have been isolated in several laboratories with the expectation that their properties would reveal both the nature of specific surface components necessary for the penetration of colicin molecules and the nature of colicin targets themselves. The properties of Tol mutants have proved to be rather complex, and progress so far has been largely restricted to their genetical characterization. However, the way is now open for their biochemical characterization and, as a first step, it is hoped that the lesions present in each mutant can be localized in either the inner or the outer membrane. Mutants of the latter class may be particularly important in determining the specificity of the interaction between the invading colicin and the inner membrane. For colicins of the E1 type, site specificity of this kind may include the target structure itself, and studies of the functional integrity *in vitro* of membrane fragments from, for example, TolC mutants, which have already been initiated, should be extremely informative. Even though knowledge of colicin action has been slow to emerge from the study of Tol mutants, looked at in the broader context of the synthesis and function of membrane proteins as a whole, the properties of the mutants appear extremely encouraging. Membrane protein operons have been identified, and analysis of their genetic fine structure has commenced. What appear to be regulatory mutations, affecting the synthesis or maturation of membrane proteins including possible alterations in the transcriptional specificity of RNA polymerase, have also been observed. In addition, colicin-tolerant mutants are frequently found to carry conditional lethal mutations affecting, in particular, the cell-division process or the ability to carry out essential energy-transducing steps. Future studies should therefore see extensive use of such mutants in the exploration of these fundamental biological processes.

XIII. Acknowledgements

I am extremely grateful to Dr. P. J. F. Henderson for permission to reproduce Fig. 4, and for his many valuable suggestions and criticisms on the subject of energy transduction in bacteria. I am also greatly indebted to the many workers in the field who have provided me with reprints and unpublished data. I am especially grateful to Dr. A. Bernstein, Dr. J. Foulds, Professor B. W. Holloway, Dr. A. M. Jetten, Dr. J. Konisky, Professor M. Nomura, Dr. P. Reeves, Dr. P. Ringrose, Dr. B. Rolfe, Dr. Linda Saxe, Dr. C. A. Schnaitman, Dr. T. F. R. Schwarz, Dr. L. W. Wendt and Dr. J. S. Wiberg for freely supplying manuscripts, and much additional data, prior to publication, thus enabling me to gain a comprehensive view of current interests in this subject. I would also like to acknowledge the assistance of my wife, Dr. É. M. Holland, to whom I am deeply grateful for her patient and critical reading of the manuscript; to Sheila Mackley and Margaret Peake for typing the manuscript, and finally to my colleagues Valerie Darby and Gordon Churchward for helpful discussions and assistance in the preparation of this review.

References

Alföldi, L., Jacob, F. and Wollman, E. L. (1957). *Compte rendu hebdomadaire des séances de l'Academie des sciences* **244**, 2974.

Almendinger, R. and Hager, L. P. (1972). *Nature, New Biology* **235**, 199.

Almendinger, R. and Hager, L. P. (1973). *Antimicrobial Agents and Chemotherapy* **4**, 167.

Amati, P. (1964). *Journal of Molecular Biology* **8**, 239.

Barry, G. T., Everhart, D. L., Abbot, V. and Graham, M. (1965). *Zentralblatt für Bakteriologie, Parasitenkunde, Infektionskrankheiten und Hygiene* (*Abteilung I*) **196**, 248.

Bayer, M. E. (1968a). *Journal of Virology* **2**, 346.

Bayer, M. E. (1968b). *Journal of General Microbiology* **53**, 395.

Ben-Gurion, R. (1970). *Biochemical and Biophysical Research Communications* **40**, 1281.

Beppu, T. and Arima, K. (1967). *Journal of Bacteriology* **93**, 80.

Beppu, T. and Arima, K. (1971). *Journal of Biochemistry* **70**, 263.

Beppu, T. and Arima, K. (1972). *Biochimica et Biophysica Acta* **262**, 453.

Beppu, T., Kawabata, K. and Arima, K. (1972). *Journal of Bacteriology* **110**, 485.

Bernstein, A. (1972). Ph.D. Thesis: University of Toronto.

Bernstein, A. (1973). *Molecular and General Genetics* **123**, 111.
Bernstein, A., Rolfe, B. and Ondera, K. (1972). *Journal of Bacteriology* **112**, 74.
Bernstein, A., Rolfe, B. and Ondera, K. (1973). *Molecular and General Genetics* **121**, 325.
Bhattacharyya, P., Wendt, L., Whitney, E. and Silver, S. (1970). *Science, New York* **168**, 998.
Birdsell, D. C. and Cota-Robles, E. H. (1967). *Journal of Bacteriology* **93**, 427.
Boon, T. (1971). *Proceedings of the National Academy of Sciences of the United States of America* **68**, 2421.
Boon, T. (1972). *Proceedings of the National Academy of Sciences of the United States of America* **69**, 549.
Bowman, C. M., (1972). *Federation of European Biochemical Societies Letters* **22**, 73.
Bowman, C. M., Dahlberg, J. E., Ikemura, T., Konisky, J. and Nomura, M. (1971a). *Proceedings of the National Academy of Sciences of the United States of America* **68**, 964.
Bowman, C. M., Sidikaro, J. and Nomura, M. (1971b). *Nature, New Biology* **234**, 133.
Bowman, C. M., Sidikaro, J. and Nomura, M. (1973). *In* "Chemistry and Functions of Colicins" (L. P. Hager, ed.), p. 87. Academic Press, New York.
Brand, L. and Gohlke, J. R. (1972). *Annual Review of Biochemistry* **41**, 843.
Braun, V., Gnirke, H., Henning, U. and Rehn, K. (1973). *Journal of Bacteriology* **114**, 1264.
Braun, V. and Rehn, K. (1969). *European Journal of Biochemistry* **10**, 426.
Braun, V. and Wolff, H. (1973). *Federation of European Biochemical Societies Letters* **34**, 77.
Burman, L. G. and Nordström, K. (1971). *Journal of Bacteriology* **106**, 1.
Buttin, G. and Wright, M. (1968). *Cold Spring Harbor Symposium on Quantitative Biology* **33**, 259.
Buxton, R. S. (1971). *Molecular and General Genetics* **113**, 154.
Buxton, R. S. (1973). Ph.D. Thesis: University of Leicester.
Buxton, R. S. and Holland, I. B. (1973). *Molecular and General Genetics* **127**, 69.
Buxton, R. S. and Holland, I. B. (1974a). *Federation of European Biochemical Societies Letters* **39**, 1.
Buxton, R. S. and Holland, I. B. (1974b). *Molecular and General Genetics* **131**, 159.
Capaldi, R. A. and Green, D. E. (1972). *Federation of European Biochemical Societies Letters* **25**, 205.
Cavard, D. and Barbu, E. (1969). *Compte rendu hebdomadaire des séances de l'Académie des sciences* **268**, 1133.
Cavard, D. and Barbu, E., (1970). *Annales de l'Institut Pasteur* **119**, 420.
Cavard, D., Marotel-Schirman, J. and Barbu, E. (1971). *Compte rendu hebdomadaire des séances de l'Académie des sciences* **273**, 1167.
Cavard, D., Rampini, C., Barbu, E. and Polonovski, J. (1968). *Bulletin de la Société de Chimie Biologique* **50**, 1455.
Chang, Y.-Y. and Hager, L. P. (1970). *Journal of Bacteriology* **104**, 1106.
Changeux, J.-P. and Thiery, J. (1967). *Journal of Theoretical Biology* **17**, 315.
Chapman, D. and Wallach, D. F. H. (1968). *In* "Biological Membranes" (D. Chapman, ed.), pp. 125–136. Academic Press, New York.
Chappell, J. B. and Crofts, A. R. (1966). *In* "Regulation of Metabolic Processes of Mitochondria", p. 582. Elsevier Publishing Company, Amsterdam.
Clowes, R. C. (1965). *Zentralblatt für Bakteriologie Parasitenkunde, Infektionskrankheiten und Hygiene (Abteilung I)* **196**, 152.

Clowes, R. C. (1972). *Bacteriological Reviews* **36**, 361.
Cramer, W. A. and Phillips, S. K. (1970). *Journal of Bacteriology* **104**, 819.
Cramer, W. A., Phillips, S. K. and Keenan, T. W. (1973). *Biochemistry, New York* **12**, 1177.
Dahlberg, A. E., Lund, E., Kjeldgaard, N. O., Bowman, C. M. and Nomura, M. (1973). *Biochemistry, New York* **12**, 948.
Dalgarno, L. and Shine, J. (1973). *Nature, New Biology* **245**, 261.
Dandeu, P., Billault, A. and Barbu, E. (1969). *Compte rendu hebdomadaire des séances de l'Académie des sciences* **269**, 2044.
Davies, J. K. and Reeves, P. (1974). In Press.
De Graaf, F. K. (1973). *Antonie van Leeuwenhoek* **39**, 109.
De Graaf, F. K., van Vught, A. M. J. J. and Stouthamer, A. H. (1973). *Antonie van Leeuwenhoek* **39**, 51.
De Petris, S. (1967). *Journal of Ultrastructure Research* **19**, 45.
De Witt, W. and Helinski, D. R. (1965). *Journal of Molecular Biology* **13**, 692.
Di Masi, D. R., White, J. C., Schnaitman, C. A. and Bradbeer, C. (1973). *Journal of Bacteriology* **115**, 506.
Dürwald, H. and Hoffmann-Berling, (1968). *Journal of Molecular Biology* **34**, 331.
Edelman, G. M. and McClure, W. D. (1968). *Accounts of Chemical Research, Washington* **1**, 65.
Ehresmann, C., Fellner, P. and Ebel, J. P. (1971). *Federation of European Biochemical Societies Letters* **13**, 325.
Endo, H., Kamiya, T. and Ishizawa, M. (1963). *Biochemical and Biophysical Research Communications* **11**, 477.
Eriksson-Grennberg, K. G. and Nordström, K. (1973). *Journal of Bacteriology* **115**, 1219.
Farrell, L. and Reiter, H. (1973). *Antimicrobial Agents and Chemotherapy* **4**, 320.
Feingold, D. (1970). *Journal of Membrane Biology* **3**, 372.
Fields, K. (1969). *Journal of Bacteriology* **97**, 78.
Fields, K. and Luria, S. E. (1969a). *Journal of Bacteriology* **97**, 57.
Fields, K. and Luria, S. E. (1969b). *Journal of Bacteriology* **97**, 64.
Fisher, H. F. (1964). *Proceedings of the National Academy of Sciences of the United States of America* **51**, 1285.
Foulds, J. (1971). *Journal of Bacteriology* **107**, 833.
Foulds, J. (1974). *Journal of Bacteriology* **117**, 1354.
Foulds, J. and Barrett, C. (1973). *Journal of Bacteriology* **116**, 885.
Foulds, J. and Shemin, D. (1969). *Journal of Bacteriology* **99**, 655.
Fredericq, P. (1949). *Compte rendu des séances de la Société de biologie* **143**, 1011.
Fredericq, P. (1957). *Annual Review of Microbiology* **11**, 7.
Fredericq, P. (1958). *Symposium of the Society for Experimental Biology* **12**, 104.
Gale, E. F. (1971). *Journal of General Microbiology* **68**, 1.
Gill, D. M., Pappenheimer, A. M. and Uchida, T. (1973). *Federation Proceedings. Federation of American Societies for Experimental Biology* **32**, 1508.
Glick, J. M., Kerr, S. J., Gold, A. M. and Shemin, D. (1972). *Biochemistry, New York* **11**, 1153.
Godson, G. N. and Sinsheimer, R. L. (1967). *Biochimica et Biophysica Acta* **149**, 489.
Goebel, W. (1962). *Proceedings of the National Academy of Sciences of the United States of America* **48**, 214.
Goebel, W. (1973). *Proceedings of the National Academy of Sciences of the United States of America* **70**, 854.

Grodzicker, T., Arditti, R. R. and Eisen, H. (1972). *Proceedings of the National Academy of Sciences of the United States of America* **69**, 366.
Guterman, S. K. (1971). *Biochemical and Biophysical Research Communications* **44**, 1149.
Guterman, S. K. and Luria, S. E. (1969). *Science, New York* **164**, 1414.
Harold, F. (1970). *Advances in Microbial Physiology* **4**, 45.
Harold, F. (1972). *Bacteriological Reviews* **36**, 172.
Harold, F. and Baarda, J. R. (1968). *Journal of Bacteriology* **94**, 53.
Hausman, C. and Clowes, R. C. (1971). *Journal of Bacteriology* **107**, 633.
Henderson, P. J. F. (1971). *Annual Reviews of Microbiology* **25**, 393.
Herschmann, H. R. and Helinski, D. R. (1967a). *Journal of Biological Chemistry* **242**, 5360.
Herschmann, H. R. and Helinski, D. R. (1967b). *Journal of Bacteriology* **94**, 691.
Hill, C. and Holland, I. B. (1967). *Journal of Bacteriology* **94**, 667.
Hirata, H., Fukui, S. and Ishikawa, S. (1969). *Journal of Biochemistry* **65**, 843.
Holland, É. M. and Holland, I. B. (1970). *Journal of General Microbiology* **64**, 223.
Holland, É. M. and Holland, I. B. (1972). *Biochimica et Biophysica Acta* **281**, 179.
Holland, I. B. (1962). *Journal of General Microbiology* **29**, 603.
Holland, I. B. (1967a). *Molecular and General Genetics* **100**, 242.
Holland, I. B. (1967b). *In* "Antibiotics" (Gottlieb and Shaw, eds.), Vol. 1, p. 688. Springer-Verlag, Berlin.
Holland, I. B. (1968). *Journal of Molecular Biology* **31**, 267.
Holland, I. B. and Darby, V. (1973). *Federation of European Biochemical Societies Letters* **33**, 106.
Holland, I. B., Threlfall, E. J., Holland, É. M., Darby, V. and Samson, A. C. R. (1970). *Journal of General Microbiology* **62**, 371.
Holland, I. B. and Tuckett, S. (1972). *Journal of Supramolecular Structure* **1**, 77.
Holloway, B. W., Rossiter, H., Burgess, D. and Dodge, J. (1974). *Genetical Research* **22**, 239.
Hull, R. R. and Reeves, P. (1971). *Journal of Virology* **8**, 355.
Hutton, J. J. and Goeble, W. (1962). *Journal of General Physiology* **45**, 125.
Isaacson, R. E. and Konisky, J. (1972). *Journal of Bacteriology* **109**, 1322.
Isaacson, R. E. and Konisky, J. (1973). *Journal of Bacteriology* **113**, 452.
Ivánovics, G., Alföldi, L. and Nagy, I. (1959). *Journal of General Microbiology* **21**, 51.
Jacob, F., Siminovitch, A. and Wollman, E. (1952). *Annales de l'Institut Pasteur* **83**, 295.
Jasper, P., Whitney, E. N. and Silver, S. (1972). *Genetical Research* **19**, 305.
Jetten, A. M. (1973). Ph.D. Thesis: University of Nijmegen, Netherlands.
Jetten, A. M. and Vogels, I. G. D. (1972a). *Journal of Bacteriology* **112**, 243.
Jetten, A. M. and Vogels, I. G. D. (1972b). *Antimicrobial Agents and Chemotherapy* **2**, 456.
Jetten, A. M. and Vogels, I. G. D. (1973). *Biochimica et Biophysica Acta* **311**, 483.
Jesaitis, M. (1970). *Journal of Experimental Medicine* **131**, 1016.
Jones, N. C. and Donachie, W. D. (1973). *Nature, New Biology* **243**, 100.
Kaback, H. R. (1972). *Biochimica et Biophysica Acta* **265**, 367.
Kadner, R. J. and Liggins, G. L. (1973). *Journal of Bacteriology* **115**, 514.
Kennedy, C. (1971). *Journal of Bacteriology* **108**, 10.
Kennedy, C., Durkacz, B. W. and Sherratt, D. J. (1973). *Society for General Microbiology Proceedings* **1**, 15.
Klein, W. L. and Boyer, P. D. (1972). *Journal of Biological Chemistry* **247**, 7257.

Konisky, J. (1972). *Journal of Biological Chemistry* **247**, 3750.
Konisky, J. (1973). *In* "Chemistry and Functions of Colicins" (L. P. Hager, ed.), p. 68. Academic Press, New York.
Konisky, J. (1974). (In Press).
Konisky, J. and Cowell, B. S. (1972). *Journal of Biological Chemistry* **247**, 6524.
Konisky, J., Cowell, B. S. and Gilchrist, M. J. (1973). *Journal of Supramolecular Structure* **1**, 208.
Konisky, J. and Nomura, M. (1967). *Journal of Molecular Biology* **26**, 181.
Konisky, J. and Richards, F. M. (1970). *Journal of Biological Chemistry* **245**, 2972.
Kulpa, C. F. and Leive, L. (1972). *In* "Membrane Research" (C. F. Fox, ed.), p. 155. Academic Press, New York.
Kutter, E. M. and Wiberg, J. S. (1968). *Journal of Molecular Biology* **38**, 395.
Leive, L. (1968). *Journal of Biological Chemistry* **243**, 2373.
Levinsohn, R. Konisky, J. and Nomura, M. (1967). *Journal of Bacteriology* **96**, 811.
Linden, C., Wright, K., McConnell, H. M. and Fox, C. F. (1974). *Proceedings of the National Academy of Sciences of the United States of America* **70**, 2271.
Lubin, M. (1964). *Federation Proceedings. Federation of American Societies for Experimental Biology* **27**, 1278.
Luria, S. E. (1964). *Annales de l'Institut Pasteur* **107**, 67.
Lusk, J. E. and Nelson, D. L. (1972). *Journal of Bacteriology* **112**, 148.
Machtiger, N. A. and Fox, C. F. (1973). *Annual Review of Biochemistry* **42**, 575.
Maeda, A. and Nomura, M. (1966). *Journal of Bacteriology* **91**, 685.
Margolin, K. and Kennedy, C. (1973). *Society for General Microbiology Proceedings* **1**, 26.
Marotel-Schirman, J., Cavard, D., Sandler, L. and Barbu, E. (1970). *Compte rendu hebdomadaire des séances de l'Académie des sciences* **270**, 230.
Mayr-Harting, A. and Shimeld, C. (1965). *Zentralblatt für Bakteriologie, Parasitenkunde, Infektionskrankheiten und Hygiene (Abteilung I)* **196**, 263.
Meyhack, B., Meyhack, I. and Apirion, D. (1973). *Proceedings of the National Academy of Sciences of the United States of America* **70**, 156.
Mitchell, P. (1972). *Journal of Bioenergetics* **3**, 5.
Mitusi, E. and Mizuno, D. (1969). *Journal of Bacteriology* **100**, 1136.
Mizushima, S. and Nomura, M. (1970). *Nature, London* **226**, 1214.
Nagel de Zwaig, R. (1969). *Journal of Bacteriology* **99**, 913.
Nagel de Zwaig, R. and Luria, S. E. (1967). *Journal of Bacteriology* **94**, 1112.
Nagel de Zwaig, R. and Luria, S. E. (1969). *Journal of Bacteriology* **99**, 78.
Nagel de Zwaig, R. and Vitelli-Flores, J. (1973). *Federation of European Biochemical Societies Letters* **29**, 318.
Nagy, I., Alföldi, L. and Ivánovics, G. (1959). *Acta Microbiologica Academiae Scientiarum Hungaricae* **6**, 327.
Nomura, M. (1963). *Cold Spring Harbor Symposia on Quantitative Biology* **28**, 315.
Nomura, M. (1964). *Proceedings of the National Academy of Sciences of the United States of America* **52**, 1514.
Nomura, M. (1967). *Annual Reviews of Microbiology* **21**, 257.
Nomura, M. and Maeda, A. (1965). *Zentralblatt für Bakteriologie, Parasitenkunde, Infektionskrankheiten und Hygiene (Abteilung I)* **196**, 216.
Nomura, M. and Nakamura, M. (1962). *Biochemical and Biophysical Research Communications* **7**, 306.
Nomura, M., Traub, P. and Bechmann, H. (1968). *Nature, London* **219**, 793.
Nomura, M. and Witten, C. (1967). *Journal of Bacteriology* **94**, 1093.
Normark, S. and Westling, B. (1971). *Journal of Bacteriology* **108**, 45.

Nose, K. and Mizuno, D. (1968). *Journal of Biochemistry* **64**, 1.
Nose, K. and Mizuno, D. (1971). *Biochimica et Biophysica Acta* **246**, 20.
Nose, K., Ono, M. and Mizuno, D. (1970). *Journal of Bacteriology* **101**, 102.
Obdržálek, V., Šmarda, J., Čeck, O. and Adler, J. (1969). *Experientia* **25**, 331.
Obinata, M. and Mizuno, D. (1970). *Biochimica et Biophysica Acta* **199**, 330.
Ochi, T., Yano, K. and Amano, T. (1971). *Bikens Journal* **14**, 423.
Osborn, M. J., Gander, J. E., Parisi, E. and Carson, J. (1972a). *Journal of Biological Chemistry* **247**, 3962.
Osborn, M. J., Gander, J. E. and Parisi, E. (1972b). *Journal of Biological Chemistry* **247**, 3973.
Overath, P., Schairer, H. V. and Stoffel, W. (1970). *Proceedings of the National Academy of Sciences of the United States of America* **67**, 606.
Ozaki, M., Higashi, Y., Saito, H., An, T. and Amano, T. (1966). *Bikens Journal* **9**, 201.
Ozeki, H., Stocker, B. A. D. and de Margerie, H. (1959). *Nature, London* **184**, 337.
Pace, N. R. (1973). *Bacteriological Reviews* **37**, 562.
Phillips, S. K. and Cramer, W. A. (1973). *Biochemistry, New York* **12**, 1170.
Plate, C. (1973). *Antimicrobiol Agents and Chemotherapy* **4**, 16.
Plate, C. and Luria, S. E. (1972). *Proceedings of the National Academy of Sciences of the United States of America* **69**, 2030.
Ptashne, M. (1972). *In* "Bacteriophage Lambda" (A. Hershey, ed.), p. 128. Cold Spring Harbor Monograph.
Reader, R. W. and Siminovitch, L. (1971). *Virology* **43**, 607.
Reeves, P. (1963). *Australian Journal of Experimental Biology and Medical Science* **41**, 163.
Reeves, P. (1965a). *Australian Journal of Experimental Biology and Medical Science* **43**, 191.
Reeves, P. (1965b). *Bacteriological Reviews* **29**, 24.
Reeves, P. (1973). The Bacteriocins. Molecular Biology Monographs 11. Springer-Verlag, Berlin.
Reynolds, B. L. and Reeves, P. (1969). *Journal of Bacteriology* **100**, 301.
Ringrose, P. S. (1970). *Biochimica et Biophysica Acta* **213**, 320.
Ringrose, P. S. (1971). Ph.D. Thesis: University of Cambridge.
Ringrose, P. S. (1972). *Federation of European Biochemical Societies Letters* **23**, 241.
Ringrose, P. S. (1973). *Biochimica et Biophysica Acta* **312**, 656.
Rolfe, B. and Ondera, K. (1971). *Biochemical and Biophysical Research Communications* **44**, 767.
Rolfe, B., Schell, J., Becker, A., Heip, J., Ondera, K. and Schell-Frederick, E. (1973). *Molecular and General Genetics* **120**, 1.
Roseman, S. (1972). *In* "The Molecular Basis of Biological Transport" (J. F. Woessner and F. Huijing, eds.), pp 181–212. Academic Press, New York.
Rothfield, L. and Romeo, D. (1971). *In* "Structure and Function of Biological Membranes". Molecular Biology Monograph. Academic Press, New York.
Sabet, S. F. and Schnaitman, C. A. (1971). *Journal of Bacteriology* **108**, 422.
Sabet, S. F. and Schnaitman, C. A. (1973). *Journal of Biological Chemistry* **248**, 1797.
Samson, A. C. R. (1970). Ph.D. Thesis: University of Leicester.
Samson, A. C. R. and Holland, I. B. (1970). *Federation of European Biochemical Societies Letters* **11**, 33.
Samson, A. C. R., Senior, B. W. and Holland, I. B. (1972). *Journal of Supramolecular Structure* **1**, 135.

Santer, M. and Santer, V. (1973). *Journal of Bacteriology* **116**, 1304.
Schnaitman, C. A. (1971a). *Journal of Bacteriology* **108**, 553.
Schnaitman, C. A. (1971b). *Journal of Bacteriology* **108**, 545.
Schwarz, T. F. R. (1972). Ph.D. Thesis: University of Dublin.
Schwartz, S. A. and Helinski, D. R. (1971). *Journal of Biological Chemistry* **246**, 6318.
Senior, B. W. (1968). Ph.D. Thesis: University of Newcastle upon Tyne.
Senior, B. W. and Holland, I. B. (1971). *Proceedings of the National Academy of Sciences of the United States of America* **68**, 959.
Senior, B. W., Kwasniak, J. and Holland, I. B. (1970). *Journal of Molecular Biology* **53**, 205.
Seto, A., Shinozawa, T. and Maeda, A. (1973). *Biochimica et Biophysica Acta* **324**, 305.
Shannon, R. and Hedges, A. J. (1973). *Journal of Bacteriology* **116**, 1136.
Singer, S. J. and Nicolson, G. L. (1972). *Science, New York* **175**, 720.
Sidikaro, J. and Nomura, M. (1973). *Federation of European Biochemical Societies Letters* **29**, 15.
Skurray, R. A. and Reeves, P. (1973). *Journal of Bacteriology* **113**, 58.
Šmarda, J. and Adler, J. (1971). *Antonie van Leeuwenhoek* **37**, 507.
Šmarda, J. and Taubeneck, U. (1968). *Journal of General Microbiology* **52**, 161.
Smilowitz, H. (1974). *Journal of Virology* **13**, 94.
Stocker, B. A. D. (1966). *Heredity* **21**, 166.
Stonington, O. G. and Pettijohn, D. E. (1971). *Proceedings of the National Academy of Sciences of the United States of America* **68**, 6.
Swift, R. L. and Wiberg, J. S. (1971). *Journal of Virology* **8**, 303.
Swift, R. L. and Wiberg, J. S. (1973a) *Journal of Molecular Biology* **80**, 743.
Swift, R. L. and Wiberg, J. S. (1973b). *Journal of Virology* **11**, 386.
Szybalski, W., Kubenski, H. and Sheldrick, W. (1966). *Cold Spring Harbor Symposia on Quantitative Biology* **31**, 123.
Takagaki, Y., Kunugita, K. and Matsuhashi, M. (1973). *Journal of Bacteriology* **113**, 42.
Taylor, A. L. (1970). *Bacteriological Reviews* **34**, 155.
Threlfall, E. J. and Holland, I. B. (1970). *Journal of General Microbiology* **62**, 383.
Timmis, K. (1972). *Journal of Bacteriology* **109**, 12.
Turnowsky, F., Drews, J., Eich, F. and Högenauer, G. (1973). *Biochemical and Biophysical Research Communications* **52**, 327.
Walker, J. R. and Pardee, A. B. (1967). *Journal of Bacteriology* **93**, 107.
Weltzien, H. V. and Jesaitis, M. A. (1971). *Journal of Experimental Medicine* **133**, 534.
Wendt, L. W. (1970). *Biochemical and Biophysical Research Communications* **40**, 489.
Whitney, E. N. (1971). *Genetics* **67**, 39.
Willetts, N. S. (1972). *Annual Reviews of Genetics* **6**, 257.
Worcel, A. and Burgi, E. (1972). *Journal of Molecular Biology* **71**, 127.
Worcel, A. and Burgi, E. (1974). *Journal of Molecular Biology* **82**, 91.
Wright, M., Buttin, G. and Hurwitz, J. (1971). *Journal of Biological Chemistry* **246**, 6543.
Zinder, N. D. (1973). *Proceedings of the National Academy of Sciences of the United States of America* **70**, 3160.

Bacterial Glycolipids and Glycophospholipids

NORMAN SHAW

Microbiological Chemistry Research Laboratory, School of Chemistry, The University of Newcastle upon Tyne, Newcastle upon Tyne, NE1 7RU, England

I. Introduction

The expansion of our knowledge of lipids has been effectively controlled by the availability of suitable experimental techniques for their investigation. The present exponential growth was initiated more than a decade ago by the introduction of gas–liquid chromatography and thin-layer chromatography, and by a recognition amongst biochemists and microbiologists that lipids, as major components of cell membranes, must play a fundamental role in cell processes. Numerous authors have attempted previously to define "lipid", and I do not wish here to engage in this particular argument, but it is indeed unfortunate that this surge of interest, which has revealed the great diversity of chemical types classified as "lipids", has also led to a proliferation of trivial nomenclature unparalleled in other areas of biochemistry. Such is the rate of progress that it is impracticable in a single essay satisfactorily to cover all aspects of bacterial lipids and, in this review, I shall attempt to assess developments in the area of bacterial glycolipids and glycophospholipids, i.e. those lipids which contain carbohydrate residues. One review on glycolipids (Shaw, 1970) and other reviews on various aspects of bacterial lipids have been published latterly (Op den Kamp *et al.*, 1969a; Lennarz, 1970; Goldfine, 1972; Shaw, 1974).

II. Glycolipids

Bacterial glycolipids have been divided into two categories, namely (i) glycosyl diglycerides, and (ii) acylated sugar derivatives (Shaw, 1970). The glycosyl diglycerides are analogous to the phosphoglycerides, i.e. they are derivatives of 1,2-diacyl-*sn*-glycerol with the carbohydrate residue glycosidically bound at the 3-position of glycerol. Acylated sugars do not contain glycerol, and have the acyl groups esterified directly to the carbohydrate residue. Methods for their isolation and purification have been discussed previously (Shaw, 1970).

A. Glycosyl Diglycerides

More than a decade has passed since the first isolation from bacteria of this type of glycolipid (Macfarlane, 1961) which was originally discovered in wheat flour and subsequently in plants. The ensuing years have established both their widespread distribution in bacteria and their structural diversity. Table 1 shows the distribution of those glycolipids

Table 1. Distribution of Glycosyl Diglycerides of Known Structure in Bacteria

Carbohydrate residue[a]	Organism	References
	MONOGLYCOSYL DIGLYCERIDES[b]	
Glcα 1 →	*Acholeplasma laidlawii*	Shaw *et al.* (1968)
	Acholeplasma modicum	Mayberry *et al.* (1974)
Glcβ 1 →	*Mycoplasma neurolyticum*	Smith (1972)
Galα 1 →	*Treponema pallidum*	Livermore and Johnson (1970)
Galβ 1 →	*Arthrobacter* sp.	Shaw and Stead (1971)
Gal*f*β 1 →	*Mycoplasma mycoides*	Plackett (1967)
	Bifidobacterium bifidum[c]	Veerkamp (1972)
GlcAα 1 →	*Pseudomonas diminuta*	Wilkinson (1969)

GlcAβ 1 →	*Pseudomonas rubescens*	Wilkinson (1968)
GlcNβ 1 →	*Bacillus megaterium*	Phizackerley *et al.* (1972)
GlcNAcα 1 →	*Streptococcus haemolyticus*	Ishizuka and Yamakawa (1968)
	DIGLYCOSYL DIGLYCERIDES	
Glcα 1 → 2 Glcα 1 →	Streptococci	Shaw (1970)
	Acholeplasma laidlawii	Shaw *et al.* (1968)
Glcβ 1 → 6 Glcβ 1 →	Staphylococci	Shaw (1970)
	Bacilli	Shaw (1970)
	Mycoplasma neurolyticum	Smith (1972)
	Pseudomonas iodinium	R. Mirauer and N. Shaw (unpublished observations)
Galβ 1 → 6 Galβ 1 →	*Arthrobacter* sp.	Shaw and Stead (1971)
Galβ 1 → 2 Galβ 1 →	*Bifidobacterium bifidum*	Veerkamp (1972)
Galfβ 1 → 2 Galfβ 1 →[c]	*Bifidobacterium bifidum*	Veerkamp (1972)
Manα 1 → 3 Manα 1 →	*Micrococcus lysodeikticus*	Lennarz and Talamo (1966)
	Microbacterium lacticum	Shaw (1968)
	Arthrobacter sp.	Shaw and Stead (1971)
Galα 1 → 2 Glcα 1 →	Pneumococci, lactobacilli	Shaw (1970)
	Listeria monocytogenes	Kosari and Carroll (1971)
Glcβ 1 → 4 GlcAα 1 →	*Pseudomonas diminuta*	Wilkinson (1969)
Glcα 1 → 4 GalAα 1 →[c]	*Streptomyces* LA7017	Bergelson *et al.* (1970)
	TRIGLYCOSYL DIGLYCERIDES	
Glcα 1 → 2 Glcα 1 → 2 Glcα 1 →	*Streptococcus haemolyticus*	Ishizuka and Yamakawa (1968)
Glcα 1 → 6 Galα 1 → 2 Glcα 1 →	*Lactobacillus casei*	Shaw *et al.* (1968)
3′-SO_3-Galβ 1 → 6 Manα 1 → 2 Glcα 1 →[d]	*Halobacterium cutirubrum*	Kates and Deroo (1973)
Galβ 1 → 2 Galβ 1 → 2 Galβ 1 →	*Bifidobacterium bifidum*	Veerkamp (1972)
	TETRAGLYCOSYL DIGLYCERIDES	
Glc 1 → 6 Glcα 1 → 6 Galα 1 → 2 Glcα 1	*Lactobacillus acidophilum*	Shaw (1970)
Galf 1 → 2 Gal 1 → 6 GlcNHCOR 1 → 2 Glc 1 →	*Flavobacterium thermophilum*	Oshima and Yamakawa (1972)

[a]For convenience the diglycerides, to which all the carbohydrate residues are linked, have been omitted. Unless otherwise indicated all the sugars are present in the pyranose form.

[b]Monoglycosyl diglycerides in these organisms are either present as principal components or have not been characterized previously.

[c]These glycolipids contain additional acyl residues.

[d]Derivative of 1,2-di-O-alkyl glycerol.

for which precise structures have been established, and Table 2 the distribution of partially characterized glycolipids. The most common structural type is that glycolipid containing a disaccharide unit linked to the diglyceride, i.e. diglycosyl diglyceride (Fig. 1) and nine different types have now been characterized (Table 1). Of these, five have only been isolated from a single genus; these are a glucosylglucuronyl diglyceride from *Pseudomonas diminuta* (Wilkinson, 1969), a glucosylgalacturonyl diglyceride from *Streptomyces* (Bergelson *et al.*, 1970) and both digalactopyranosyl and digalactofuranosyl diglycerides from *Bifidobacterium bifidum* (Veerkamp, 1972). The remaining five diglycosyl diglycerides,

TABLE 2. Distribution in Bacteria of Partially Characterized Glycosyl Diglycerides

Constituent Sugars	Organism	References
	MONOGLYCOSYL DIGLYCERIDES[a]	
Glucose	*Treponema zuelzerae*	Meyer and Meyer (1971)
Glucuronic acid	Unclassified halotolerant bacterium	Peleg and Tietz (1971)
	DIGLYCOSYL DIGLYCERIDES	
Mannose	*Microbacterium thermosphactum*	Shaw and Stead (1970)
	Corynebacterium aquaticum	Khuller and Brennan (1972a)
Glucose	*Yersinia pseudotuberculosis*	Tornabene (1973)
	Nocardia polychromogenes	Khuller and Brennan (1972b)
	Thermus aquaticum	Ray *et al.* (1971)
Galactose	*Vibrio fetus*	Tornabene and Ogg (1971)
	Butyrivibrio fibrisolvens	Kunsman (1970)
	Mycoplasma pneumoniae	Plackett *et al.* (1969)
Glucose, mannose	*Chromatium* strain D	Steiner *et al.* (1969)
Glucose, galactose	*Mycoplasma pneumoniae*	Plackett *et al.* (1969)
	Sulfolobus acidocaldarius	Langworthy *et al.* (1974)
	TRIGLYCOSYL DIGLYCERIDES	
Mannose, mannose, glucose	*Chromatium* strain D	Steiner *et al.* (1969)
Galactose, rhamnose and unidentified	*Chloropseudomonas ethylicum*	Constantopoulos and Bloch (1967)
Galactose	*Mycoplasma pneumoniae*	Plackett *et al.* (1969)
Galactose, glucose	*Mycoplasma pneumoniae*	Plackett *et al.* (1969)
	PENTAGLYCOSYL DIGLYCERIDES	
Galactose, galactose, glucose, glucose, mannoheptose	*Acholeplasma modicum*	Mayberry *et al.* (1974)

[a]In these organisms, monoglycosyl diglycerides are principal components.

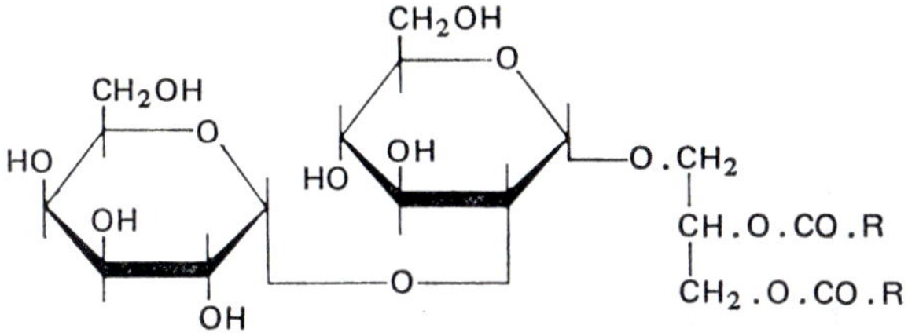

FIG. 1. The diglycosyl diglyceride isolated from pneumococci and lactobacilli, namely 3-[*O*-α-D-galactopyranosyl-(1-2)-*O*-α-D-glucopyranosyl]-*sn*-1,2-diglyceride. Nine different types of diglycosyl diglyceride have been characterized (Table 1).

α-diglucosyl, β-diglucosyl, digalactosyl, dimannosyl and galactosylglucosyl are the most prevalent, and they have been isolated from organisms in several genera.

Monoglycosyl diglycerides, although precursors in the biosynthesis of diglycosyl diglycerides, do not usually accumulate in significant quantities, but, occasionally, larger amounts are observed. Of the nine diglycosyl diglycerides, only one of the corresponding monoglycosyl diglycerides, namely galacturonosyl diglyceride, has yet to be observed. Diglycosyl diglycerides containing glucosamine have not yet been found but two glucosaminyl diglycerides have been characterized. N-Acetyl α-glucosaminyl diglyceride has been isolated as a minor component (less than 3% total glycolipid) of *Streptococcus haemolyticus* (Ishizuka and Yamakawa, 1969), and the de-N-acetylated derivative of opposite anomeric configuration is the sole glycolipid component of *Bacillus megaterium* (Phizackerley *et al.*, 1972). This organism also has glycophospholipids containing glucosamine (see Section IIIB, p. 151).

Tri- and tetraglycosyl diglycerides have been found in a few bacteria. The thermophile *Flavobacterium thermophilum* has, as its major lipid, an unusual tetraglycosyl diglyceride containing galactose, glucose and glucosamine, the last residue having an amide-linked fatty acid residue (Oshima and Yamakawa, 1972). The largest glycolipid of this type so far reported is a pentaglycosyl diglyceride from *Acholeplasma modicum* (Mayberry *et al.*, 1974). This glycolipid is interesting not only for its size but also for its carbohydrate components which include D-mannoheptose. The latter, although a common component of lipopolysaccharides (see Section IVA, p. 155), has not previously been found in glycolipids. The problems associated with the extraction of these larger glycosyl diglycerides are discussed in Section IVC (p. 157).

Biosynthesis of diglycosyl diglycerides has been investigated in several bacteria, and the details discussed previously (Shaw, 1970). The synthesis proceeds through sequential transfer of sugars from the appropriate sugar nucleotides first to 1,2 diacyl-*sn*-glycerol and then to the newly synthesized monoglycosyl diglyceride. An enzyme preparation from *Mycobacterium smegmatis* catalyses transfer of galactose and glucose from the appropriate sugar nucleotides into lipid products characterized as mono- and diglycosyl diglycerides, but no evidence was presented for their occurrence in whole organisms (Schultz and Elbein, 1974). The biosynthetic pathway to the larger glycosyl diglycerides is still unknown. Chemical syntheses of monoglucosyl-, galactosyl-, mannosyl- and diglucosyl diglycerides have been reported (Wehrli and Pomeranz, 1969; Shvets *et al.*, 1973).

B. Acylated Sugar Derivatives

This type of glycolipid, although present in many bacteria, has received much less attention than glycosyl diglycerides, and the number of fully characterized structures is comparatively small (Table 3, Fig. 2). The most familiar member is diacyl treholase, or "cord-factor", present

TABLE 3. Distribution of Acylated Sugar Derivatives in Bacteria

Lipid	Organism	References
6-*O*-Mycolyl glucose	*Corynebacterium diphtheriae*, mycobacteria, *Brevibacterium thiogenitalis*	Brennan *et al.* (1970) Okazaki *et al.* (1969)
Acylated glucoses, hexoses	*Mycoplasma* strain J	Smith and Mayberry (1968)
	Streptococcus faecalis	Welsh *et al.* (1968)
	Escherichia coli, *Aerobacter aerogenes*,	Brennan *et al.* (1970)
	Pseudomonas fluorescens	Welsh *et al.* (1968)
	Nocardia, corynebacteria and mycobacteria	Brennan *et al.* (1970) Khuller and Brennan (1970 a, b)
Dirhamnosyl β-hydroxydecanoyl-β-hydroxydecanoate ("rhamnolipid")	*Pseudomonas aeruginosa*	Edwards and Hayashi (1965)
Diacylinositol mannoside	*Propionibacterium* sp.	Prottey and Ballou (1968) Shaw and Dinglinger (1969)
Diacyl trehalose ("cord-factor")	Corynebacteria, nocardia mycobacteria	Lederer (1967); Yano *et al.* (1971); Ioneda *et al.* (1970); Khuller and Brennan (1972)
Esters of "phleic acids" and trehalose	*Mycobacterium phlei*	Asselineau *et al.* (1972)

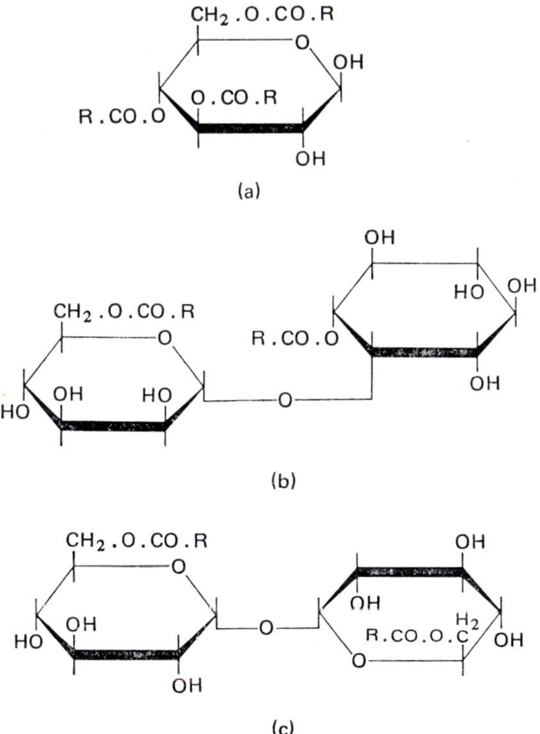

FIG. 2. Some glycolipids of the acylated sugar type: (a) Triacyl glucose from *Mycoplasma*, strain J; (b) Diacyl inositol mannoside from *Propionibacterium* sp.; (c) Diacyl trehalose, cord-factor, from *Corynebacterium* sp.

in mycobacteria and related organisms (Lederer, 1967). A recent development in this area is the isolation from *Mycobacterium phlei* of trehalose derivatives containing "phleic acids", the latter being a homologous series of polyunsaturated fatty acids, the main member of which is hexatriaconta-4,8,12,16,20-pentaenoic acid (Asselineau *et al.*, 1972). Propionic acid bacteria contain a diacyl inositol mannoside (Prottey and Ballou, 1968; Shaw and Dinglinger, 1969) which so far is the only glycolipid containing inositol, and a possible relationship to the phosphatidylinositol mannosides is apparent (see Section IIIA, p. 148). The simplest lipids in this class are the acylated glucoses which have been isolated with varying degrees of acylation from many different bacteria.

Biosynthesis of the rhamnolipid of *Pseudomonas aeruginosa* occurs by a route analogous to that of the diglycosyl diglycerides (Burger *et al.*, 1963). Two rhamnose units are transferred sequentially from TDP-rhamnose to β-hydroxydecanoyl-β-hydroxydecanoate. Prottey and Ballou

(1968) suggested that the diacyl inositol mannoside might arise from the action of phospholipase D on a diacyl phosphatidylinositol mannoside. However, phosphatidic acid, the other product from the reaction, is not present in propionic-acid bacteria nor has any phospholipase activity been demonstrated.

III. Glycophospholipids

The scheme of nomenclature suggested by Shaw and Stead (1972) will be used. The term "glycophospholipid" may be applied to any lipid containing carbohydrate and phosphate residues, the latter not necessarily implying the presence of a phosphatidyl group. The term "phosphoglycolipid" will only be used for those lipids which are derived from glycolipids by the addition of a phosphate-containing residue.

A. Phosphatidylinositol Mannosides

This founder member of the glycophospholipids was discovered by Anderson and his colleagues in the late 1930s during their intensive investigations into the lipids of mycobacteria but it was not until 25 years later that complete structures were proposed (Lee and Ballou, 1965). A family of phosphatidylinositol mannosides exists containing from one up to five mannose units (Fig. 3). The first mannose unit is glycosidically linked to the hydroxyl at C-2 of the inositol ring and subsequent mannose units are added sequentially to the hydroxyl at C-6 to form a series of mono-, di-, tri- and tetrasaccharides. However, the phosphatidylinositol dimannoside usually predominates. From *Mycobacterium tuberculosis*, Pangborn and McKinney (1966) isolated a series of phosphatidylinositol dimannosides containing a total of two, three and four acyl residues. The exact location of the additional acyl residues was not determined, and their significance remains unknown. However, this phenomenon is becoming increasingly prevalent; a similar situation has been found in the phosphoglycolipid in *Streptococcus faecalis* (See Section IIID, p. 153) and acyl derivatives of phosphatidylglycerol and phosphatidylinositol have been described (Olsen and Ballou, 1971; Brennan, 1968).

Amongst the Actinomycetales, mycobacteria, nocardia, streptomycetes and microbispora all contain this type of glycophospholipid. *Micromonospora* sp. F3 (Tabaud *et al.*, 1971) and *Streptomyces griseus* (Kataoka and Nojima, 1967) both contain a monomannoside, and *Nocardia coeliaca* and *Nocardia polychromogenes* a dimannoside (Khuller and Brennan, 1972b). Brennan and Lehane (1971) have examined a number of corynebacteria, and identified the dimannoside (or acylated derivatives) as major components in *C. diphtheria*, *C. xerosis*, *C. equi* and *C. ovis*.

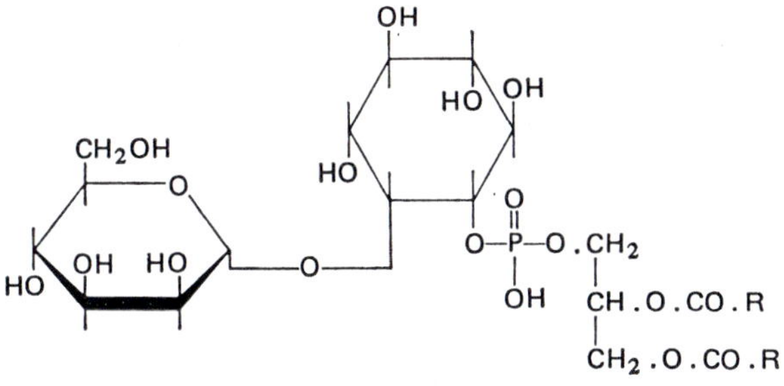

(a)

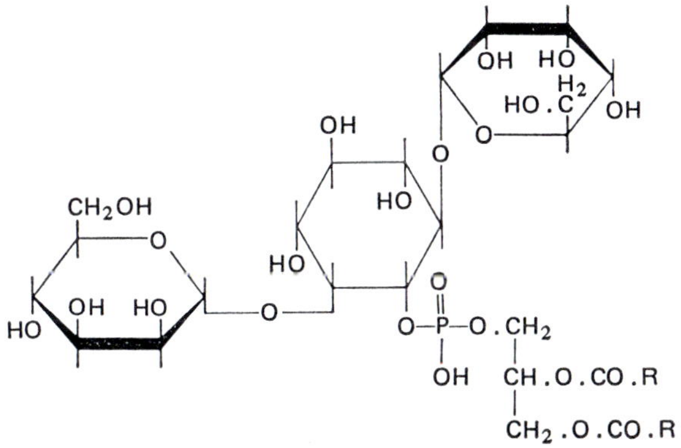

(b)

Fig. 3. (a) Phosphatidylinositol monomannoside; (b) Phosphatidylinositol dimannoside. The tri- and tetramannosides have the additional mannose units joined by α(1-6) linkages to the mannose located at position-6 of the inositol ring. In the pentamannoside, the terminal mannose has an α(1-2) linkage.

The monomannoside was predominant in *C. aquaticum* (Khuller and Brennan, 1972a). Although precise chemical structures have been elucidated for the phosphatidylinositol mannosides from mycobacteria, a rigorous chemical identification for many of the lipids isolated from the other organisms has not been obtained. Structures have been proposed on the basis of the identification of mannose and inositol in hydrolysates and/or paper chromatographic comparison of the phosphate esters obtained by deacylation with those of the authentic mycobacterial lipids. Such methods would probably not distinguish, for example, any differences in the location of the mannose residue on the inositol residue or in the anomeric configuration of the glycosidic linkage. Such differences have been observed in other isomeric glycophospholipids (see Section IIIB, p. 151).

Phosphatidylinositol mannosides have not yet been reported in any other members of the Corynebacteriaceae. Phosphatidylinositol is present in *Arthrobacter* species but the mannosyl derivatives could not be detected (Shaw and Stead, 1971). The major inositol lipid in *Propionibacterium* species is a glycolipid, namely diacylinositol mannoside (Shaw and Dinglinger, 1969; Prottey and Ballou, 1968) and, although these organisms are capable of synthesizing phosphatidylinositol monomannoside from exogenous phosphatidylinositol *in vitro* (Brennan and Ballou, 1968b), a rigorous identification of the glycophospholipids as *in vivo* constituents has not yet been reported.

The biosynthesis of phosphatidylinositol mannosides has received considerable attention, but a number of features still require further elaboration. In *M. phlei*, GDP-mannose acts as sugar donor and phosphatidylinositol the acceptor to yield as major product the acylated phosphatidylinositol dimannoside (Brennan and Ballou, 1967). The point at which the additional acyl residues are introduced was not determined although an enzyme system capable of catalysing transfer of acyl groups to the dimannoside was subsequently identified (Brennan and Ballou, 1968a). Alternatively, acylation of the presumed precursor of the dimannoside, the monomannoside, might be necessary prior to transfer of the second mannose unit. A measure of support for this proposal has been provided by Takayama and Goldman (1969) who demonstrated, in *M. tuberculosis*, that there is some restriction on the sequential addition of the second mannose unit to newly synthesized phosphatidylinositol monomannoside. A cell-free system, utilizing endogenous acceptors and CDP[^{14}C]mannose, synthesized both phosphatidylinositol[^{14}C]monomannoside and phosphatidylinositol[^{14}C]dimannoside but, in the latter, only the mannose at position 6 was labelled. Thus, newly synthesized monomannoside could not act as acceptor for the second mannose unit without further modification (e.g. acylation?) which the cell-free system was unable to carry out. An alternative hypothesis was presented by Shaw and Dinglinger (1969) following their isolation of the diacylinositol mannoside from a *Propionibacterium* species. Transfer of the phosphatidyl group from CDP-diglyceride to a diacyl inositol mannoside would give directly a diacyl phosphatidylinositol monomannoside which could then act as acceptor for further mannose units. Circumstantial support for this proposal has come from studies on the biosynthesis of phosphatidylinositol monomannoside, the principal glycophospholipid in *Corynebacterium aquaticum* (Khuller and Brennan, 1972a). An enzyme system prepared by methods used successfully with mycobacteria did not catalyse transfer of mannose from GDP-mannose to either endogenous acceptor or added phosphatidylinositol. Thus, the precise sequence of events leading to synthesis of the phosphatidylinositol monomannoside and dimannoside is still unknown as indeed is the synthesis of the higher

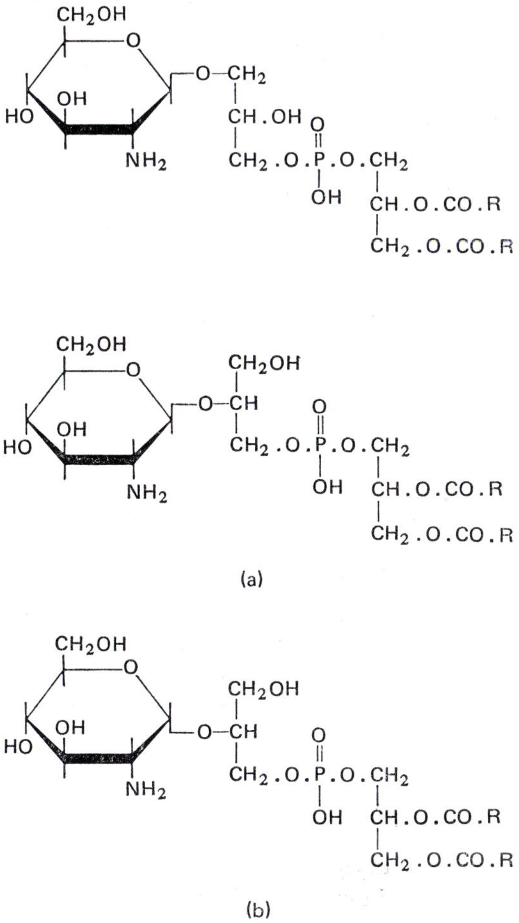

(a)

(b)

FIG. 4. (a) The two isomers of glucosaminyl phosphatidylglycerol from *Bacillus megaterium*. (b) Glucosaminyl phosphatidylglycerol from *Pseudomonas ovalis*.

mannosides. Brennan and Ballou (1967) reported very poor incorporation (less than 1% of the lipid-bound [^{14}C]mannose) into the tetra- and pentamannosides.

B. PHOSPHATIDYLGLYCEROL GLYCOSIDES

These glycophospholipids are also related to a known phospholipid, namely phosphatidylglycerol, and the first example is a glucosaminyl derivative thereof. At least three isomers are known, their structures varying in the anomeric configuration of the glucosamine residue and its position of attachment to the glycerol. *Bacillus megaterium* contains two isomers, both β-glucosaminides (Fig. 4) but, in one isomer, the glycosidic linkage is to the hydroxyl at C-2 of the glycerol residue and in the other to the hydroxyl at C-3 of the glycerol (MacDougall and Phizackerley, 1969). *Pseudomonas ovalis* contains a third isomer (Fig. 4) in which the glucosamine residue is attached by an α-glycosidic linkage to the hydroxyl group at C-2 of the glycerol residue (Op den Kamp *et al.*, 1969; MacDougall and Phizackerley, 1969). The occurrence of a non-acetylated glucosamine residue is extremely unusual and enables the lipid to exist as a zwitterion, as in phosphatidylethanolamine. No biosynthetic evidence has been reported, but a possible route would be the transfer of glucosamine to phosphatidylglycerol from the appropriate sugar nucleotide precursor, UDP-glucosamine. This last compound has, however, to the author's knowledge, never been isolated, and an alternative pathway would involve formation of an N-acetylglucosaminylphosphatidylglycerol followed by enzymic N-deacetylation. Some bacilli are known to have peptidoglycans containing non-acetylated glucosamine residues, and an N-deacetylating enzyme has been reported (Araki *et al.*, 1971).

The glucose analogue of glucosaminylphosphatidylglycerol has been isolated from an unclassified halotolerant bacterium (Peleg and Tietz, 1971). Hydrolysis with phospholipase C gave a diglyceride and a water-soluble product which liberated glycerol phosphate on acid hydrolysis. This result is consistent with the proposed structure, but additional evidence is clearly desirable, in particular the precise nature of the glucose linkage to glycerol.

C. Phosphatidylglucose

The name "phosphatidylglucose" implies a structure containing a phosphatidyl residue joined through a phosphodiester linkage to a hydroxyl group of glucose, and a chemical synthesis of one of the several isomers possible has been carried out (Verheij *et al.*, 1970). In addition, the water-soluble deacylation products of three isomeric phosphatidylglucoses have also been synthesized (Shaw *et al.*, 1970). These synthetic experiments were expected to confirm the structure of a glucose-containing phospholipid isolated from *Acholeplasma laidlawii* and for which a phosphatidylglucose structure was proposed (Smith and Henrikson, 1965). However these experiments and further structural analysis showed the structure to be incorrect, and an alternative glycerylphosphoryldiglucosyl diglyceride structure has now been established (see Section IIID, p. 153; Shaw *et al.*, 1970). A similar structure is also probable for another lipid to which a phosphatidylglucose structure has been ascribed (Short and White, 1970) but further evidence is desirable (Shaw and Stead, 1972). Thus far phosphatidylglucose remains elusive.

D. Phosphoglycolipids: Glycerylphosphoryl- and Phosphatidylglycosyl Diglycerides

Unlike the lipids discussed so far in this section, these carbohydrate-containing lipids are structurally related to known glycolipids, hence the generic term "phosphoglycolipids", and they are proving to be prevalent in Gram-positive bacteria. The first example was reported in 1965 by

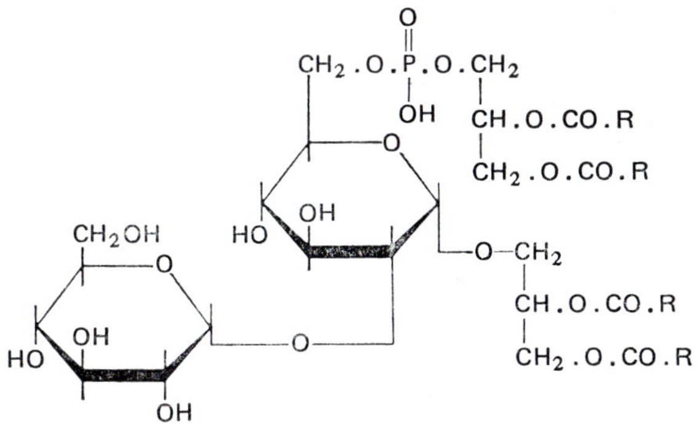

Fig. 5. Phosphatidyldiglucosyl diglyceride from *Streptococcus faecalis.*

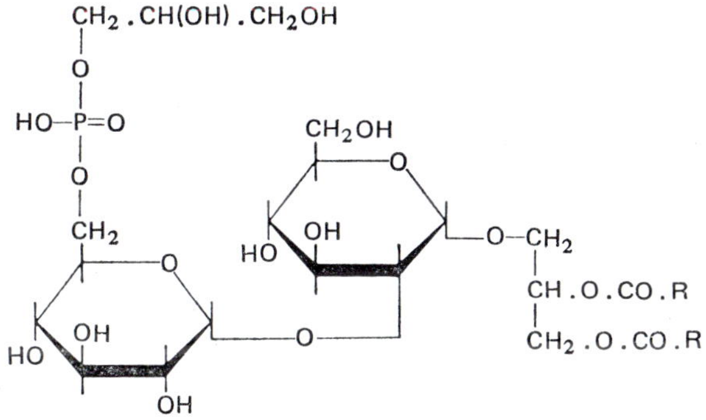

Fig. 6. Glycerylphosphoryldiglucosyl diglyceride from *Streptococcus faecalis.*

Smith and Henrikson as the major phospholipid in *Acholeplasma laidlawii* although it was several years later that the basic structure was established as a glycerophosphate derivative of a diglucosyl diglyceride (Shaw *et al.*, 1970). Meanwhile Ambron and Pieringer (1971) and Fischer (1970) independently reported the isolation of a phosphoglycolipid from several streptococci in which a phosphatidyl residue is linked to the hydroxyl group at C-6 of the internal glucose in a diglucosyl diglyceride (Fig. 5), the latter possessing a structure identical with the glycolipid found in the same organism. Subsequently Fischer *et al.* (1973) isolated from the same organisms a glycerylphosphoryldiglucosyl diglyceride analogous to the phosphoglycolipid from *A. laidlawii* in which the glycerophosphate group is linked to the hydroxyl at C-6 of the terminal glucose in diglucosyl diglyceride (Fig. 6). The occurrence of both diacyl and tetra-acyl derivatives is reminiscent of the acylated phosphatidylinositol mannosides (see Section IIIA, p. 148), but here the situation is more complex as other structural differences exist, namely the location of the phosphate residue. This excludes the possibility of any acylation-deacylation interconversion. A phosphatidyldiglucosyl diglyceride has been isolated from *A. laidlawii* (Smith, 1972) although, in contrast to *Strep. faecalis*, it is the minor phosphoglycolipid component. An important structural difference between the phosphoglycolipids of *A. laidlawii* and *Strep. faecalis* is the stereochemistry of the glycerophosphate unit. Both the glycerophosphate and the phosphatidyl groups in the two phosphoglycolipids from *A. laidlawii* have the *sn*-glycerol 3-phosphate configuration. In the lipid from *Strep. faecalis*, however, whilst the phosphatidyl group has the anticipated *sn*-glycerol 3-phosphate configuration, the glycerophosphate of the glycerylphosphoryldiglucosyl diglyceride is *sn*-glycerol 1-phosphate. The biosynthetic implications of these configurations are discussed below.

The structural relationship between phosphoglycolipids and diglycosyl diglycerides suggests that the former may be as widely distributed as the latter. Glycerophosphate derivatives of the β-diglucosyl diglyceride have been isolated from two *Cellulomonas* species, as have derivatives of the galactosylglucosyl diglyceride from *Leuconostoc mesenteroides* and *Listeria monocytogenes* (Shaw and Stead, 1972). Monoglycosyl diglycerides are the biosynthetic precursors of the diglycosyl diglycerides and, although they do not usually accumulate in significant quantities, the occurrence of related phosphoglycolipids would not be unexpected. *Pseudomonas diminuta* contains a phosphatidylglucosyl diglyceride, and this represents the first isolation from a Gram-negative organism (Wilkinson and Bell, 1971). The major phospholipid of *Thermoplasma acidophilum*, a thermophilic, acidophilic member of the Mycoplasmatales, is a diether analogue of a glycerylphosphoryl monoglycosyl diglyceride (Langworthy *et al.*, 1972). In this unusual phosphoglycolipid, the normal acyl residues are replaced by C_{40} isopranols in ether linkage and the sugar, as yet unidentified, is not a normal hexose. The alkyl ethers are also present in the other lipids of this organism and also in a lipoglycan (see Section IVC, p. 157). Alkyl ethers derived from mevalonate have been found in the lipids of extremely halophilic bacteria (Kates *et al.*, 1966).

The phosphatidyldiglucosyl diglyceride from *Strep. faecalis* is formed directly from diglucosyldiglyceride, and the donor is either phosphatidylglycerol or bisphosphatidylglycerol (Pieringer, 1972). Pieringer was unable to distinguish conclusively between these two phospholipids as the crude enzyme system catalysed their interconversion. The source of glycerophosphate in glycerylphosphoryldiglycosyl diglycerides is unknown, but transfer of the glycerophosphate residue from phosphatidylglycerol would lead to the required *sn*-glycerol 1-phosphate configuration. This route could not operate in *A. laidlawii* where the enantiomeric configuration is found. The involvement of a preformed phospholipid rather than a nucleotide intermediate as phosphatidyl donor has also been demonstrated in the biosynthesis of bisphosphatidylglycerol (Hirschberg and Kennedy, 1972).

IV. Lipid-Polysaccharide Complexes

Many organisms, even after exhaustive extraction with organic solvents, still contain bound fatty acids which are only released after acid or alkaline hydrolysis. The nature of many of these components is still unknown, but a rapidly increasing number of lipid-polysaccharide complexes are being isolated and characterized in which the lipid moiety is recognizable as a glycolipid or glycophospholipid. Undoubtedly the most familiar examples are the endotoxins of the Gram-negative envelope, and they are widely distributed in strains of Enterobacteriaceae, Neisseriaceae and Pseudomonadaceae. The generic term "lipopolysaccharide" has been given to this class of polymers whose chemical and immunological properties have been extensively investigated. More recently other lipid-polysaccharide complexes have been isolated to which the name "lipopolysaccharide" could have been applied but, as they show many structural differences from the endotoxins, it would be inappropriate to use this term which has now acquired a particular connotation. Accordingly the term "lipoglycan" will be used to describe any polymeric material which contains both carbohydrate and lipid moieties.

A. Lipopolysaccharides of Gram-Negative Cell Walls

Extraction of the Gam-negative cell wall with aqueous phenol produces a high molecular-weight water-soluble material called lipopolysaccharide. Characteristic components of the lipopolysaccharide are mannoheptose, 2 keto-2-deoxyoctanoic acid and lipid A. The last compound is covalently bound to the polysaccharide and is only extractable into chloroform after acid hydrolysis. Lipid A, so called to distinguish it from the unbound lipid present in many lipopolysaccharide preparations and which is now known to be predominantly phosphatidylethanolamine, contains phosphate, glucosamine and fatty acids. Its complete structure is not yet known but the available evidence shows it to be dissimilar to any known glycophospholipid. In particular it does not contain glycerol, and the fatty acids contain high proportions of hydroxyacids which are not found in the free lipids. The basic unit of the lipid A from *Salmonella* is a β-(1 $\rightarrow$ 6) linked disaccharide of glucosamine containing β-hydroxymyristic acid joined in amide linkage to the amino group of both glucosamine residues. Several disaccharide units are bridged by phosphate groups, and the remaining hydroxyl groups esterified with fatty acids. For a comprehensive review on all aspects of lipopolysaccharide structure see Luderitz *et al.* (1973). These major structural features are probably common to lipid A preparations from most Gram-negative bacteria. However it should be stressed that, as lipid A is obtained from an acidic hydrolysis, its structure may have been modified from that present in the original lipopolysaccharide and, in turn, the original structure of the latter almost certainly will have been affected by the phenolic extraction procedure. Nowotny (1971a) has shown that lipid A preparations are extremely heterogeneous, and this is of particular importance when considering their biological properties. The isolation of mutants in which the lipopolysaccharide lacks most of the polysaccharide backbone has enabled endotoxin preparations to be isolated which consist primarily of the lipid moiety; such preparations exhibit full endotoxicity. The primary effect is pyrogenicity although the overall toxicity is comparatively low. Thus the present evidence suggests that the lipid, besides linking the lipopolysaccharide to the cell wall, plays a significant role in the biological properties of endotoxins (Nowotny, 1971b).

B. Lipoteichoic Acids of Gram-Positive Bacteria

The molecular architecture of the Gram-positive cell wall is considerably less complex than that of Gram-negative organisms. The majority of walls in Gram-positive bacteria do not contain any lipid, either bound or unbound, and are predominantly composed of peptidoglycan and teichoic acids. The latter polymers may be considered the equivalent of lipopolysaccharides but are linked to the peptidoglycan through phosphate residues (Baddiley, 1972). Glycerol teichoic acids have also been found associated with the outer surface of the cytoplasmic membrane (Hay *et al.*, 1963) but, until recently, the nature of this association was unknown. Using isolation procedures similar to those used for the preparation of lipopolysaccharides, Wicken and Knox (1970) isolated a membrane teichoic acid preparation from *Lactobacillus fermenti* which, from analysis of the products of alkaline hydrolysis, it was concluded contained covalently bound glycolipid and phospholipid and accordingly was named "lipoteichoic acid". Toon *et al.* (1972) purified a lipoteichoic acid from *Streptococcus faecalis* and, by treatment with hydrogen fluoride, isolated a diglucosyldiglyceride identical in structure with that present in the membrane lipids. In addition, the presence of diglycerides amongst the products of hydrolysis led them to suggest that the teichoic acid was joined through its terminal phosphate group to a phosphatidyldiglucosyl diglyceride, the latter also being present in the membrane lipids. Lipoteichoic acids have also been isolated from *Lactobacillus plantarum*, *Lactobacillus buchneri* and *Staphylococcus aureus* and, in each organism, diglycosyl diglycerides have been obtained by treatment with hydrogen fluoride but the presence of diglyceride has not been conclusively established (Coley *et al.*, 1972). Also the available evidence cannot exclude glycerylphosphoryldiglycosyl diglyceride as the lipid moiety because hydrolysis of such a lipoteichoic acid would also give diglycosyl diglyceride. Thus it appears that, in membrane-associated teichoic acids, the glycerophosphate polymer is covalently linked to either glycolipid or phosphoglycolipid and that this lipid, as a constituent of the lipoprotein layer, anchors the polymer to the membrane. The lipoteichoic acids from several lactobacilli are antigenic when injected together with Freund's adjuvant into rabbits, whereas the lipid-free teichoic acids are not (Wicken and Knox, 1971). Toon *et al.* (1972) estimated that approximately 12% of the total diglycosyl diglyceride (either as glycolipid or phosphoglycolipid) was linked to teichoic acid, and, if further work establishes that the lipid moiety is glycolipid, then it would seem unlikely that this is the sole or even major function of these compounds in view of the small amounts involved. However, if the lipid moiety proves to be phosphoglycolipid, this may indeed represent their major function as these unbound lipids constitute only a small fraction of the total cellular lipid. However, it is unlikely that these unbound phosphoglycolipids have arisen through chemical or enzymic degradation during isolation procedures. Phosphoglycolipids are major lipid constituents in *Acholeplasma laidlawii* and, although a polymer of N-acetylglucosamine and N-acetylgalactosamine has been isolated from this cell wall-less organism (Gilliam and Morowitz, 1972), demonstration of covalent bonding to lipids has not been reported.

C. Lipoglycans

The unusual nature of the lipids from *Thermoplasma acidophilum*, an acidophilic, thermophilic organism, has already been discussed (Section IIID, p. 153). Using the phenol extraction procedure, a lipoglycan has been isolated from this organism which also contains C_{40} isopranol units in ether linkage (Mayberry-Carson *et al.*, 1974). The lipid moiety is a diether analogue of glucosyl diglyceride to which is bound, through the reducing group, a polysaccharide of 24 mannose units. Although uncharged, this lipoglycan is of a similar size to the lipoteichoic acid from *Streptococcus faecalis* and readily forms high molecular-weight micelles in solution. An interesting point of nomenclature now arises. This lipoglycan might also be categorized as a glycolipid, namely a glycosyl diglyceride with a hydrophilic moiety of 25 hexose units. When does a glycolipid become a lipoglycan or *vice versa*? As discussed in Section IIA (p. 142) the largest glycolipid of this type so far reported is a pentaglycosyl diglyceride which is soluble in chloroform-methanol mixtures. Thereafter a point will soon be reached when solubility in organic solvents is minimal, and it seems probable that isolation of decaglycosyl diglycerides and higher homologues by conventional solvent extraction methods will not be possible, and the term "glycolipid" will be inappropriate.

Micrococcus lysodeikticus does not possess a conventional lipoteichoic acid but contains instead an acylated mannan (Powell *et al.*, 1974). Glycosyl diglycerides or phosphoglycolipids are not present in the complex but fatty acids are released on alkaline hydrolysis. This complex is therefore a polymeric derivative of the acylated sugar type of glycolipid. Residues of glycerol or phosphate are not present but it is still negatively charged due to an unidentified acidic residue. The physical properties of this lipoglycan are therefore similar to those of lipoteichoic acids, and it may have a similar function, e.g. controlling the availability of divalent cations to the membrane (Hughes *et al.*, 1971). The structural analogy to lipoteichoic acids requires that the acyl residues are located at one end of the molecule in order to anchor it to the membrane, and that the acidic residues are distributed along the remainder of the polymer. Both of these features have yet to be established.

An acidic lipoglycan of the acylated sugar type has been isolated from several strains of mycobacteria (Saier and Ballou, 1968). Glucose, 6-*O*-methyl glucose and 3-*O*-methyl glucose make up the polysaccharide which is glycosidically bound to D-glyceric acid. Distributed along the polymer are eight short-chain acyl residues including two half-esters of succinic acid which contribute to the acidity of the molecule (Gray and Ballou, 1973). Clearly this lipoglycan is structurally very different from those described previously. First, the acyl residues are much shorter and, second, they are distributed throughout the molecule rather than being concentrated in a specific area. This would preclude any association with membrane; indeed it is found as a soluble component of the disrupted cells and is released into the extracellular fluid during all stages of the growth cycle. This lipoglycan is apparently unique to mycobacteria. A search for its presence in closely related organisms, namely species of *Corynebacterium*, *Nocardia* and *Propionibacterium*, gave negative results.

V. Distribution and Taxonomy

The first attempt to correlate lipid composition with taxonomic classification was made by Abel *et al.* (1963) and, in the ensuing years as more information on lipids became available, the expediency of this approach to chemotaxonomy has become established. As two extensive reviews on this subject have appeared recently (Goldfine, 1972; Shaw, 1974) it will be necessary here only to summarize the results.

Lipids offer several advantages as a chemotaxonomic method. They are present in all bacteria; they are easily and specifically extracted; they are convenient to analyse; and finally, and probably most important of all, the great variety of lipids usually present allows the comparison of several variables.

Recognition of the widespread distribution of glycosyl diglycerides and their taxonomic potential was first discussed by Shaw and Baddiley (1968). The increasing number of glycolipids characterized since then has substantiated in large measure their original conclusions. They are found predominantly, although not exclusively, in Gram-positive bacteria and it is usually the diglycosyl diglyceride which is the sole or major glycolipid. Whereas the same glycosyl diglyceride may be found in organisms belonging to different families, members of the same genus normally contain the same glycolipid. For example, streptococci contain the α-diglucosyl diglyceride, and lactobacilli the galactosylglucosyl diglyceride. A notable exception to this conclusion is the family Mycoplasmatales. *Acholeplasma laidlawii* contains the same diglucosyl diglyceride as in the streptococci (Shaw *et al.*, 1968), *M. gallinarum* Strain J contains a triacyl glucose similar to that found in *Streptococcus faecalis* (Smith and Mayberry, 1968), and *M. neurolyticum* has the same diglucosyl diglyceride as *Staphylococcus aureus* (Smith, 1972). Other mycoplasmas contain unique glycolipids such as the pentaglycosyl diglyceride in *Acholeplasma modicum* (Mayberry *et al.*, 1974). Thus, whilst there is an overall similarity between this group of cell wall-less organisms and the Gram-positive bacteria, which may be circumstantial or suggest some evolutionary relationship, there is great diversity within the family. This is also apparent from other criteria such as DNA composition and nutritional requirements; colonial morphology remains the primary criterion for identification.

Arthrobacter remains the only genus to contain two unrelated diglycosyl diglycerides. *Arthrobacter globiformis*, *A. pascens* and *A. crystallopoietes* all contain both the digalactosyl and the dimannosyl diglycerides. The reason for this is unknown but it is not related to the rod-sphere morphogenesis exhibited by these organisms (Shaw and Stead, 1971). A digalactosyl diglyceride of different structure is found in *Bifidobacterium bifidum* together with a digalactofuranosyl diglyceride. The latter compound carries an additional acyl residue on the galactose as does the corresponding monogalactofuranosyl diglyceride found in the same organism (Veerkamp, 1972).

There is insufficient evidence concerning the distribution of acylated sugar derivatives to enable many significant conclusions to be reached. The diacyl inositol mannoside appears to be diagnostic for *Propionibacterium* species, and its absence from *P. acnes* (N. Shaw and A. Stead, unpublished observations), an organism of uncertain classification (Johnson and Cummins, 1972), supports its exclusion from this genus.

Phosphatidylinositol mannosides have been found in corynebacteria, mycobacteria, nocardia, streptomycetes and microbispora, and diacyl trehalose (cord-factor) is also present in members of the first three genera. This similarity in lipid composition supports the proposition of a close phylogenetic relationship between these genera, but they may be conveniently distinguished by their fatty-acid residues, including mycolic acids, nocardic acids and corynemycolic acids (Shaw, 1974).

The structural relationship between glycosyl diglycerides and glycerylphosphoryl- and phosphatidylglycosyl diglycerides suggests that the latter compounds might also be present in those bacteria where the former are located. Phosphoglycolipids derived from three of the diglycosyl diglycerides have been isolated (Section IIID, p. 153) and certainly other examples will be found. However it is doubtful if they will be of taxonomic importance as the related glycolipids are usually present in larger concentrations and are easier to identify.

VI. Location and Function

The ease with which lipids may be specifically extracted from intact microbial cells has resulted in comparatively few studies which rigorously establish their location. It is generally assumed that walls of Gram-positive bacteria are devoid of lipids, and that the latter are present in the cytoplasmic membrane. This has been confirmed for several organisms where pure membrane preparations have been analysed, including *Streptococcus faecalis* (Vorbeck and Marinetti, 1965) and *Bacillus subtilis* (Bishop *et al.*, 1967), and it is reasonable to suppose this is true for most Gram-positive bacteria. However Huis in't Veld and Villiers (1973) have reported that cell-wall preparations from streptococci of groups Fs and Z_3, even after extensive purification, still contain appreciable quantities of diglucosyl diglyceride; the absence of phospholipids excludes contamination with cytoplasmic membrane. By contrast, walls of Group-A streptococci are typical in that they do not contain lipid, and a possible correlation between the presence of glycolipids in cell walls and the type III antigen was suggested.

Having established that glycolipids are usually present in the cytoplasmic membrane, are they part of a unit membrane or are they located at a specific area in the membrane? Chapman and Urbina (1971), from their results of differential scanning calorimetry on membranes from *Acholeplasma laidlawii*, speculated that one possible interpretation is that the glycolipids are arranged together in the membrane, leaving the remaining lipid to be organized in different configurations with lipid-protein interactions. Reinert and Steim (1970), however, have interpreted their results from similar experiments as supporting an extended regular bilayer configuration. There are obviously many difficulties in extending observations from isolated individual lipids *via* artificial membranes to natural membranes. In this respect, it is pertinent to note that most of the physicochemical studies on artificial membranes and black lipid films have conveniently ignored the existence and possible effects of glycolipids as membrane substituents.

The status of the mesosome with respect to the plasma membrane and its functional role in the cell are still a matter for conjecture, but a role in membrane synthesis has been suggested (Fitz-James, 1968). Although the isolation and preliminary chemical analysis of several mesosomal preparations have been reported, a detailed lipid analysis is available only for mesosomes from *Micrococcus lysodeikticus* (Thomas and Ellar, 1973). Their results indicated a preferential accumulation of glycolipid in the mesosomal fraction (15·9% of total lipid) as compared with 5·7% in the plasma membrane, but they did not support the proposal that mesosomes are preferential sites for lipid synthesis.

The problem of location is more complicated in both Gram-negative organisms and the mycobacteria and related organisms, where differentiation between the multilayered cell wall and cytoplasmic membrane is not easily established. The acylated sugar derivatives described by Shaw *et al.* (1968) were contaminants of lipopolysaccharide preparations, and are probably located in the cell envelope. The glucuronic acid-containing glycolipids in *Pseudomonas diminuta* and *Pseudomonas rubescens* were also isolated from cell-envelope preparations (Wilkinson, 1968, 1969); the phosphatidylinositol mannosides from mycobacteria are similarly located (Akamatsu *et al.*, 1967). The phleic acid derivatives of trehalose from *Mycobacterium phlei* are also located in the outer parts of the cell as they can be extracted by washing the bacteria with hexane, a process which does not destroy the viability of the cells (Asselineau *et al.*, 1972). A specific role for these surface lipids is suggested by the observation that, when the medium is supplemented with a surfactant to prevent bacterial aggregation, the amount of these lipids present decreases dramatically.

The lipids of mycoplasma, L-forms and other cell wall-less organisms must of necessity be in the plasma membrane. The differences in lipid composition of L-forms and their parent organisms, and analogies with the mycoplasmas, have been discussed previously (Shaw, 1970). There is a significant increase in glycolipid concentration in L-forms, comparable to the high concentrations in mycoplasmas, suggesting an attempt by these organisms to stabilize their membranes and maintain the integrity of the cell. Brundish *et al.* (1967) have suggested that one possible structural function for glycolipids is to form pores in the membrane, regions of hydrophilicity through which small molecules might pass. An examination of the molecular shape of most glycolipids, irrespective of the nature of the carbohydrate residue, shows that they can adopt a conformation in which all of the hydroxyl groups lie on one side of the molecule and the lipophilic groups lie on the other. The presence of larger glycosyl diglycerides may represent an attempt to regulate the size of such pores. The location of these pores near the membrane surface would allow an involvement in binding intracellular polymers. The presence of either glycolipids or phosphoglycolipids, as covalently bound units, in lipoteichoic acids has already been described (Section IVB, p. 156). This might also imply an asymmetric distribution of these lipids within the membrane. It cannot however represent their major function, particularly for glycolipids, as the major proportion is unbound and freely extractable.

Earlier suggestions for a possible role for glycolipids as intermediates in the biosynthesis of polysaccharides may now be discounted in view of the recognition of such a role for the glycosyl phosphate derivatives of polyisoprenols (Rothfield and Romeo, 1971). Moreover no evidence has been presented for the rapid metabolism of glycosyl diglycerides which such a role would require. The acylated sugar derivatives in *Mycobacterium smegmatis* do show a metabolic turnover of about ten times per generation (Winder *et al.*, 1972). This somewhat slow turnover would suggest that any metabolic role is not a major one in the bacterial economy, and would also exclude these glycolipids as a mechanism for sugar transport across membranes.

Synthesis of acyl glucoses in corynebacteria is dependent upon the presence of glucose in the medium (Brennan *et al.*, 1970a) and they disappear when the glucose in the medium is replaced by glycerol. The effects of culture conditions upon bacterial composition are well established, and significant changes in wall composition have been observed under certain growth-limiting conditions (Ellwood, 1970). Changes in lipid composition under similar conditions have led to certain proposals for interrelationships between particular lipids. *Bacillus subtilis* W23 and *Bacillus cereus* T, when grown under conditions of apparent phosphate limitation, accumulate diglucosyl diglyceride with a corresponding decrease in the proportion of phosphatidylethanolamine. *Bacillus cereus* T also produces small amounts of an acidic glycolipid in the stationary phase with a corresponding fall in the proportion of the acidic phospholipids (Minnikin *et al.*, 1971). Thus, an interchangeability is proposed first, between the neutral lipids phosphatidylethanolamine and glycosyl diglycerides, and second between acidic phospholipids and acidic glycolipids. The latter proposition had previously been proposed by Wilkinson (1968) following his investigations on the lipids of *Pseudomonas diminuta* which were low in phospholipid, phosphatidylethanolamine being completely absent, but high in acidic glycolipids when grown on a solid medium or in submerged culture. By the use of continuous cultures in a chemostat, Minnikin *et al.* (1974) examined this phenomenon further in *P. diminuta*, and presented additional evidence which, in their view, supports the acidic phospholipid-acidic glycolipid interchangeability hypothesis. A critical examination of their data, however, reveals some discrepancies which do not support their contention. When grown under conditions of magnesium limitation, the total phospholipid content is 29% of which 22% is phosphatidylglycerol and the acidic glycolipids constitute 48% of the total lipid. Under conditions of phosphate limitation, although the phospholipid content falls practically to zero (0·3%), the acidic glycolipids only increased to 57% of the total lipid. Thus, whilst the phospholipids virtually disappeared from the membranes, an exceedingly interesting observation in itself, the proportion of acidic glycolipids is only marginally increased. Further, the authors did not comment upon the most obvious change, that is the proportion of neutral glucosyl diglyceride increased from 18% to 40% of the total lipid. Virtually all of the phospholipids therefore were replaced not by acidic lipids as inferred, but by neutral glycolipids. *Pseudomonas diminuta* is exceptional in its inability to synthesize phosphatidylethanolamine, an almost universal constituent of Gram-negative bacteria. Beebe (1971) isolated a mutant of *Bacillus subtilis* which is also deficient in phosphatidylethanolamine. An analysis of the lipid composition of the parent organism with that of the mutant, when grown in a synthetic medium, shows that as the phosphatidylethanolamine content falls from 7·9% in the parent to 0·2% in the mutant the glycolipid content increases only slightly from 11·9% to 13·8%. Under these conditions, phosphatidylethanolamine is not replaced by glycolipid. Stern and Tietz (1973), during studies on the effects of culture composition on the acidic glycolipids of an unclassified halotolerant bacterium, were unable to demonstrate any relationship between acidic glycolipids and phospholipids under phosphate-limiting conditions. The number of bacteria in which phosphatidylethanolamine and glycosyl diglycerides occur together is comparatively small; it is unlikely that the small concentrations of glycolipids present in most Gram-positive bacteria fulfil a similar role to the large concentrations of phosphatidylethanolamine present in most Gram-negative bacteria. Although similarities have been observed between these two types of lipids in ion-permeability studies (Hopfer *et al.*, 1970), phosphatidylethanolamine is a zwitterion whereas glycosyl diglycerides are neutral uncharged molecules. Undoubtedly growth conditions can affect membrane composition, and it will be interesting to see if the changes shown in some bacilli can be demonstrated in other bacteria, but at present the interpretation of such results primarily in terms of lipid functionality will require more substantial evidence.

Studies on the serological properties of glycolipids have mainly been restricted to those from mycoplasmas, and this is not too surprising in view of the importance of their membrane surfaces. In *Mycoplasma pneumoniae*, hapten activity, measured in complement-fixation tests with both rabbit and human antisera, was found to be associated with the glycolipid fraction (Plackett *et al.*, 1969). Cross-reactivity has been observed between antisera to *M. pneumoniae* and purified diglucosyl diglycerides from a *Streptococcus* sp. (Plackett and Shaw, 1967). Streptococcal L-forms are immunogenic, and evidence has been presented that the antigenic determinants are glycolipids (Feinman *et al.*, 1973). The importance of the lipid moiety in the serological properties of lipoteichoic acids has been discussed previously (Section IVB, p. 157). The antigenic identity established between nervous tissue and trepenoma is due to the presence in the trepenoma of a galactosyl diglyceride hapten which cross reacts with a cerebroside present in the nervous tissue (Dupouey, 1972). The glycosyl diglycerides are proving to be the bacterial counterparts of the ubiquitous and immunologically important cerebrosides in animal cells (Brady, 1966).

References

Abel, K., de Schmertzing, H. and Peterson, J. L. (1963). *Journal of Bacteriology* **85**, 1039.

Akamatsu, Y., Ono, Y. and Nojima, S. (1967). *Journal of Biochemistry* **61**, 96.

Ambron, R. T. and Pieringer, R. A. (1971). *Journal of Biological Chemistry* **246**, 4216.

Araki, Y., Fukuoka, S., Oba, S. and Ito, E. (1971). *Biochemical and Biophysical Research Communications* **45**, 751.

Asselineau, C. P., Montrozier, H. L., Promé, J. C., Savagnac, A. M. and Welet, M. (1972). *European Journal of Biochemistry* **28**, 102.

Baddiley, J. (1972) *Essays in Biochemistry* **8**, 35.

Beebe, J. L. (1971). *Journal of Bacteriology* **107**, 704.

Bergelson, L. D., Batrakov, S. G. and Pilipenko, T. V. (1970). *Chemistry and Physics of Lipids* **4**, 181.

Bishop, D. G., Rutberg, L. and Samuelson, B. (1967). *European Journal of Biochemistry* **2**, 448.

Brady, R. O. (1966). *Journal of the American Oil Chemists Society* **43**, 67.

Brennan, P. J. (1968). *Biochemical Journal* **109**, 158.

Brennan, P. J. and Ballou, C. E. (1967). *Journal of Biological Chemistry* **242**, 3046.

Brennan, P. J. and Ballou, C. E. (1968a). *Journal of Biological Chemistry* **243**, 2975.

Brennan, P. J. and Ballou, C. E. (1968b). *Biochemical and Biophysical Research Communications* **30**, 69.

Brennan, P. J. and Lehane, D. P. (1971). *Lipids* **6**, 401.

Brennan, P. J., Flynn, P. and Griffin, P. F. S. (1970a). *Federation of European Biochemical Societies Letters* **8**, 322.

Brennan, P. J., Lehane, D. P. and Thomas, D. W. (1970b). *European Journal of Biochemistry* **13**, 117.

Brundish, D. E., Shaw, N. and Baddiley, J. (1967). *Biochemical Journal* **105**, 885.

Burger, M. M., Glaser, L. and Burton, R. M. (1963). *Journal of Biological Chemistry* **238**, 2595.

Chapman, D. and Urbina, J. (1971). *Federation of European Biochemical Societies Letters* **12**, 169.

Constantopoulos, G. and Bloch, K. (1967). *Journal of Bacteriology* **93**, 1788.

Coley, J., Duckworth, M. and Baddiley, J. (1972). *Journal of General Microbiology* **73**, 587.

Dupouey, P. (1972). *Journal of Immunology* **109**, 146.

Edwards, J. R. and Hayashi, J. A. (1965). *Archives of Biochemistry and Biophysics* **111**, 415.

Ellwood, D. C. (1970). *Biochemical Journal* **118**, 367.

Feinmann, S. B., Prescott, B. and Cole, R. M. (1973). *Infection and Immunology* **8**, 752.

Fitz-James, P. C. (1968). *In* "Microbial Protoplasts, Spheroplasts and L-forms", (L. B. Guze, ed.), p. 124. Williams and Wilkins, Baltimore.

Fischer, W. (1970). *Biochemical and Biophysical Research Communications* **41**, 731.

Fischer, W., Ishizuka, I., Landgraf, H. R. and Herrmann, J. (1973). *Biochimica et Biophysica Acta* **296**, 527.

Gilliam, J. M. and Morowitz, H. J. (1972). *Biochimica et Biophysica Acta* **274**, 353.

Goldfine, H. (1972). *Advances in Microbial Physiology* **8**, 1.

Gray, G. R. and Ballou, C. E. (1973). *Journal of Biological Chemistry* **247**, 8129.

Hay, J. B., Wicken, A. J. and Baddiley, J. (1963). *Biochimica et Biophysica Acta* **71**, 188.

Hirschberg, C. B. and Kennedy, E. P. (1972). *Proceedings of the National Academy of Sciences of the United States of America* **69**, 640.

Hopfer, U., Lehninger, A. L. and Lennarz, W. J. (1970). *Journal of Membrane Biology* **2**, 41.

Huis in't Veld, J. H. J. and Villiers, J. M. N. (1973). *Antonie van Leeuwenhoek* **39**, 281.

Hughes, A. H., Stow, M., Hancock, I. C. and Baddiley, J. (1971). *Nature New Biology* **229**, 53.

Ishizuka, I. and Yamakawa, T. (1969). *Journal of Biochemistry* **64**, 13.

Ioneda, I., Lederer, E. and Rozanis, J. (1970). *Chemistry and Physics of Lipids* **4**, 375.

Johnson, J. L. and Cummins, C. S. (1972). *Journal of Bacteriology* **109**, 1047.

Kataoka, T. and Nojima, S. (1967). *Biochimica et Biophysica Acta* **144**, 681.
Kates, M. and Deroo, P. W. (1973). *Journal of Lipid Research* **14**, 438.
Kates, M., Wassef, M. K., Krushner, D. J. and Gibbons, N. E. (1966). *Biochemistry, New York* **5**, 4092.
Khuller, G. K. and Brennan, P. J. (1972a). *Biochemical Journal* **127**, 369.
Khuller, G. K. and Brennan, P. J. (1972b). *Journal of General Microbiology* **73**, 409.
Kosaric, N. and Carroll, K. K. (1971). *Biochimica et Biophysica Acta* **239**, 428.
Kunsman, J. E. (1970). *Journal of Bacteriology* **103**, 104.
Langworthy, T. A., Smith, P. F. and Mayberry, W. R. (1972). *Journal of Bacteriology* **112**, 1193.
Langworthy, T. A., Mayberry, W. R. and Smith, P. F. (1974). *Journal of Bacteriology* **119**, 106.
Lederer, E. (1967). *Chemistry and Physics of Lipids* **1**, 294.
Lee, Y. C. and Ballou, C. E. (1965). *Biochemistry, New York* **4**, 1395.
Lennarz, W. J. (1970). *In* "Lipid Metabolism", (S. J. Wakil, ed.), p. 155. Academic Press, New York.
Lennarz, W. J. and Talamo, B. (1966). *Journal of Biological Chemistry* **241**, 2207.
Livermore, B. P. and Johnson, R. C. (1970). *Biochimica et Biophysica Acta* **210**, 315.
Luderitz, O., Galanos, C., Lehmann, V., Nurminen, M., Rietschel, E. T., Rosenfelder, G., Simon, M. and Westphal, O. *Journal of Infectious Diseases* **128**, 517.
MacDougall, J. C. and Phizackerley, P. J. R. (1969). *Biochemical Journal* **114**, 361.
Macfarlane, M. G. (1961). *Biochemical Journal* **80**, 45.
Mayberry, W. R., Smith, P. F. and Langworthy, T. A. (1974). *Journal of Bacteriology* **118**, 898.
Mayberry-Carson, K. J., Langworthy, T. A., Mayberry, W. R. and Smith, P. F. (1974). *Biochimica et Biophysica Acta* **360**, 217.
Meyer, H. and Meyer, F. (1971). *Biochimica et Biophysica Acta* **231**, 93.
Minnikin, D. E., Abdolrahimzadeh, H. and Baddiley, J. (1971). *Biochimica et Biophysica Acta* **249**, 651.
Minnikin, D. E., Abdolrahimzadeh, H. and Baddiley, J. (1974). *Nature, London* **249**, 268.
Nowotny, A. (1971a). *Naturwissenschaften* **58**, 397.
Nowotny, A. (1971b). *Microbial Toxins* **4**, 309.
Okazaki, H., Sugino, H., Kanzaki, T. and Fukuda, H. (1969). *Agricultural and Biological Chemistry* **33**, 764.
Olsen, R. W. and Ballou, C. E. (1971). *Journal of Biological Chemistry* **246**, 3305.
Op den Kamp, J. A. F., Van Deenen, L. L. M. and Tomasi, V. (1969a). *In* "Structural and Functional Aspects of Lipoproteins in Living Systems", (E. Tria and A. M. Scance, eds.), p. 227. Academic Press, London & New York.
Op den Kamp, J. A. F., Bonsen, P. P. M. and Van Deenen, L. L. M. (1969b). *Biochimica et Biophysica Acta* **176**, 298.
Oshima, M. and Yamakawa, T. (1972). *Biochemical and Biophysical Research Communications* **49**, 185.
Pangborn, M. C. and McKinney, J. A. (1966). *Journal of Lipid Research* **7**, 627.
Peleg, E. and Tietz, A. (1971). *Federation of European Biochemical Societies Letters* **15**, 309.
Pieringer, R. A. (1972). *Biochemical and Biophysical Research Communications* **49**, 502.
Phizackerley, P. J. R., MacDougall, J. C. and Moore, R. A. (1972). *Biochemical Journal* **126**, 499.
Plackett, P. (1967). *Biochemistry, New York* **6**, 2746.

Plackett, P. and Shaw, E. J. (1967). *Biochemical Journal* **104**, 61.

Plackett, P., Marmion, B. P., Shaw, E. J. and Lemcke, R. M. (1969). *Australian Journal of Experimental Medical Science* **47**, 171.

Powell, D., Duckworth, M. and Baddiley, J. (1974). *Federation of European Biochemical Societies Letters* **41**, 259.

Prottey, C. E. and Ballou C. E. (1968). *Journal of Biological Chemistry* **243**, 6196.

Ray, P. H., White, D. C. and Brock, T. D. (1971). *Journal of Bacteriology* **108**, 227.

Reinert, J. C. and Steim, J. M. (1970). *Science, New York* **168**, 1580.

Rothfield, L. and Romeo, D. (1971). *Bacteriological Reviews* **35**, 14.

Saier, M. H. and Ballou, C. E. (1968). *Journal of Biological Chemistry* **243**, 4332.

Shaw, N. (1968). *Biochimica et Biophysica Acta* **152**, 427.

Shaw, N. (1970). *Bacteriological Reviews* **34**, 365.

Shaw, N. (1974). *Advances in Applied Microbiology* **17**, 63.

Shaw, N. and Baddiley, J. (1968). *Nature, London* **217**, 142.

Shaw, N. and Dinglinger, F. (1969). *Biochemical Journal* **112**, 769.

Shaw, N. and Stead, D. (1970). *Journal of Applied Bacteriology* **33**, 470.

Shaw, N. and Stead, D. (1971). *Journal of Bacteriology* **107**, 130.

Shaw, N. and Stead, D. (1972). *Federation of European Biochemical Societies Letters* **21**, 249.

Shaw, N., Smith, P. F. and Koostra, W. L. (1968). *Biochemical Journal* **107**, 329.

Shaw, N., Heatherington, K. and Baddiley, J. (1968). *Biochemical Journal* **107**, 491.

Shaw, N., Smith, P. F. and Verheij, H. M. (1970). *Biochemical Journal* **129**, 167.

Short, S. A. and White, D. C. (1970). *Journal of Bacteriology* **104**, 126.

Shultz, J. C. and Elbein, A. D. (1974). *Journal of Bacteriology* **117**, 107.

Shvets, V. I., Bashakova, A. I. and Evstigreeva, R. P. (1973). *Chemistry and Physics of Lipids* **10**, 267.

Smith, P. F. (1972). *Journal of Bacteriology* **112**, 554.

Smith, P. F. and Henrikson, C. V. (1965). *Journal of Lipid Research* **6**, 106.

Smith, P. F. and Mayberry, W. R. (1968). *Biochemistry, New York* **7**, 2706.

Steiner, S., Conti, S. F. and Lester, R. L. (1969). *Journal of Bacteriology* **98**, 10.

Stern, N. and Tietz, A. (1972). *Biochimica et Biophysica Acta* **296**, 130.

Takayama, K. and Goldman, D. S. (1969). *Biochimica et Biophysica Acta* **176**, 196.

Tabaud, H., Tisnovska, H. and Vilkas, G. (1971). *Biochimie* **53**, 55.

Thomas, T. D. and Ellar, D. J. (1973). *Biochimica et Biophysica Acta* **316**, 180.

Toon, P., Brown, P. E. and Baddiley, J. (1972). *Biochemical Journal*, **127**, 399.

Tornabene, T. G. (1973). *Biochimica et Biophysica Acta* **306**, 173.

Tornabene, T. G. and Ogg, J. E. (1971). *Biochimica et Biophysica Acta* **239**, 133.

Veerkamp, J. H. (1972). *Biochimica et Biophysica Acta* **273**, 359.

Verheij, H. M., Smith, P. F., Bonsen, P. P. M. and Van Deenen, L. L. M. (1970). *Biochimica et Biophysica Acta* **218**, 97.

Vorbeck, M. L. and Marinetti, G. V. (1965). *Journal of Lipid Research*, **6**, 3.

Wehrli, H. and Pomeranz, Y. (1969). *Chemistry and Physics of Lipids* **3**, 357.

Welsh, K., Shaw, N. and Baddiley, J. (1968). *Biochemical Journal* **107**, 313.

Wicken, A. J. and Knox, K. W. (1970). *Journal of General Microbiology* **60**, 293.

Wicken, A. J. and Knox, K. W. (1971). *Journal of General Microbiology* **67**, 251.

Wilkinson, S. G. (1968). *Biochimica et Biophysica Acta* **164**, 148.

Wilkinson, S. G. (1969). *Biochimica et Biophysica Acta* **187**, 492.

Wilkinson, S. G. and Bell, M. E. (1971). *Biochimica et Biophysica Acta* **248**, 293.

Winder, F. G., Tighe, J. J. and Brennan, P. J. (1970). *Journal of General Microbiology* **73**, 539.

Yano, I., Furakawa, Y. and Kusunose, M. (1971). *Journal of General and Applied Microbiology, Tokyo* **17**, 329.

The Physiology of Obligate Anaerobiosis

J. G. MORRIS

Department of Botany & Microbiology, School of Biological Sciences, The University College of Wales, Aberystwyth, SY23 3DA, Wales

I. Introduction

Pasteur, in 1861, found to his considerable surprise that the causative microbes of a saccharolytic butyric fermentation were actually growing in the total absence of air. He was further intrigued by the fact that, although these micro-organisms could dispense with air, they were not indifferent to it since on aeration they ceased their movements and metabolic activities, and very quickly died. So far from benefiting from oxygen, these microbes were evidently poisoned by it. It is instructive to read both Pasteur's account of his observations and the ensuing comments of his contemporaries (Vallery-Radot, 1901), if only to recapture some of the initial incredulity which greeted his conclusion that life could be sustained in the absence of air, and that oxygen could prove lethal to some living creatures. After all, it was only in the preceding year (1860) that a scientific journalist writing in *La Presse*, had said of his refutation of spontaneous generation, "I am afraid that the experiments you quote M. Pasteur will turn against you . . . the world into which you wish to take us is really too fantastic". Yet Pasteur's vision of an oxygen-independent biosphere populated by obligately anaerobic micro-organisms has subsequently been fully vindicated.

In his *Dictionary of Microbial Taxonomic Usage*, Cowan (1968) defined an *anaerobe* as "an organism incapable of growing in air, receiving its essential oxygen from a source other than the atmosphere". Organisms that normally grow anaerobically, but are able to grow to a small extent in the presence of air, he termed *aerotolerant*. Others (for example, Stanier *et al.*, 1971) have distinguished between the *obligate anaerobe*, for which molecular oxygen is a toxic substance which kills the organism or inhibits its growth, and the *facultative anaerobe* which can grow either in the absence or presence of molecular oxygen. Some bacteriologists see further merit in subdividing facultative anaerobes into two sub-groups: (a) those that in response to the presence or absence of oxygen can switch between anaerobic and aerobic-respiratory modes of energy-yielding metabolism; and (b) those that even when growing in the presence of oxygen continue to employ a wholly anaerobic mode of energy-yielding metabolism. Distinguishable from all of these are the *microaerophilic* organisms—obligate aerobes which grow best at partial pressures of oxygen considerably lower than that present in air.

Using enzymological criteria, Decker *et al.* (1970) proposed that a useful distinction could be made between anaerobes and facultative aerobes. While anaerobes are organisms that do not possess either cytochrome oxidase or oxygenases, facultative aerobes, though they have an oxygenase-independent metabolism, are nevertheless able to obtain energy both by oxygen-independent and by oxygen-dependent (cytochrome oxidase-linked) redox processes.

Whatever scheme of classification is adopted, an obligate anaerobe emerges as an organism which: (i) generates energy and synthesizes its substance without recourse to molecular oxygen; and (ii) demonstrates a singular degree of adverse oxygen-sensitivity which renders it unable to grow under an atmosphere of air. As we shall see, even truly obligate anaerobes display a spectrum of tolerance to oxygen. The most oxygen-sensitive organisms are generally spoken of as the very strict, extreme or fastidious anaerobes; this imprecise, but convenient, terminology serving to distinguish them from the more oxygen-tolerant, less exacting or moderate anaerobes.

II. Nature and Distribution of Obligate Anaerobes

Though facultatively anaerobic eukaryotes are not uncommon, obligate anaerobiosis is such a rarity amongst eukaryotes that in these "higher organisms" it would appear to be a secondarily acquired character, selected for by retrogressive evolution in specialized oxygen-free habitats (Stanier *et al.*, 1971). Thus the relatively few species of protozoa reported to be obligately anaerobic have been discovered in the digestive tract of termites and other metazoa (Cleveland, 1925), in the rumens of herbivores (Hungate, 1966) and alongside oxygen-consuming aerotolerant species in sapropelic or bottom mud, aquatic habitats (Noland and Gojdics, 1967). It remains the case that no obligately anaerobic fungus is known, for *Aqualinderella fermentans*, cited by Stanier (1970) as the "one strictly anaerobic fungus discovered to date", alas does not qualify for this distinction. It has proved in fact to be an oxygen-indifferent, facultative anaerobe which, like some other obligately fermentative water moulds (e.g. *Blastocladia pringsheimii*), will grow in carbon dioxide-enriched air even though it lacks the capacity to metabolize oxygen (Emerson and Held, 1969).

In contrast to the dearth of oxygen-sensitive eukaryotes, there exist very many groups of eubacteria and spirochetes whose marked intolerance of oxygen proclaims their right to be considered as truly obligate anaerobes. Yet even amongst prokaryotes the strictly anaerobic species are outnumbered by those that are facultatively or necessarily aerobic—especially since all cyanobacteria (blue-green algae) generate molecular oxygen during their photosynthesis. Though strictly anaerobic bacteria display great versatility in their exploitation of a variety of growth substrates, their distribution in nature is dictated by the availability of those organic and inorganic electron donors and acceptors that they utilize for energy generation (p. 174). Fermentative species are likely, therefore, to be particularly numerous in organic nutrient-rich anaerobic locations. Thus over 97% of the human and animal faecal flora is composed of obligately anaerobic bacteria (Moore and Holdeman, 1972), whilst the rumen operates virtually as an anaerobic continuous culture fermenter (Hungate, 1966; Hobson, 1971). By the use of stringent anaerobic procedures, both Gram-positive and Gram-negative rods and cocci have been isolated from such situations and have been assigned to such genera as *Acidaminococcus*, *Bacteroides*, *Butyrivibrio*, *Fusobacterium*, *Megasphaera*, *Peptococcus*, *Ruminococcus*, *Selenomonas*, *Sphaerophorus*, *Succinivibrio* and *Veillonella* (Barnes, 1969; Hungate, 1966;

Holdeman and Moore, 1972). The methane-generating bacteria in digestive tract contents, sewage sludge and organic sediments and muds have come in for particular scrutiny in recent years because of their novel biochemistry and potential industrial importance (Stadtman, 1967; Kirsch and Sykes, 1971; Wolfe, 1971). Several genera have been described, for example *Methanobacterium*, *Methanococcus*, *Methanosarcina* and *Methanospirillum* (Wolfe, 1971), and have acquired the reputation of being amongst the most rigorous of obligate anaerobes (Paynter and Hungate, 1968). Though methane production is limited to a relatively small group of bacteria, these are widespread in nature, being readily found in anaerobic locations where organic materials are being actively decomposed. Here they play the part of terminal organisms in anaerobic "food chains", for example, that which stems from cellulose breakdown in the black muds of lakes and swamps (Wolfe, 1971). However, by reason of their clinical significance or one-time economic importance as industrial fermentation agents, the most comprehensively studied of all obligately anaerobic bacteria are various members of the genus *Clostridium*. These Gram-positive, sporulating, rod-shaped bacteria are notable for their catholic distribution, and for the diversity of organic substrates that different species can ferment, including carbohydrates, amino-acids and purines (Barker, 1956, 1961; Stadtman, 1973). Potentially pathogenic species, and their production of exotoxins, have naturally merited particular study (Willis, 1964) as has the dinitrogen-fixing ability of *Clostridium pasteurianum* (Dalton and Mortenson, 1972; Dalton, 1974). The considerable differences in oxygen-tolerance that can be displayed by quite closely related species of obligately anaerobic bacteria is well illustrated within this genus whose members cover the whole range of oxygen-sensitivity from near aerotolerance to extreme intolerance of low concentrations of molecular oxygen (Morris and O'Brien, 1971).

Of the photosynthetic bacteria, all are strict anaerobes save the purple nonsulphur bacteria (Athiorhodaceae). Thus both the green photosynthetic bacteria (*Chlorobium*, *Pelodictyon*) and the purple sulphur bacteria (*Amoebobacter*, *Chromatium*, *Ectothiorhodospira*, *Lamprocystis*, *Thiocapsa*, *Thiocystis*, *Thiospirillum*, *Rhodothece*) are to be found in anaerobic illuminated situations, where sulphide (the common photosynthetic electron donor) is freely available. Being aquatic organisms, they flourish in natural sulphur springs or in shallow organically polluted ponds, where they populate a somewhat dimly lit de-oxygenated layer (Stanier *et al.*, 1971). The sulphide that is oxidized by such photosynthetic anaerobes might very well have originated in the activities of another class of obligately anaerobic bacterium, viz. sulphate reducers living in the bottom muds. Both fresh water and marine species of sulphate-reducing bacteria are known, and these are very widely distributed in sulphate-rich anaerobic muds. Two main genera have been described, viz. *Desulfovibrio* and *Desulfotomaculum*, the latter consisting of sporulating rod-shaped organisms liable to be confused with *Clostridium* (Le Gall and Postgate, 1973). These sulphate reducers contain high concentrations of cytochromes which are employed in the anaerobic respiration for which the sulphate acts as terminal electron acceptor (p. 173). Even so, these organisms, like other exacting anaerobes, only thrive in locations of low redox potential (Connell and Patrick, 1968) where, because of their somewhat limited substrate specificity, they may be sustained by syntrophy with other bacteria (Le Gall and Postgate, 1973). Examples of the natural ecosystem based on sulphate reduction (i.e. a sulphuretum) are still available for study (Durner *et al.*, 1965; Fenchel and Riedl, 1970) though this might have been much more important as a primary-producing ecosystem in those primitive times when the earth's atmosphere was non-oxidizing (Postgate, 1968).

It should be obvious even from this cursory survey of the major types of present-day obligately anaerobic bacteria that, although they obviously have in common a super-sensitivity to oxygen (which means that all must operate some oxygen-independent metabolism), they are in other respects likely to differ as widely as are the members of any quite arbitrary collection of bacteria. However, in view of their oxygen sensitivity it is rather unexpected to find that such obligate anaerobes are by no means restricted to intestinal tracts, rich muds and airless composts, but are also prevalent in seemingly exposed aerobic locations which, at first sight, would seem little suited to any oxygen-intolerant species. The widespread distribution of species of *Clostridium* in soils and airborne dust could possibly be explained by their possession of oxygen-insensitive spores. Yet many asporogenous, strictly anaerobic bacteria are similarly widely dispersed. For example, on the surface of normal human skin, anaerobes outnumber aerobic and facultatively anaerobic organisms by 10:1, whilst on the mucous membrane of the mouth the ratio (30:1) is even more in favour of the strict anaerobes (Rosebury, 1962). As pointed out by Smith (1967), the mere fact that axenic new born animals quickly establish the same sort of intestinal flora that is characteristic of the adult animal must mean that even the most fastidious anaerobes amongst this population must have withstood the exposure to air that accompanies its transmission from one individual to another. It would seem that the reason why anaerobes that are extremely sensitive to oxygen in pure culture can apparently tolerate long term contact with air under "natural conditions" is to be sought in the fact that, in their native situation, they are constituents of a more complex microbial flora which contains many oxygen consuming species. These possibly create, and help to maintain, the local oxygen-depleted low-potential conditions required by their fellow anaerobes. Furthermore, on mucous membranes lining some of the body cavities, such anaerobes could live in a state of commensalism, not only with other bacteria but also with the epithelial cells on which they grow (Smith, 1967). Perhaps even dead cells of the same, or companion, species could aid in the protection against oxygen that is obviously afforded to many anaerobes in the natural state. But so little is known about such matters that present speculation must give way to future experimentation. However, the undoubted reality of such protective agencies must constantly remind us that we should not seek to interpret the normal "mixed flora" situation solely in terms of findings made with pure cultures of its anaerobic components.

III. How Obligate Anaerobes Contrive to Dispense with Molecular Oxygen

Though by far the greater part of the molecular oxygen that is consumed by a heterotrophic, aerobic bacterium is utilized solely as an electron acceptor in resipiration, a small amount is directly incorporated into the substance of the organism as a consequence of the activities of certain oxygenases and mixed-function hydroxylases. This means that not only must a strict anaerobe exploit an oxygen-independent mode of energy generation (i.e. neither aerobic respiration nor green plant-type photophosphorylation), but it must also either dispense with those materials whose fabrication in the aerobe is achieved with fixation of molecular oxygen, or it must avail itself of anoxic biosynthetic routes to the same products.

A. Energy-Yielding Metabolism

The three classes of ATP-generating oxidation-reduction processes exploited by strict anaerobes are: (1) fermentation; (2) anaerobic (bacterial) photosynthesis; and (3) anaerobic respiration.

1. Fermentation

A fermentation is an energy-yielding sequence of oxidation-reduction reactions in which organic compounds serve as primary electron donor(s) and terminal electron acceptor(s). It follows that all fermentations must lead to the accumulation of quite large quantities of those reduced organic compounds that are the products of reduction of the terminal electron acceptors. In certain situations, other partially oxidized organic compounds (the products of oxidation of the primary electron donors) may also accumulate. The production and accumulation of these partially oxidized/reduced substances must make fermentation (on a –J

mol^{-1} basis) an energetically less rewarding process than that total combustion of organic compounds to carbon dioxide and water which is generally accomplished by aerobic respiration. Under conditions so contrived that the substance of the growing anaerobe is wholly synthesized from additionally supplied nutrients, the quantities of fermentable substrate(s) utilized during growth are wholly recoverable in the culture medium in the form of characteristic organic fermentation products whose net oxidation level will be identical with that of the substrate(s) fermented.

In some cases, separate sources of electron donor and electron acceptor have to be supplied, for example in the Stickland fermentation of suitable pairs of amino acids that is accomplished by *Clostridium sporogenes* (Barker, 1961) or in the fermentation by *Clostridium kluyveri* of acetate plus ethanol (Thauer *et al.*, 1968). In other instances, the compounds that perform these functions are metabolites formed by the organism from a single fermentable substrate, for example in the homolactic fermentation of glucose (Wood, 1961) or the fermentation of alanine as effected by *Clostridium propionicum* (Cardon and Barker, 1946). In such "intramolecular" redox processes, the need to produce the electron acceptor from the electron donor places a new constraint on the efficiency of the whole fermentation process, though some flexibility is generally ensured by such devices as: (a) production of hydrogen gas as a means of discharging varying amounts of reducing power; (b) "over-reduction" of certain organic electron acceptors as an alternative means of dealing with excess reducing power; or (c) accumulation of less reduced fermentation products when reducing power is diverted to biosynthetic ends.

The versatility of obligately anaerobic bacteria is reflected in the tremendous range of organic compounds that they can ferment (e.g. carbohydrates, amino acids, purines and pyrimidines). In their excellent review of anaerobic energy-yielding processes, Decker *et al.* (1970) listed over fifty primary electron-donating reactions, and a similar number of electron-accepting organic couples, exploited by anaerobic bacteria. These include carboxylic, α-oxo-, hydroxy- and α,β-Δ, unsaturated acids, aldehydes, alcohols and amino acids. It would seem that some classes of substance (e.g. alkanes and aromatic compounds) are not amenable to anaerobic fermentation for thermodynamic and biochemical reasons—which is perhaps just as well, else we would have been deprived of our deposits of fossil fuels. Some organisms are more restricted than others in the range of their fermentable substrates. Thus, in the genus *Clostridium*, we find species as different as *Cl. kluyveri*, which can ferment only: (a) crotonate; or (b) a mixture of ethanol and acetate (or propionate), *Cl. butyricum* which can ferment a wide range of sugars and sugar alcohols, *Cl. tetanomorphum* which can ferment amino acids or sugars, and *Cl. acidi-urici* which is only able to ferment certain purines. The chief purpose served by all these fermentations is the concurrent net synthesis of ATP which is accomplished by substrate-level phosphorylation (SLP) reactions. It is therefore particularly interesting to find that only a relatively small number of SLP reactions serve all fermentation processes (Table 1). The SLP reaction most generally exploited by obligate anaerobes is probably that catalysed by acylate kinases; i.e.,

$$\text{acyl phosphate} + \text{ADP} = \text{acylate} + \text{ATP} \quad (\Delta G^{0\prime} = -10{\cdot}46\,\text{kJ mol}^{-1})$$

It is possibly significant that, unlike some other enzymes that catalyse SLP reactions (e.g. N^{10}-formyl tetrahydrofolate synthetase, pyruvate kinase and phosphoglycerate kinase), acylate kinases are not found in strictly aerobic organisms.

The necessity to produce one or more of the high-energy compounds that are the substrates of these ATP-yielding reactions whilst ensuring internal balance of the key electron-donating and accepting processes, frequently contrives to create metabolic pathways the like of which are not found in aerobic bacteria—for example, the fermentations recently discussed in detail by Wolfe (1971) or Stadtman (1973).

The potential span (E_0') between certain of the electron donating and accepting couples encountered in fermentations is theoretically quite sufficient to effect ATP production in a thermodynamically spontaneous coupled reaction. It is therefore not surprising that on several occasions it has been suggested that even in obligate anaerobes electron transport-linked phosphorylation could in some instances supplement ATP production by substrate-level phosphorylation. Determinations of maximum growth yield (*g* cells produced per mol of substrate fermented), taken in conjunction with calculated values of Y_{ATP}^{MAX} (*g* cells produced per mol of ATP utilized), have suggested the yield of ATP which a given anaerobe is able to "extract" per mol of substrate fermented (Gunsalus and Shuster, 1961; Stouthamer, 1969, 1973; Stouthamer and Bettenhausen, 1973). Whenever this value has seemed in excess of that explicable by the supposed route of fermentation, electron-transport (oxidative) phosphorylation has been invoked as a possibility. Thus it was at one time suggested that during fermentation by *Clostridium kluyveri* of acetate and ethanol, ATP might be generated by "oxidative phosphorylation" coupled with transfer of electrons via flavoprotein between reduced nicotinamide nucleotide and crotonyl-CoA (Barker, 1956; Shuster and Gunsalus, 1958). However, when the fully determined route of this fermentation was re-examined for its complement of SLP reactions, it was discovered that these alone were quite sufficient to account for the observed growth yield (Thauer *et al.*, 1968). Another possible candidate

TABLE 1. The "High-Energy" Substrates of the Key ATP-Synthesizing Enzymes of Substrate-Level Phosphorylation (from Decker *et al*., 1970)

The two classes of substrate-level phosphorylation reactions are:

(i) $ADP + Substrate \sim P \rightleftharpoons ATP + Substrate$

(ii) $ADP + P_i + substrate \sim X \rightleftharpoons ATP + substrate + X$

Type of Compound	Substrate ~ P or substrate ~ X	$\Delta G^{\circ\prime}$ kJ mol^{-1}	Enzyme
Phospho-acyl anhydride	Acetyl phosphate	−44·8	Acetate kinase (E.C.2.7.2.1)
	Carbamyl phosphate	−42·7	Carbamate kinase (E.C.2.7.2.2)
	3-Phosphoglyceryl 1-phosphate	−62·3	3-Phosphoglycerate kinase (E.C.2.7.2.3)
	Propionyl phosphate	−44·8	Propionate kinase (E.C.2.7)
	Butyryl phosphate	−44·8	Butyrate kinase (E.C.2.7)
Phospho-enol ester	Phosphoenol pyruvate	−58·2	Pyruvate kinase (E.C.2.7.1.40)
Acyl thioester	Succinyl Coenzyme A	−37·7	Succinate thiokinase (E.C.6.2.1.4)
Acyl anilide	N^{10}-Formyltetrahydrofolate	−26·0	N^{10}-Formyltetrahydrofolate synthetase (E.C.6.3.4.3)

for anaerobic electron-transport-phosphorylation, supplementary to fermentation, is the fermentation of molecular hydrogen plus carbon dioxide by species of *Methanobacterium* (Roberton and Wolfe, 1970; Wolfe, 1971). Yet at this time there seems to be no good evidence to support most of these hypotheses, and fermentation in obligate anaerobes is still generally considered a process which generates ATP solely by substrate-level phosphorylation. On the other hand, there is some evidence that in certain cytochrome-containing anaerobes (e.g. *Selenomonas ruminantium*, *Anaerovibrio lipolytica*, *Veillonella alcalescens* and propionic acid bacteria) some ATP may be formed by oxidative phosphorylation when fumarate acts as terminal electron acceptor in anaerobic electron transport (de Vries *et al.*, 1974). It is somewhat paradoxical that it may be oxygen's inhibition of cytochrome synthesis in the propionic-acid bacteria, with the consequent loss of ability to undertake oxidative phosphorylation, that is the cause of their inability to grow aerobically on the surface of agar-solidified media (de Vries *et al.*, 1972).

2. *Bacterial Photosynthesis*

The most significant feature of the photosynthesis practised by bacteria other than cyanobacteria is that it is a wholly anaerobic process. Although a variety of reduced organic and inorganic compounds can be used as primary electron donors, hydroxyl ion (from water) is not utilized and oxygen is therefore not evolved. It would seem, therefore, that the anaerobic photosynthetic bacteria differ from cyanobacteria, algae and green plants in possessing a relatively simple photosynthetic electron transport system with only one type of photochemical reaction centre (Evans and Whatley, 1970). Supplied as they are with organic or inorganic electron donors which can effect reduction of nicotinamide nucleotides without the intervention of light energy, such organisms are primarily reliant on their photosynthetic apparatus for ATP (generated by cyclic photophosphorylation).

Some species of purple non-sulphur bacteria are not obligately photosynthetic. These species are oxygen-tolerant and can grow aerobically and heterotrophically in the dark. However, oxygen at quite low concentrations specifically inhibits the synthesis of their photopigments so that members of the Athiorhodaceae (e.g. *Rhodopseudomonas spheroides* or *Rhodospirillum rubrum*), grown aerobically, are virtually non-pigmented (Cohen-Bazire *et al.*, 1957). This very dramatic effect of oxygen is achieved in *Rps. spheroides* by a complex series of direct and indirect inhibitory controls of key steps in the synthesis of bacteriochlorophyll and carotenoids (Lascelles and Altschuler, 1969; Davies *et al.*, 1973). Similar inhibition of photopigment synthesis is also observed as one consequence of aeration of cultures of the obligately anaerobic green sulphur and purple sulphur species (Pfennig, 1967). It has indeed been suggested in the case of both *Rhodospirillum molischianum* and *Chromatium* D that inhibition by oxygen of bacteriochlorophyll synthesis, in the absence of any mechanism for obtaining ATP by aerobic respiration, is sufficient to account for the organism's strictly anaerobic mode of life (Sistrom, 1965; Hurlbert, 1967).

3. *Sulphate Dissimilation* (*Obligately Anaerobic Respiration*)

Those obligately anaerobic bacteria that can obtain energy for growth by reduction of sulphate, or certain other inorganic sulphur compounds, fall into two distinct genera (p. 173). The five "authentic" species of *Desulfovibrio* are non-sporing organisms containing a cytochrome c_3 and the distinctive pigment desulphoviridin, whilst the three species of *Desulfotomaculum* are spore-forming rods possessing a *b*-type cytochrome and no desulphoviridin. The physiology of these organisms has recently been expertly reviewed (Le Gall and Postgate, 1973) so that it is unnecessary to do more than highlight one or two of the more intriguing features of their energy-generating metabolism. Though this would seem to resemble the similar "anaerobic respiration" practised by facultatively anaerobic, denitrifying bacteria, *all* strains of *Desulfovibrio* and *Desulfotomaculum* are strict anaerobes. Sulphate is reduced to hydrogen sulphide at the expense of oxidation of some organic electron donor (e.g. lactate or malate). Only incomplete oxidation is accomplished, so that fatty acids (generally acetic acid) and carbon dioxide are produced. Gaseous hydrogen can also be utilized to reduce sulphate and to provide reducing power for biosyntheses, and some species, by fermenting compounds such as pyruvate, fumarate or choline, can grow anaerobically in the absence of sulphate.

Surprisingly little detail is known of the electron-transport chain in sulphate-reducing bacteria though a bewildering variety of components have been isolated from the various species. These include multiple *c*-type cytochromes, *b*- and *d*-type cytochromes, non-haem iron proteins, flavoproteins and even quinones. The roles of ferredoxin, flavodoxin, rubredoxin, and of the menaquinone MK-6, have yet to be elucidated. However, the presence of such low-potential electron carriers is consistent with the low redox potential of the ecological niches occupied by the sulphate-reducing bacteria, the capacity of several strains to undertake dinitrogen fixation, and their active hydrogenases.

Dissimilatory reduction of sulphate is an ATP-dependent process involving the intermediary formation of adenylyl sulphate (APS) and its subsequent reduction to sulphite and AMP by the following three-step enzyme-catalysed process:

$$SO_4^{2-} + ATP \rightleftharpoons APS + PP_i \text{ (ATP sulphurylase)}$$
$$H_2O + PP_i \rightleftharpoons 2P_i \text{ (pyrophosphatase)}$$
$$APS + 2e \rightleftharpoons AMP + SO_3^{2-} \text{ (APS reductase)}$$

The further direct reduction of sulphite to sulphide is a six electron transfer reaction:

$$3H_2 + SO_3^{2-} \rightleftharpoons S^{2-} + 3H_2O$$

possibly accomplished via a variety of routes, one of which might involve reduction of a recycled sulphite pool with trithionate and thiosulphate as intermediates:

$$3HSO_3^- + 3H^+ + 2e \rightleftharpoons S_3O_6^{2-} + 3H_2O \text{ (bisulphite reductase)}$$
$$S_3O_6^{2-} + 2e \rightleftharpoons S_2O_3^{2-} + SO_3^{2-} \text{ (trithionate reductase)}$$
$$S_2O_3^{2-} + 2e \rightleftharpoons S^{2-} + SO_2^{3-} \text{ (thiosulphate reductase)}$$

The sulphate-reducing machinery is spatially separated from the biosynthetic machinery of the cell (Sorokin, 1966) and considerable interest has been aroused by the finding that all three distinct bisulphite reductases that are present in various species (viz, desulfoviridin, desulforubidin and P_{582}) are novel haemoproteins (Murphy *et al.*, 1973, 1974).

From the probable reactions involved in the lactate-sulphate or ethanol-sulphate anaerobic respiration, it appears that the net yield of ATP from substrate level phosphorylation must be zero:

$$2 \text{ ethanol} \rightarrow 2 \text{ acetaldehyde} + [4H]$$
$$2 \text{ acetaldehyde} + 2P_i \rightarrow 2 \text{ acetyl phosphate} + [4H]$$
$$2 \text{ acetyl phosphate} + AMP + 2H^+ \rightarrow 2 \text{ acetate} + ATP$$
$$SO_4^{2-} + ATP + [8H] \rightarrow S^{2-} + 2H_2O + AMP + 2P_i + 2H^+$$

It appears therefore that oxidative phosphorylation is necessary for growth on these (and other) substrates (Peck, 1974). In a particle-containing cell-free extract of *Desulfovibrio gigas*, ATP was formed during reduction of sulphite by molecular hydrogen (Peck, 1966). Ferredoxin was manifestly involved in this process, and the phosphorylation was uncoupled by 2,4-dinitrophenol, pentachlorophenol and gramicidin, but not by oligomycin. Production of ATP also accompanied fumarate reduction by either molecular hydrogen or L(+)-lactate (Barton *et al.*, 1970; Barton and Peck, 1971). A phosphorylation yield (P:2e) of less than 0·5 was estimated from the ratios of growth yields with pyruvate and lactate, at various concentrations of their substrates (Vosjan, 1970). Thus, though it is the possession of haem pigments (cytochromes) by sulphate-reducers that is usually remarked upon as being particularly interesting, their ability to obtain energy via oxidative phosphorylation is equally worthy of note. Yet in this they may not be unique amongst obligate anaerobes, for it is suspected that certain other cytochrome-containing anaerobes that utilize fumarate (or possibly nitrate) as terminal electron acceptor may similarly indulge in oxidative phosphorylation (see p. 178, and de Vries *et al.*, 1974).

B. Catabolism and Biosynthesis

The utilization of molecular oxygen in catabolic, detoxification, or biosynthetic processes is accomplished by the action of specific enzymes known as oxygenases. Dioxygenases catalyse reactions in which both atoms of each oxygen molecule are inserted into the substrate (Hayaishi and Nozaki, 1969). Mono-oxygenases (hydroxylases) catalyse reactions in which one atom only of each oxygen molecule is transferred to the substrate (Katagiri and Takemori, 1973). A great many of these hydroxylases require to be provided with an additional reluctant (e.g. $NAD(P)H_2$ or ascorbate); these are known as mixed-function oxygenases, for whilst the one oxygen atom is incorporated into the substrate (as an OH group) the other atom is utilized to oxidize the co-reductant, e.g.

$$AH + NADPH_2 + O_2 \rightleftharpoons AOH + NADP + H_2O$$

Figure 1 illustrates a few of the many reactions catalysed in aerobic bacteria by such dioxygenases and mixed function oxygenases.

Yet, seemingly similar reactions can be accomplished by strictly anaerobic bacteria, when the oxygen that is incorporated into the substate must have originated not in molecular oxygen but in some other compound (e.g. water). Hydroxylation reactions can after all be accomplished by wholly anaerobic means by a combination of dehydrogenation and hydration reactions as in the "usual" routes of synthesis of malate from succinate or of β-hydroxybutyryl-CoA from butyryl-CoA (Fig. 2). We can discover several more such examples in the catabolic and biosynthetic metabolism of strict anaerobes where key oxygenation steps are either circumvented or accomplished by an oxygen-independent mechanism.

1. *Degradation of Aromatic Compounds*

While aerobic degradation of aromatic compounds by species of *Pseudomonas* involves such oxygen-dependent steps as those catalysed by *p*-hydroxybenzoate hydroxylase or catechol 1,2-oxygenase, an alternative reductive pathway is used to accomplish catabolic cleavage of the benzene ring when *Rhodopseudomonas palustris* grows anaerobically and photosynthetically on benzoate (Dutton and Evans, 1969; Guyer and Hegeman, 1969), or when *Pseudomonas* PN-1 grows anaerobically on *p*-hydroxybenzoate with nitrate as terminal electron acceptor (Taylor *et*

(1) **Dioxygenases**

(a) *Catechol 1,2-oxygenase*

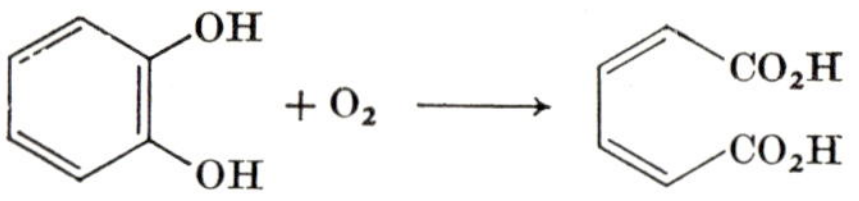

Catechol $+ O_2 \longrightarrow$ *Cis, Cis*-muconic acid

(b) *Tryptophan pyrrolase*

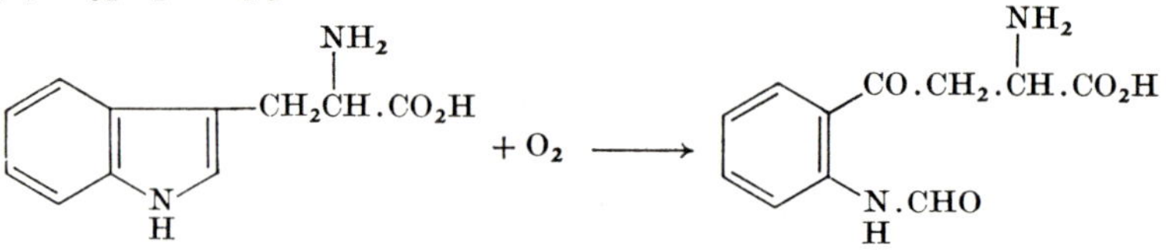

tryptophan $+ O_2 \longrightarrow$ N-formylkynurenine

(2) **Mono-oxygenases** (i.e. mixed function oxygenases)

(c) *p-Hydroxybenzoate hydroxylase* ($NADPH_2$-dependent)

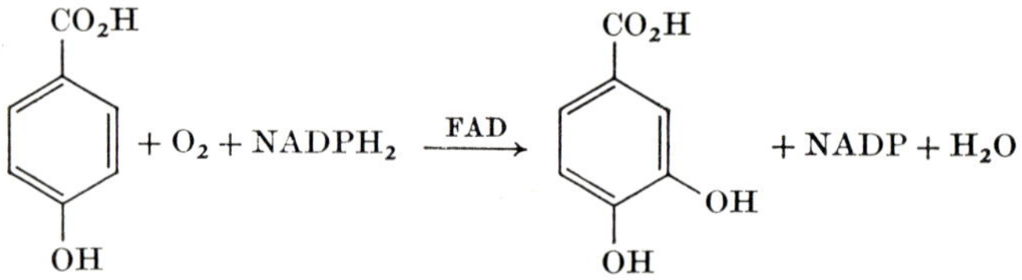

p-hydroxybenzoic acid $+ O_2 + NADPH_2 \xrightarrow{FAD}$ protocatechuic acid $+ NADP + H_2O$

(d) *Salicylate hydroxylase* ($NADH_2$-dependent)

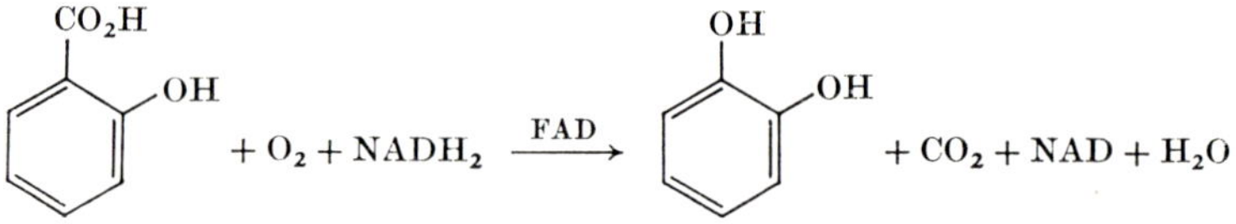

Salicylic acid $+ O_2 + NADH_2 \xrightarrow{FAD}$ catechol $+ CO_2 + NAD + H_2O$

FIG. 1. Examples of (1) di-oxygenases, and (2) mono-oxygenases (mixed function oxygenases), produced by species of *Pseudomonas* sp. during aerobic growth on various aromatic substrates.

al., 1970). These pathways (illustrated in Fig. 3) are discussed in greater detail by Dagley (1971).

2. *Biosyntheses of Essential Metabolites*

In this context, Goldfine (1965) instanced syntheses of mono-unsaturated fatty acids, tyrosine and nicotinic acid.

(a) *Mono-unsaturated fatty acids.* In some bacteria, as also in yeasts, plants and animals, desaturation of long-chain fatty acids is accomplished by the action of $NADPH_2$-requiring mixed-function oxidases.

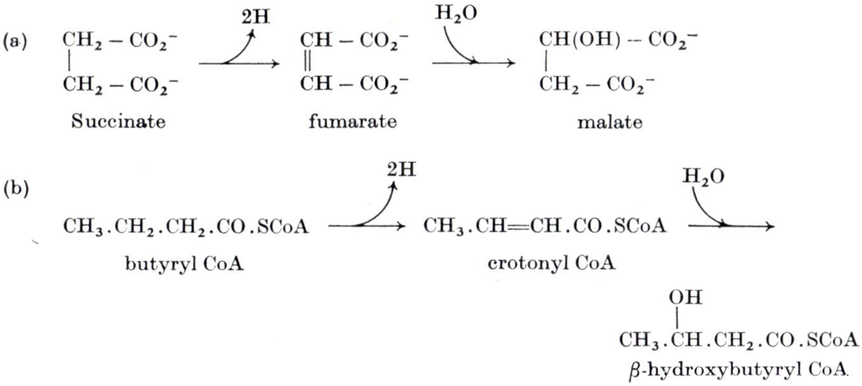

(a) $CH_2-CO_2^-$ / $CH_2-CO_2^-$ Succinate $\xrightarrow{2H}$ $CH-CO_2^-$ ‖ $CH-CO_2^-$ fumarate $\xrightarrow{H_2O}$ $CH(OH)-CO_2^-$ / $CH_2-CO_2^-$ malate

(b) $CH_3.CH_2.CH_2.CO.SCoA$ butyryl CoA $\xrightarrow{2H}$ $CH_3.CH{=}CH.CO.SCoA$ crotonyl CoA $\xrightarrow{H_2O}$

$CH_3.CH(OH).CH_2.CO.SCoA$ β-hydroxybutyryl CoA.

FIG. 2. "Usual" routes for the formation of (a) malate from succinate, and (b) β-hydroxybutyryl-CoA from butyryl-CoA, two examples of an "anaerobic" hydroxylation mechanism.

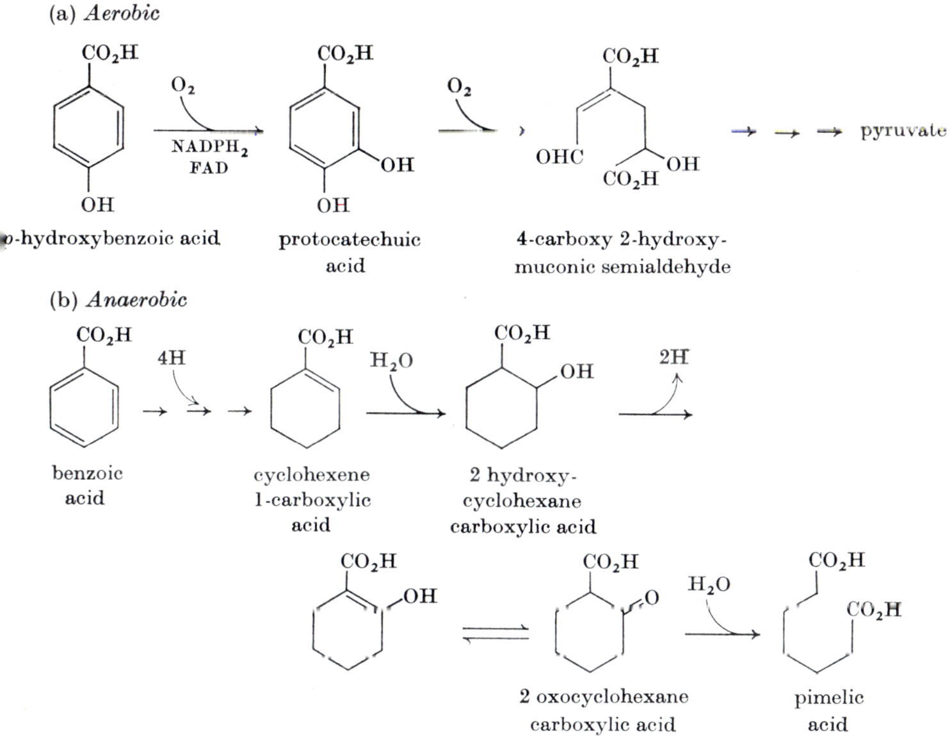

(a) *Aerobic*

CO_2H, OH p-hydroxybenzoic acid $\xrightarrow[NADPH_2,\ FAD]{O_2}$ CO_2H, OH, OH protocatechuic acid $\xrightarrow{O_2}$ CO_2H, OHC, OH, CO_2H 4-carboxy 2-hydroxy-muconic semialdehyde → → → pyruvate

(b) *Anaerobic*

CO_2H benzoic acid → $\xrightarrow{4H}$ → CO_2H cyclohexene 1-carboxylic acid $\xrightarrow{H_2O}$ CO_2H, OH 2 hydroxy-cyclohexane carboxylic acid $\xrightarrow{2H}$

CO_2H, OH ⇌ CO_2H, O 2 oxocyclohexane carboxylic acid $\xrightarrow{H_2O}$ CO_2H, CO_2H pimelic acid

FIG. 3. Initial steps in catabolism of aromatic compounds by *Rhodopseudomonas palustris* (a) Oxidative route employed during aerobic growth in the dark on p-hydroxybenzoate (Hegeman, 1967). (b) Reductive route employed during anaerobic growth in the light on benzoate (Dutton and Evans, 1969; Guyer and Hegeman, 1969).

In this way, the C_{16} saturated acid (palmitic) gives rise to the Δ^9-C_{16} mono-unsaturated acid (palmitoleic), whilst the Δ^9-C_{18} acid (oleic) is derived from the corresponding saturated C_{18} compound (stearic acid). In obligately anaerobic bacteria, and also in some facultatively anaerobic and aerobic species, production of mono-unsaturated fatty acids does not involve the homologous saturated acids. Thus *Escherichia coli*, *Lactobacillus plantarum* and *Clostridium kluyveri* were shown not to form their Δ^9-C_{16} (palmitoleic) and Δ^{11}-C_{18} (*cis* vaccenic) acids from palmitic and stearic acids (Bloch *et al.*, 1961). Instead, in *E. coli* at least, synthesis of palmitoleic acid starts with β-hydroxydecanoate. The enzyme β-hydroxydecanoyl thioester dehydrase catalyses dehydration of β-hydroxydecanoyl-S-acyl carrier protein to yield $\Delta^{3,4}$-decenoyl-S-ACP. Three further C_2 units (as malonyl-CoA) are then added to the carboxyl end of this unsaturated C_{20}-acyl-S-ACP to yield the acyl carrier protein (ACP) ester of palmitoleic acid. This would suggest that anaerobic bacteria similarly circumvent their inability to introduce a double bond directly into a preformed fatty acid by "building in" this double bond during the process of chain lengthening, using specific dehydrases acting on C_8 and C_{10} β-hydroxyacyl intermediates. In these organisms, the pathways for synthesis of saturated and unsaturated fatty acids will therefore diverge at the C_{10} (or possibly C_8) level.

(b) *Tyrosine*. In mammals for which phenylalanine is an essential amino acid, tyrosine is formed from dietary phenylalanine by a hydroxylation reaction catalysed by phenylalanine hydroxylase (a mixed-function oxygenase requiring $NADPH_2$ as co-reductant, and dihydrobiopterin as cofactor). Those anaerobic bacteria which do not show a nutritional requirement for tyrosine are presumably able to employ the oxygen-independent route for *de novo* tyrosine synthesis from prephenic acid which has been demonstrated in organisms such as *E. coli*, *Klebsiella aerogenes* and *Bacillus subtilis* (Gibson and Pittard, 1968).

(c) *Nicotinic acid*. In some eukaryotes (i.e. *Neurospora*, yeast and mammals) nicotinic acid is derived from 3-hydroxyanthranilate which, in turn, is formed from tryptophan via a route employing two oxygen-dependent reactions catalysed by tryptophan pyrrolase (dioxygenase) and kynurenine 3-hydroxylase (a mixed-function, $NADPH_2$-utilizing oxygenase). This route is evidently not available to strict anaerobes, many of which are nevertheless able to synthesize nicotinic acid. Though this alternative anaerobic pathway has not been fully elucidated, work with whole cells, and extracts, of *Clostridium butylicum* suggests that it is indeed a novel route that does not proceed via tryptophan. Isquith and Moat (1965, 1966) showed that aspartate and formate were required for production of nicotinic acid by *Cl. butylicum*, and proposed that quinolinate was an intermediate in this synthesis. More recently, N-formyl aspartate has been implicated as a key intermediate in this anaerobic biosynthesis of nicotinate, crude extracts of *Cl. butylicum* being able to obtain the remaining two carbon atoms from acetate or pyruvate or glutamine (Scott *et al.*, 1969). Related, though possibly not identical, pathways for the oxygen-independent synthesis of nicotinic acid from C_4 and C_3 precursors are operative in some plants (Yang and Waller, 1965), *Mycobacterium tuberculosis* (Gross *et al.*, 1965, 1967), *Serratia marcescens* (Scott and Hussey, 1965) and *Escherichia coli* (Chandler *et al.*, 1970; Chandler and Gholson, 1972).

In all of these instances, as in several others that could just as well have been described, the obligate anaerobe is at no great biosynthetic disadvantage through its inability to exploit molecular oxygen. Indeed, it is of considerable interest, and of possible phylogenetic significance, that these "anaerobic pathways" have been conserved in some species that have the option of living aerobically.

The position is somewhat different in the case of some special cell components which are only found in aerobic organisms and whose formation is an oxygen-dependent process. Thus polyunsaturated fatty acids and sterols that are important constituents of plant and animal cell membranes are not produced by any alternative anaerobic routes. Yet, since polyunsaturated fatty acids are seemingly absent from all, and sterols from many, aerobic bacteria, it can hardly be claimed that lack of these substances is a penalty exclusively exacted from obligate anaerobes.

IV. Culture E_h Values and the Growth of Obligate Anaerobes

Ever since Potter (1910) discovered that, when a platinum electrode was immersed in a bacterial culture it registered a more negative potential than in the original uninoculated medium, the hope has been sustained that the easily measurable redox potential (E_h) of a microbial culture could prove to be one of its more informative properties. This expectation was encouraged by the finding that different bacteria, when grown under similar conditions, established characteristic redox potentials in their cultures (Cannan *et al.*, 1926), and the further report that a very low (i.e. strongly negative) potential could be elevated by aeration (Gillespie, 1920). Such early findings were the precursors of the very many more detailed studies of the significance of culture E_h measurements that have been comprehensively reviewed by Hewitt (1950), Rabotnowa (1963) and Jacob (1970).

A. Difficulties in Interpreting Culture E_h Values

Though it is a simple enough matter to measure the E_h value of a culture using a platinum and reference electrode pair, what this measured potential actually represents has been a source of contention for many years past—especially as the value obtained will depend not only on the composition of the culture medium but even on the manner in which the measuring electrode is prepared (e.g. whether, and in what way, the platinum is polished; Jacob, 1970). One source of confusion has been the mistaken tendency of some microbiologists to view E_h as a measure of "electron concentration"—an error which probably stems from their being more used to dealing with pH value which they rightly comprehend to be a convenient index of "proton concentration (activity)" in the medium. But in fact the proper comparison is between E_h of a redox couple and the acid strength of a proton-exchanging (acid-conjugate base) couple (since both E° and pK_a values are proportional to ΔG° mol^{-1} of "particles" transferred—electrons and protons, respectively). The E_h value of a redox couple is therefore a measure of its electron-transfer potential and is an exact statement of a thermodynamically meaningful characteristic (Morris, 1974). On the other hand, to talk of the E_h value of a bacterial culture, which doubtless contains a multitude of different redox couples, is conceptually the equivalent of discussing the strength of a complex mixture of different acids, bases and buffers. Should all the redox couples in the culture be freely reversible contributors to a rapidly established equilibrium, and if the terminal (dominant) couple be perfectly electromotively active, one could argue that what is measured is the E_h value of this end couple, and that this reflects the overall reduction/oxidation tendency of the culture. But should the culture contain sluggish, irreversible and/or electromotively inactive redox components, the potential registered at the immersed platinum electrode could be misleadingly unrepresentative of the states of individual redox couples which might nevertheless be of crucial significance to the microbe's well being. Furthermore, the bacterial culture is an open system which may achieve and maintain a steady state but not a "true" equilibrium. What is imprecisely called the E_h value of a bacterial culture is therefore the resultant of a possibly imperfect interaction between many contributory couples and must not be invested with too great theoretical significance (see also, Morris and Stumm, 1967; Harrison, 1972). Even so, it is a real and measurable property of the culture which we cannot afford to ignore. Indeed, the potential difference between a bacterial culture and its uninoculated medium, or between one culture and another, can be exploited as a source of energy (i.e. as a biological fuel cell; Potter, 1911; Sisler, 1971). Therefore, whatever suspicions we may have as to its precise significance, by monitoring the E_h value of a bacterial culture we can (pragmatically) hope to follow changes in the magnitudes of the contending reducing and oxidizing agencies that are its prime determinants. Clark (1924) was very aware of the difficulties of interpretation posed by E_h measurements on bacterial cultures, but when specifically referring to cultures of anaerobes he circumvented the problem thus: "While there is still some doubt regarding the interpretation of certain observed electrode potentials, there can be no doubt that certain anaerobic cultures generate a hydrogen overvoltage. It is extremely difficult to conceive of molecular oxygen playing any part in the activity of a cell that is producing a hydrogen overvoltage and tearing to pieces by reductive action materials which resist strong chemical reducing agents."

B. E_h Values of Aerated Cultures

Though there are no secure grounds for the expectation that the measured "extracellular" E_h value in a culture mirrors the intracellular E_h value of the bacteria, such measurements as have been made suggest that the intracellular E_h value will prove to be close to, if a little lower than, the electrometrically measured culture potential (Leman, 1965; Jacob, 1970). There are similar "practical" grounds for hope that measurement of E_h value in aerated bacterial cultures can be employed to assay dissolved oxygen in concentrations below those detectable by polarimetric means. Thus, Squires and Hosler (1958) found that the redox potential of aerated nutrient medium changed in proportion to the logarithm of the prevailing pO_2, and measurement of E_h value has been successfully employed as a means of assessing and controlling the low oxygen concentrations required to give optimum yields of desirable bacterial growth products (Tengerdy, 1961; Lengyel and Nyiri, 1965). It is certainly possible, by following changes in culture E_h value, to reproduce required degrees of oxygenation in pure cultures of a bacterium growing in a simple defined medium (Wimpenny, 1969, 1970; O'Brien and Morris, 1971).

C. E_h Values of Cultures of Obligate Anaerobes

A proper appreciation of the "meaning" of culture E_h value is especially required of those working with obligately anaerobic bacteria since great significance is generally attached to the finding that their cultures register particularly low E_h values and that the "ceiling" E_h limits for initiation of their growth in a given medium are generally much lower than those displayed by facultative anaerobes. Consequently, to obtain growth of an obligate anaerobe in any medium it is usually necessary to expel all dissolved oxygen (thereby lowering the E_h value) and to add some suitable reducing agent (cysteine, thioglycollate, dithionite or ascorbate) to poise the culture E_h value at a still lower value (e.g. -200 to -350 mV at pH 7). Alternatively, growth may often be achieved in a medium whose E_h value is initially unpropitiously high by employing a large and compact inoculum which will quickly reduce its immediate environs and then progressively invade the remainder of the medium as the zone of reduction is extended. An E_h gradient may therefore be established in such cultures, as the anaerobic organisms strive to establish and maintain an agreeably reducing environment. The major advances in techniques of isolating, purifying and handling obligate anaerobes have therefore consisted, in the main, of developments in the incubation and transfer of cultures in oxygen-free atmospheres, and in the selection of effective poising agents. Particularly helpful accounts of current anaerobic microbiological procedures have been given by Willis (1964), Moore (1966), Dowell and Hawkins (1968), Hungate (1969), Barnes (1969), Cato *et al.* (1970), Shapton and Board (1971), Dowell (1972), Holdeman and Moore (1972), Miraglia (1974).

The relation between the E_h values of their cultures and the growth of various species of *Clostridium* has recently been reviewed in an article which lists both the minimum E_h values reportedly established in growing cultures of these anaerobes, and the upper limits of E_h value compatible with their growth (Morris and O'Brien, 1971). The following conclusions were drawn from this survey: (a) although a low E_h value is not absolutely essential for initiation of growth of some species of *Clostridium*, the growth of all is invariably accompanied by the creation and maintenance of a negative culture E_h value attributable to the organisms anaerobic metabolism; (b) addition to the medium of substances which lower its E_h value facilitates the initiation of growth, especially from small inocula; (c) though there may be a limiting culture E_h value above which a given species of *Clostridium* will not grow, its value is not constant, being determined by several factors, for example the size of inoculum, richness of the medium, whether or not the E_h value is artificially maintained, and if so, by what agency.

It seems likely that such behaviour is common to all obligate anaerobes, with the more stringent of them requiring, and thereafter maintaining, the most negative E_h values. We shall return to this subject when we discuss the hypothesis that part at least of the deleterious action of molecular oxygen on obligate anaerobes may be mediated through its tendency to raise the culture E_h value (p. 206).

V. Effects of Oxygen on Obligate Anaerobes

Since all obligate anaerobes are primarily distinguished by their sensitivity to molecular oxygen, any account of their physiology must seek to explain the basis of this oxygen hypertoxicity. Effects of oxygen on various groups of strictly anaerobic bacteria have been surveyed in several comprehensive reviews (Grunberg, 1948, 1949; Smith, 1967; Morris and O'Brien, 1971), as has the possibly related phenomenon of the toxicity of hyperbaric oxygen to aerobes (Haugaard, 1968; Gottlieb, 1971). Other recent accounts of the regulatory action of oxygen on the growth and metabolism of facultatively anaerobic and aerobic bacteria are also available (Wimpenny, 1969; Hughes and Wimpenny, 1969; Harrison, 1972, 1973). It is for this reason that I have thought it best in the present essay to give possibly undue prominence to the more recent advances, being encouraged in this course by the novelty of some of the currently held views of oxygen toxicity. But, to appreciate these, the microbiologist has first to make the acquaintance of possible by-products of oxygen's metabolism that have not as yet been accorded the *imprimatur* of inclusion in textbooks of bacteriology. The reader who is already well versed in the chemistry of oxygen will appreciate the need for some explanation of the unique properties of this very reactive molecule and will, I trust, condone (even if he chooses to ignore) my attempt to introduce novitiates to the mysteries of singlet oxygen, superoxide anion and hydroxyl free radical.

A. Chemistry of Oxygen and Some Derivatives

1. Oxygen

The O atom (proton number 8, molecular orbital arrangement $1s^2\,2s^2\,2p^4$) possesses six valence electrons. Thus, in the expectation that the octet rule is followed, one would presume a tendency to gain two electrons in order to complete the outer valence shell. Accordingly, it might be assumed that the diatomic molecule of oxygen would have the structure $:\ddot{O}=\underset{\cdot\cdot}{O}:$. Yet the oxygen molecule is found to be paramagnetic and its chemistry to be dominated by a diradical character deriving from its possession of two unpaired valence electrons (Ardon, 1965; Taube, 1965). To explain the paramagnetism of oxygen in terms of the valence bond theory, Pauling (1931) suggested the structure :O$\underset{\cdots}{\overset{\cdots}{-}}$O: which contained one single bond and two 3-electron bonds. Each 3-electron bond would have approximately half the energy of the single bond, and each 3-electron bond would possess an unpaired electron. These could interact by pairing (to produce a singlet) or by remaining in parallel (to form a triplet). Since the triplet state is more stable than the singlet (Wheland, 1937) the structure suggested by Pauling might be expected to be more stable in its triplet state than the "more obvious" double bonded molecule. Application of the molecular orbital theory predicts that the valence electron distribution in the oxygen molecule is as shown in Fig. 4, i.e. $[\sigma(2s)]^2$, $[\sigma^*(2s)]^2$, $[\sigma(2p_x)]^2$, $[\pi(2p_z)]^2$, $[\pi(2p_y)]^2$, $[\pi^*(2p_z)]^1$, $[\pi^*(2p_y)]^1$. The paramagnetism of the molecule thus arises from the unpaired electrons of the

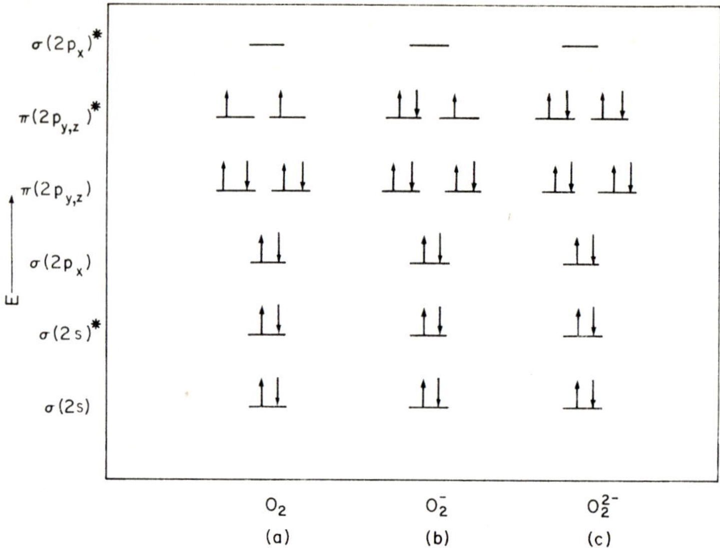

FIG. 4. Approximate energy-level diagram for some binuclear oxygen species, viz: (a) oxygen (in its triplet ground state $^3\Sigma_g^-$, (b) superoxide anion (a free radical), (c) peroxide anion. From Lagowski (1973).

double degenerate (i.e. equal energy) π^* orbitals. With two unpaired electrons, the molecule therefore behaves as a diradical, even though it possesses overall an even number of valence electrons.

It is the ability of oxygen to behave as a diradical that makes it a potent reagent in initiation and/or addition reactions. Demopoulos (1973a) instanced the following reactions of possible biological significance, wherein oxygen is symbolized as $\dot{O}$—$\dot{O}$ to emphasize its diradical nature:

```
                                  alkyl radical
      H  H    H  H               H  H    H  H              H  H    H  H
(1) —C—C = C—C—   →   —C—C = C—C—   →   —C—C = C—C—
     (H)          H               •            H               |            H
      ↓                                   •  H                 O—OH
     Ȯ—Ȯ                                 O—O
                                  perhydroxyl radical

      H  H    H  H               H  H    H  H
(2) —C—C = C—C—   →   —C—C — C—C—
      H           H              H  |      |  H
                                        O — O
           Ȯ—Ȯ
```

Radical reactions of this type may occur in lipids (e.g. peroxidation) and can be initiated and catalysed by heavy metals like copper and iron which are particularly effective when complexed (e.g. iron in haem; see also p. 196).

A universally recognized property of oxygen is its ability to act as a powerful oxidizing agent, the two redox couples of greatest traditional interest to the biologist being those inwhich the reduced species are water and hydrogen peroxide, respectively, i.e.:

$$O_2 + 4H^+ + 4e^- \rightleftharpoons 2H_2O \qquad E_0' \text{ at pH } 7 = +0{\cdot}815 \text{ V}$$
$$O_2 + 2H^+ + 2e^- \rightleftharpoons H_2O_2 \qquad E_0' \text{ at pH } 7 = +0{\cdot}270 \text{ V}$$

Oxygen is present in air at a partial pressure of 0·209 atm. (i.e. 159 mm Hg or 21·12 k*Pa*) and is only moderately soluble in water—about 9 mg dm^{-3} at 20°C under 1 atm of air (Montgomery *et al.*, 1964). It is somewhat less soluble in salt solutions than in water, and a value of 5·9 mg dm^{-3} was reported as the oxygen concentration at 30°C in a nutrient medium containing 5% glucose and several dissolved salts (Brown, 1970); a more usual assumption is that the concentration of oxygen in nutrient media in equilibrium with air at 30°C will be about 8 mg dm^{-3} (i.e. 250 μmol dm^{-3}). Though polarimetric methods are generally employed to measure oxygen activity in aqueous solutions (Beechey and Ribbons, 1972) the autoclavable membrane probes generally employed in work with microbial cultures are ineffective at low tensions of dissolved oxygen (Harrison, 1972). The significantly greater solubility of oxygen in organic solvents (generally some seven to eight-times that in water) has attracted less attention than it warrants, though this property has led to the use of inert fluorocarbons as "super-effective" oxygen reservoirs for O_2-transfer to submerged living cells (Clark and Gollan, 1966; Mazia and Ruby, 1967).

The "normal" molecular oxygen, whose chemistry we have just considered, is the most stable (ground) state of oxygen. This lowest energy form of oxygen, in which the two unpaired electrons are of parallel spin (↑) (↑), is known as the triplet state and is given the signature $^3\Sigma_g^-$. When the two unpaired electrons are anti-parallel in identical or separate orbits, more highly energized singlet states of oxygen prevail.

2. *Singlet Oxygen*

The form of singlet oxygen ($^1O_2{}^*$) which has its two antiparallel unpaired electrons in identical orbit (⇅) ( ) is given the signature $^1\Delta_g$, whilst that form in which the antiparallel unpaired electrons are located in separate orbits (↑) (↓) is represented as $^1\Sigma_g^+$. Molecular oxygen in its triplet state ($^3\Sigma_g^-$) must be energized to yield either of the singlet states:

$$^3\Sigma_g^- \xrightarrow{96 \text{ kJ mol}^{-1}} {}^1\Delta_g$$

$$^3\Sigma_g^- \xrightarrow{134 \text{ kJ mol}^{-1}} {}^1\Sigma_g^+$$

Because the unpaired electrons in singlet oxygen are antiparallel, both singlet states are diamagnetic, a property which facilitates their detection in the presence of the paramagnetic triplet state.

Singlet oxygen is present as an atmospheric pollutant; in so-called "smog" conditions some 3×10^6 $^1O_2^*$ molecules cm^{-3} have been detected (Dowty *et al.*, 1973). In the laboratory though, singlet oxygen is generally produced from triplet oxygen either by a microwave discharge procedure (Politzer *et al.*, 1971) or by dye-sensitized photo-excitation of (ground state) oxygen dissolved in water or some other solvent. Sensitizers such as porphyrins, polycyclic aromatic hydrocarbons, fluorescein derivatives and methylene blue have all proved effective. Using a low-energy sensitizer dye (rose bengal), the lower energy, longer lived, singlet oxygen $^1\Delta_g$ is formed whilst, using a high-energy sensitizer dye (eosin), it is claimed that $^1\Sigma_g^+$ singlet oxygen is produced. The use of insolubilized sensitizers, produced by immobilization of such dyes on suitable supporting materials, has much to commend it (Nilsson and Kearns, 1974).

One classical chemical means of making singlet oxygen is the reaction of a hypohalite ion (e.g. hypochlorite) with hydrogen peroxide (Maugh, 1973). Finding that singlet oxygen ($^1\Sigma_g^+$) was produced from potassium superoxide in dimethylsulphoxide, Khan (1970) was concerned lest sufficient $^1O_2^*$ could survive immediate quenching to make the use of the reaction of potassium superoxide with water a particular health hazard (for this reaction is exploited as a portable chemical source of oxygen for emergency breathing purposes in hospitals, mines, submarines and space craft). It has since been demonstrated that singlet oxygen is indeed produced in aqueous solution by the rapid dismutation of superoxide anion (Rotilio *et al.*, 1973a).

$$O_2^{\cdot-} + O_2^{\cdot-} \longrightarrow O_2^{2-} + {}^1O_2^*$$

$$O_2^{2-} \xrightarrow{2H^+} H_2O_2 \qquad k \text{ at pH } 7{\cdot}12 = 2{\cdot}1 \times 10^5 \text{ dm}^3 \text{ mol}^{-1} \text{s}^{-1}$$

To be able to distinguish between effects due to $O_2^{\cdot-}$ and those due to $^1O_2^*$, it is sometimes necessary to have recourse to a "dark" chemical reaction which forms $^1O_2^*$ by a mechanism which does not also produce superoxide anions. Such a facility is provided by the interaction of CrO_8^{3-} with water at pH 7, which generates $^1\Delta_g$ singlet oxygen (Peters *et al.*, 1972).

The excited singlet oxygen molecule is, of course, highly unstable and by the same token very reactive, with a tendency to release energy by any of several means. The most obvious of these is by direct relaxation to the triplet ground state, the energy so lost appearing as light (fluorescence and/or chemiluminescence). It has been proposed that transitory dimers formed by the collision of two singlet oxygen molecules can produce photons with a characteristic emission peak. By this route, both singlet molecules simultaneously return to the ground state with the production of a single photon (Stauff, see Arneson, 1970). Whatever the mechanism, this chemiluminescence could in practice be a useful form of "self display", enabling the experimenter to detect production of the diamagnetic singlet states of oxygen from the paramagnetic triplet state, not only by electron paramagnetic resonance spectroscopy but also by the light emission that would accompany relaxation of the excited state. Thus, instantaneous light emission from $^1\Delta_g$ singlet oxygen at 1270 nm can, for assay purposes, be conveniently "shifted" into the visible region by the use of luminol (Paschen and Weser, 1973). On the other hand, Kearns (1971) has pointed out that, since the mechanism of chemiluminescence in singlet oxygen reactions is not well understood, it is not an infallible criterion to use for the presence of $^1O_2$*.

Singlet oxygen reacts with regions of high electron density (*pi* systems) in organic structures to form substituted dioxetanes (i.e. molecules with a four membered ring comprised of two adjacent oxygen atoms with two carbon atoms). These dioxetanes are generally very labile, dissociating to form electronically excited carbonyl groups that relax by chemiluminescence (Allen *et al.*, 1972; Maugh, 1973). Several examples of the chemical reactivity of $^1O_2$* are given by Kasha and Khan (1970), Politzer *et al.* (1971) and Stephenson *et al.* (1973), who between them document examples of photo-oxygenation reactions, "ene" and "diene" reactions, hydrogen abstraction mechanisms and interactions of singlet oxygen with N- and S-containing substrates. Singlet oxygen "traps" include 2,5-diphenylfuran (Porter and Ingraham, 1974) and 1,3-diphenylisobenzofuran, whose bleaching by $^1O_2$* can be followed at 415 nm (Nilsson and Kearns, 1974). An interesting finding that has important implications in studies of the mechanisms of $^1O_2$* reactions is that the lifetime of singlet oxygen in D_2O is ten times greater than in H_2O (Merkel and Kearns, 1972).

3. *Superoxide Anion*

If the oxygen molecule were to accept electrons in a stepwise manner, its reduction would proceed via superoxide anion (a free radical) and peroxy anion:

$$\underset{\text{oxygen}}{O_2} \xrightarrow{1e} \underset{\substack{\text{superoxide}\\ \text{anion}}}{O_2^{\cdot-}} \xrightarrow{1e} \underset{\substack{\text{peroxy}\\ \text{anion}}}{O_2^{2-}}$$

These single electron reductions result in a lengthening of the O—O bond from 121 pm in the oxygen molecule to 128 pm in the superoxide anion and to 148 pm in the peroxy anion, thereby mirroring the marked weakening of the bond.

The free radical, paramagnetic nature of the superoxide anion (deriving from its possession of a single unpaired electron) is evident from its molecular orbital diagram (Fig. 4). Its protonated form $HO^{\cdot-}$ is of little consequence at neutral pH values since the pK_a of its dissociation is 4·8 (Behar *et al.*, 1970). Superoxides of sodium or potassium ($K^+O_2^{\cdot-}$) are strong oxidizing agents that vigorously react with water to yield oxygen (Czapski, 1972). This is the same proton-dependent dismutation of superoxide anions mentioned previously (p. 192) as yielding singlet oxygen and hydrogen peroxide. The reaction proceeds most rapidly at pH 4·8 which would suggest that the most rapid mechanism is:

$$O_2^{\cdot-} + HO_2\cdot + H^+ \rightarrow H_2O_2 + O_2$$

Since superoxide anions can further react with hydrogen peroxide to yield hydroxyl radicals:

$$O_2^{\cdot-} + H_2O_2 + H^+ \rightarrow OH\cdot + O_2 + H_2O$$

reaction of superoxide anions with water can indirectly give rise to two further very reactive species (viz. singlet oxygen and OH·). Even so, superoxide anions can be generated in aerated aqueous solution (however transitorily) by: (i) ultrasonication (Anbar and Pecht, 1964; Lippitt *et al.*, 1972); (ii) pulse radiolysis (Rotilio *et al.*, 1972a; Klug *et al.*, 1972); (iii) photo-illumination of a number of dyes, including isoalloxazine derivatives, reduced flavins and flavoproteins, in the presence of oxygen and oxidizable compounds (Massey *et al.*, 1969; Balny and Douzou, 1974); and (iv) scavenging of hydroxyl free radicals with formate (O'Donnell and Sangster, 1970; Lippitt *et al.*, 1972). Other useful methods of preparation of $O_2^{\cdot-}$ include oxidation of hydrogen peroxide with periodate, one-electron reduction of molecular oxygen with ferrous ions in the presence of a suitable ligand such as phosphate (Michelson, 1973), and electrolytic reduction of molecular oxygen (McCord and Fridovich, 1969), preferably carried out in dry acetonitrile with tetrabutylammonium bromide as carrier electrolyte (Fee and Hildenbrand, 1974).

The superoxide free radical anion is an effective reducing agent (Rao and Hayon, 1973) and can quite easily be detected and assayed by its reduction of: (a) tetranitromethane to nitroform; (b) nitroblue tetrazolium to its formazan; or (c) ferricytochrome *c* to its ferro- form. However, it may also act as an oxidizing agent (e.g. in the oxidation at pH 10·2 of epinephrine to adrenochrome; Misra and Fridovich, 1972b), and as an initiator (e.g. of the free-radical chain reaction which accomplishes the aerobic oxidation of sulphite; Fridovich, 1972). In the presence of ascorbate or of Mn^{2+} ions, superoxide anions are reduced to hydrogen peroxide with no simultaneous formation of oxygen (Epel and Neumann, 1973).

It has been proposed that superoxide anions might be involved in oxygen-dependent hydroxylation reactions. Thus Strickland and Massey (1973) showed that $O_2^{\overline{\cdot}}$, generated by photo-illumination of lumiflavin acetic acid, accomplished the conversion of *p*-hydroxybenzoate to polyhydroxybenzoates. Whether the $O_2^{\overline{\cdot}}$ is acting via singlet oxygen production in this and related reactions has yet to be determined.

4. *Peroxide Anion*

The peroxide anion (O_2^{2-}) is diamagnetic and is not a free radical (see Fig. 4). While acidification of cold aqueous solutions of metal peroxides gives a solution of hydrogen peroxide;

$$O_2^{2-} + 2H^+ \rightarrow H_2O_2,$$

simple solution of O_2^{2-} in water yields oxygen (via the hydroperoxyl anion):

$$O_2^{2-} + H_2O \rightarrow HO_2^- + OH^-$$
$$2HO_2^- \rightarrow O_2 + 2OH^-$$

Peroxides are good oxidizing agents towards a variety of substances, but towards stronger oxidizing agents (permanganate or oxygen) O_2^{2-} can act as a reducing agent.

The peroxide anion is, of course, a diacidic base:

$$\underset{\text{peroxide anion}}{O_2^{2-}} + H^+ \rightleftharpoons \underset{\text{hydroperoxyl anion}}{HO_2^-} + H^+ \rightleftharpoons \underset{\text{hydrogen peroxide}}{H_2O_2}$$

and hydrogen peroxide in dilute aqueous solution is more acidic than water ($K_a = 1{\cdot}78 \times 10^{-12}$ at 20°C). However, in solution at high or low pH values it acts as an excellent oxidizing agent. In alkaline solution, homogeneous decomposition of hydrogen peroxide takes place (i.e. catalysed by OH^-) to yield oxygen, and innumerable other catalysts, both homogeneous and heterogeneous, are known to accelerate hydrogen peroxide breakdown to form oxygen. Transition metal ions and their complexes are particularly effective catalysts of hydrogen peroxide breakdown. Free radicals (e.g. OH·) can also be formed from hydrogen peroxide, as during Fe^{2+} oxidation in Fenton's reagent:

$$Fe^{2+} + H_2O_2 \rightarrow Fe^{3+} + OH\cdot + OH^-$$

5. *Hydroxyl Free Radical*

Produced by (i) thermal homolysis, (ii) high-energy radiation and photolysis, or (iii) one electron redox reactions, free radicals such as OH· are exceedingly reactive and can ordinarily exist in aqueous solutions only in low concentrations. In a biological context, OH· is generally encountered in discussions of the effect of ionizing radiation on living cells, for OH· (together with H· and solvated electrons) are produced by radiation of water, and have all been shown to be capable of causing radiation damage to many types of biopolymer (Powers, 1972; Pryor, 1973).

Interaction of hydroxyl radicals with adenine residues probably gives rise to adenine-7,N-oxide (Pryor, 1973), whilst OH· may also undergo addition reactions with molecules containing double bonds (Myers, 1973):

$$R_2C=CR_2 + OH\cdot \rightarrow R_2C(OH)-\dot{C}R_2$$

Again, interaction of OH· with RH (where RH is some organic cellular component) could occur with the abstraction of H and the formation of the organic free radical R·:

$$RH + OH\cdot \rightarrow R\cdot + H_2O$$

The organic free radical, R·, may then interact with itself, resulting in functionally damaging cross linking:

$$R\cdot + R\cdot \rightarrow R{-}R,$$

or, alternatively, it might interact with molecular oxygen:

$$R\cdot + O_2 \rightarrow RO_2\cdot$$

The product is an unstable and highly reactive organic peroxide free radical ROO· (Ingold, 1969). Spontaneous degeneration of these peroxide free radicals produces additional radicals and could prove to be structurally devastating. Such free radicals also arise by the schism of organic peroxides or hydroperoxides that is catalysed by iron or copper ions (Demopoulos, 1973a):

$$Fe^{3+} + -\underset{OOH}{\overset{H}{C}}-\overset{H}{C}=\overset{H}{C}-\underset{H}{\overset{H}{C}}- \rightarrow -\underset{OO\cdot}{\overset{H}{C}}-\overset{H}{C}=\overset{H}{C}-\underset{H}{\overset{H}{C}}- + H^+ + Fe^{2+}$$

Even saturated fatty acids, especially in a non-polar environment, can undergo peroxidation initiated and catalysed by Cu^{2+} or Fe^{3+}. Thus any attempt to mitigate damage (e.g. radiation injury) instigated in the first instance by OH· radicals might, with profit, concentrate on: (1) rapid "neutralization" of the organic free radicals; and (2) removal of transition metal cations. This is the rationale behind the use of the so-called "radioprotective" sulphydryl compounds (cysteine, cysteamine and D-penicillamine). These thiols are generally thought to remove the organic free radicals by a rapid repair process which reconstitutes the original substrate (RH) at the expense of the formation of a new thienyl radical (RS·) which is much less reactive, and is innocuously removed (terminated) by interaction with a fellow RS· radical to form a disulphide (—S—S—) i.e.:

$$R\cdot + R_1SH \rightarrow RH + R_1S\cdot$$
$$R_1S\cdot + R_1S\cdot \rightarrow R_1S - SR_1$$

Furthermore, cysteine, cysteamine and penicillamine have the additional helpful attributes of being both anti-oxidants and metal chelators.

6. Ozone

The triatomic molecule of ozone can be synthesized from molecular oxygen by: (i) silent electric discharge; or (ii) irradiation with ultraviolet light (<210 nm). It can also be produced by electrolysis of water (aqueous sulphuric acid) and in some reactions where molecular oxygen is formed (e.g. the action of fluorine on water). Ozone is a minor constituent of the normal atmosphere, but its concentration increases with altitude, achieving its maximum value in the so-called "ozone layer" at a height of about 25 km. This ozone layer efficiently absorbs solar ultraviolet-radiation (<300 nm), becoming heated in the process and thus being able to play some part in the temperature regulation of the lower atmosphere whilst shielding the earth's surface from biologically harmful short ultraviolet rays (p. 224).

Ozone is a particularly potent oxidizing agent that is often employed in place of chlorine for purposes of controlled bleaching and water purification. It has been reported that singlet oxygen is produced from a number of ozone oxidation reactions (Murray and Lin, 1970), and some of its effects on micro-organisms have recently been reviewed (Schilpkoter and Bruch, 1973).

B. Biochemistry of Oxygen and Some Derivatives

1. Molecular Oxygen

As we have noted (p. 174), the most obvious role played by oxygen in biological systems is to function as a terminal electron acceptor for aerobic respiration. In aerobic eukaryotes (wherein respiration proceeds in the mitochondria), as in those prokaryotic organisms whose cytochrome-based electron-transport system is located in the cell membrane, terminal reduction of molecular oxygen is catalysed by a haem protein (cytochrome oxidase), and the product is water (Smith, 1961; Meyer and Jones, 1973). In many bacteria, cytochrome-independent electron transport, catalysed by soluble enzymes, may also bring about the reduction of oxygen. Flavoprotein systems are generally the major link between the substrate and oxygen, some yielding water and others hydrogen peroxide. The location and functions of hydrogen peroxide-yielding oxidases, in the peroxisomes of eukaryotic cells, have been reviewed (de Duve and Baudhuin, 1966; Tolbert, 1971), whilst Dolin (1961b) described many of the flavoprotein oxidases and peroxidases to be found in bacteria (including anaerobes). In photoluminescent bacteria, a luciferase catalyses oxidation of $FMNH_2$ by oxygen, as part of the process (also involving a long chain aldehyde) whereby visible light is emitted when an organic molecule in an excited state relaxes to its ground state (McElroy and Seliger, 1962; Hastings *et al.*, 1973). The biosynthetic utilization of molecular oxygen in hydroxylations catalysed by dioxygenases and mixed function oxygenases has previously been mentioned (p. 181); some bacteria are also able to employ such reactions for catabolic ends, as in "ring-opening" steps in the aerobic degradation of some aromatic compounds (Dagley, 1971). The transfer of facultatively anaerobic micro-organisms from anaerobic to aerobic conditions is very often accompanied by a sophisticated and well-regulated "switch over" to aerobic respiration, which might involve repression of synthesis of specific components of the anaerobic energy-yielding metabolic machinery with concurrent acquisition of new enzymes and cofactors, even to the extent of fabrication of new sub-cellular organelles (Gray *et al.*, 1966, b; Hughes and Wimpenny, 1969; Wimpenny and Necklen, 1971; Harrison, 1972, 1973). It is in this context that we find reported the oxygen inhibition of synthesis of photopigments in photosynthetic bacteria (Cohen-Bazire *et al.*, 1957) and the oxygen-provoked biosynthesis of sterols in yeast (Klein, 1955; Adams and Parks, 1969), of haem in *Staphylococcus epidermidis* (Jacobs *et al.*, 1969), of mitochondrial DNA in yeast (Rabinowitz *et al.*, 1969) and of the mitochondria themselves (Lloyd, 1969). Though several essential cell components and cofactors are subject to "autoxidation" by molecular oxygen, it is probably best if discussion of this were deferred until the possible mechanisms of oxygen-toxicity are examined in greater detail (p. 205).

2. *Singlet Oxygen*

Following the suggestion that the antibacterial activity of raw milk and saliva could be in part due to the interaction of a peroxidase with thiocyanate ions and hydrogen peroxide (Reiter *et al.*, 1964; Mickelson, 1966; Klebanoff *et al.*, 1966), there have been several reports that the post-engulfment killing of bacteria by polymorphonucleocytes may also be mediated by myeloperoxidase plus hydrogen peroxide and halide ions (McRipley and Sbarra, 1967a, b; Paul *et al.*, 1973). It seems possible that these systems could be bactericidal by virtue of their acting as generators of singlet oxygen (Allen *et al.*, 1972); spontaneous dismutation of superoxide anions could be the more usual source of singlet oxygen in most living cells (Finazzi-Agro *et al.*, 1972; Maugh, 1973). It has further been proposed that the concurrent presence of hydrogen peroxide could promote the formation of singlet oxygen from superoxide anions by participating in the generation of OH· radicals (Haber and Weiss, 1934):

$$O_2^{\bar{\cdot}} + H_2O_2 + H^+ \rightarrow O_2 + H_2O + OH\cdot$$

Singlet oxygen would then be produced by the interaction of OH· with $O_2^{\bar{\cdot}}$ (Arneson, 1970):

$$O_2^{\bar{\cdot}} + OH\cdot \rightarrow OH^- + {}^1O_2^*$$

If this is so, then any biological system which produces both $O_2^-\cdot$ and H_2O_2 should be a good source of singlet oxygen (e.g. some NADPH oxidases; Howes and Steele, 1972). However, there are additional oxygen-consuming systems that produce singlet oxygen in the absence of any intermediary formation of superoxide anion (e.g. the lipoxidase-linoleate system; Chan, 1971; Finazzi-Agro *et al.*, 1972; Faria Oliveira *et al.*, 1974), and many natural products (e.g. chlorophylls) are very efficient sensitizers for photo-oxygenation reactions which probably proceed via a singlet oxygen mechanism (Politzer *et al.*, 1971).

In all, there seems little doubt that singlet oxygen *can* be produced in living cells that are exposed to oxygen, sometimes in controlled quantities as a biosynthetic reagent, but doubtless on other occasions in potentially hazardous amounts in sensitive locations (Goda *et al.*, 1973). For example, if ${}^1O_2^*$ were to be formed in a hydrophobic lipid/protein region (e.g. within a cell membrane), it would not be solvated and could be longer lasting and hence more damaging than in aqueous media (Maugh, 1973). Actual evidence of damage wrought within the cell by singlet oxygen is hard to find, though it has been suggested that ${}^1O_2^*$ could inactivate enzymes by destruction of key histidine residues (Foote, 1968), be involved in lipid hydroperoxidations (Rawls and van Santen, 1970; Howes and Steele, 1972; Pederson and Aust, 1973), readily oxidize cholesterol (Lamolo *et al.*, 1973), inactivate DNA (Foote, 1968) and damage mitochondrial membranes (Goda *et al.*, 1973). Exposure of pine pollen to artificially generated singlet oxygen decreased the content of unsaturated fatty acids in surface, or near-surface, lipids (Dowty *et al.*, 1973), whilst it was shown *in vitro* that methyl linoleate reacts at least 1450-times faster with ${}^1O_2^*$ than with oxygen in its ground state (Rawls and van Santen, 1970). A role for singlet oxygen in ageing processes, and even in carcinogenesis, has been proposed (Cusachs and Steele, 1967; Khan and Kasha, 1970).

Many living cells do, however, contain substances which, by acting as quenchers of singlet oxygen, could minimize the structural and other damage otherwise likely to be caused by this species. Thus it has been suggested that carotenoids could protect photosynthetic organisms against the lethal effects of sensitization by chlorophyll (Foote *et al.*, 1970a), for β-carotene will quench artificially generated singlet oxygen (Foote *et al.*, 1970b; Farmilio and Wilkinson, 1973). It has even been suggested that erythrocuprein quenches the excited states of singlet oxygen, whether these are produced by dismutation of $O_2^{\overline{\cdot}}$ or by other means (Arneson, 1970; Finazzi-Agro *et al.*, 1972; Weser and Paschen, 1972)*. Indeed, it was concluded by Paschen and Weser (1973) that the scavenging of singlet oxygen species is actually the main physiological function of "cupreins" and (as yet by inference only) those other mangano- and iron-proteins that are identifiable as superoxide dismutases. Discovering that simple copper-amino-acid chelates, which possessed substantial superoxide dismutase activity, were virtually unable to quench singlet oxygen, they concluded that this ability to react with $^1O_2^*$ was a consequence of the entatic state of the metal ion in the cupreins (the entatic state is a catalytically-poised state intrinsic to the active site; Vallee and Williams, 1968).

3. Hydrogen Peroxide

Hydrogen peroxide production in biological systems is the consequence of the two-electron reduction of molecular oxygen that can be catalysed by a number of enzymes (generally flavoproteins). Some hydrogen peroxide is doubtless produced by all aerobes, and by facultative anaerobes growing under aerated conditions. In eukaryotes, hydrogen peroxide is formed in peroxisomes and glyoxisomes (Tolbert, 1971), but may even by produced in controlled quantities by mitochondria (Loschen *et al.*, 1973). The flavoprotein enzymes responsible for hydrogen peroxide generation may primarily produce $O_2^{\overline{\cdot}}$ (e.g. aldehyde oxidase and xanthine oxidase; Arneson, 1970) or may form hydrogen peroxide without prior production of superoxide anion (e.g. D-amino-acid oxidase, diamine oxidase, glycollate oxidase and urate oxidase; Rotilio *et al.*, 1973a).

In most aerobic bacteria, hydrogen peroxide is quickly removed by the activity of the haem-containing enzyme catalase. This catalyses the dismutation of hydrogen peroxide to yield molecular (ground state) oxygen and water. Alternatively, a range of peroxidases catalyse oxidation by hydrogen peroxide of various organic reductants (Saunders *et al.*, 1964). A particularly interesting example of such a "protective peroxi-

*But see *Note added in proof*.

dase" is afforded by the seleno-enzyme, glutathione peroxidase (Flohe *et al.*, 1973), which in eukaryotic cells can prevent lipid peroxidation either by its elimination of low-molecular weight hydroperoxides, or by reduction of lipid hydroperoxides. It is worthy of note that some unconventional catalase-like activity can even be displayed by several facultatively anaerobic organisms (e.g. lactic acid bacteria) that do not contain the conventional enzyme (Jones *et al.*, 1964; Johnston and Delwiche, 1965). Furthermore, peroxidation mechanisms which contribute to the removal of accumulated quantities of hydrogen peroxide may actually be developed by some bacteria in response to elevated oxygen concentrations (Seeley and Vandemark, 1951).

Even so, bacterial secretion and accumulation of hydrogen peroxide can be particularly marked on occasions, with varying degrees of ill-effect being experienced by the producer organism. Thus, the haemolysin secreted by *Mycoplasma pneumoniae* was identified as hydrogen peroxide (Somerson *et al.*, 1965; Cohen and Somerson, 1969), possibly produced by a FMN-dependent NADH oxidase operating in the absence of a catalase (Low *et al.*, 1968; Low and Zimcus, 1973). Similarly, the bacteriocin-like activity of *Streptococcus sanguis* against other bacteria has been attributed to its production of hydrogen peroxide (Holmberg and Hollander, 1973). There is no doubt that hydrogen peroxide can be toxic to very many organisms, though how it acts is a matter for speculation (p. 214). In one study, with *Serratia marcescens*, it seemed to cause a repairable injury which affected cell division (Campbell and Dimmick, 1966), whilst its inhibition of key enzymes of energy metabolism has also been reported (de Vries and Stouthamer, 1969). On the other hand, the adverse effect of hydrogen peroxide on the growth of *Salmonella typhimurium* was attributed to its forming toxic adducts with components of the culture media, including carbonyls, amino acids and thymine (Watson and Schubert, 1969). Even catalase-containing and/or peroxidase-containing organisms may not be fully protected against the consequences of hydrogen peroxide production, for there have been reports of $^1O_2^*$ production by peroxide-catalase and peroxide-peroxidase systems (Kasha and Khan, 1970), though Porter and Ingraham (1974) found that only an exceedingly small fraction, if any, of the total molecular oxygen released in the hydrogen peroxide-catalase reaction was present as singlet oxygen.

4. *Superoxide anion*

Superoxide free radicals ($O_2^{-\cdot}$) are generated in the course of interaction with molecular oxygen of various cellular constituents including reduced flavins, flavoproteins, quinones, thiols and iron-sulphur proteins (White and Dearman, 1965; Tollin and Fox, 1967; Orme-Johnson and

Beinert, 1969; Massey *et al.*, 1969; Misra and Fridovich, 1971, 1972a; Misra, 1974). They are also produced during reactions catalysed by enzymes such as aldehyde oxidase, xanthine oxidase (Knowles *et al.*, 1969; Fridovich, 1970; Fried *et al.*, 1973) and D-galactose oxidase (Hamilton *et al.*, 1973; Kwiatkowski and Kosman, 1973). Other reported sources of $O_2^{\overline{\cdot}}$ include autoxidation of mammalian oxyhaemoglobin (Wever *et al.*, 1973), reversal of the usual "biological" route of detoxification (i.e. $O_2^{\overline{\cdot}}$ production by superoxide dismutase acting on hydrogen peroxide plus O_2; Hodgson and Fridovich, 1973a), and the operation of a few "specialist" NAD(P)H-oxidizing systems (e.g. NADPH-cytochrome P_{450} reductase; Pederson and Aust, 1972; Maugh, 1973).

Superoxide anions pose a particular threat to organisms if only because in aqueous media they are the longest lived of all oxygen-derived free radicals (Rabani and Njelsen, 1969); they may be sufficiently long-lasting for exogenously produced $O_2^{\overline{\cdot}}$ to be able to penetrate the bacterial cell envelope (Gregory *et al.*, 1973). Under controlled conditions, superoxide anions may be exploited in hydroxylation reactions carried out by cytochrome P_{450}-containing systems, for example in the fatty acid, hydrocarbon and drug hydroxylations accomplished by liver microsomes (Coon *et al.*, 1972; Kumar *et al.*, 1972), in the catabolic aryl hydroxylations perpetrated by some aerobic bacteria (Gunsalus *et al.*, 1972), or in the epoxidation and hydroxylation reactions catalysed by the ω-hydroxylation system of *Pseudomonas oleovorans* (May and Abbott, 1973; May *et al.*, 1973). Yet uncontrolled generation and accumulation of superoxide anions could prove very damaging, since these radicals can participate in lipid peroxidation, oxidize—SH groups (Jocelyn, 1970), cause destruction of tryptophan residues in proteins (Green and Cruzov, 1968; Michelson, 1973), damage DNA (Lagnado and Sourks, 1958), and react with enzyme-bound $NADH_2$ (Bielski and Chan, 1973). Indeed, testimony to the potential lethality of superoxide radicals is supplied by the suspicion that the dramatic phytotoxic action of bipyridyl herbicides (e.g. paraquat) may be attributable to the generation of $O_2^{\overline{\cdot}}$ (Farrington *et al.*, 1973; Autor, 1974). Furthermore, as we have seen (p. 199), the toxicity of $O_2^{\overline{\cdot}}$ could very well be amplified by secondary reactions leading to the production of hydroxyl readicals and singlet oxygen, and indeed there is a growing body of opinion which believes that much of the biological damage previously attributed to $O_2^{\overline{\cdot}}$ is, in fact, caused by $^1O_2^*$ and $OH\cdot$ (Paschen and Weser, 1973; Gregory and Fridovich, 1974). It is not surprising, therefore, to find that aerobic cells contain a specific defence against $O_2^{\overline{\cdot}}$ in the form of an enzyme (superoxide dismutase) which scavenges these free radicals by catalysing their dismutation:

$$O_2^{\overline{\cdot}} + O_2^{\overline{\cdot}} + 2H^+ \rightleftharpoons H_2O_2 + O_2$$

Even when uncatalysed, this reaction proceeds at a significantly rapid rate ($k = 2{\cdot}1 \times 10^5$ dm^3 mol^{-1} s^{-1}, at pH 7·12) to produce hydrogen peroxide and singlet oxygen (p. 192). But when catalysed by a superoxide dismutase, not only is the rate of destruction of $O_2^{\overline{\cdot}}$ greatly accelerated ($k = 2{\cdot}3 \times 10^9$ dm^3 mol^{-1} s^{-1}, at 25°C, and is largely independent of pH value in the range 4·8 to 9·7; Forman and Fridovich, 1973a), but there is concurrent quenching of the singlet state of the product oxygen. In other words, the enzyme-contrived dismutation of $O_2^{\overline{\cdot}}$ effects rapid scavenging of these free radicals with production of innocuous ground-state oxygen in place of the singlet oxygen which would otherwise be generated (Finazzi-Agro *et al.*, 1972).

Superoxide dismutase is currently assayed by its ability to abort reactions whose accomplishment is dependent on a continuing supply of superoxide anions. The enzyme, for example, will specifically inhibit the following $O_2^{\overline{\cdot}}$-provoked reactions: (i) chemiluminescence of luminol (Greenlee *et al.*, 1962; Hodgson and Fridovich, 1973b); (ii) reduction of ferricytochrome *c* (McCord and Fridovich, 1968; Lippitt *et al.*, 1972), (iii) autoxidation of epinephrine (Misra and Fridovich, 1972b); and (iv) reduction of nitroblue tetrazolium to its insoluble formazan (Beauchamp and Fridovich, 1971). The inhibition of nitroblue tetrazolium reduction by bands of superoxide dismutases, separated on disc-gel electrophoretograms, suggests that the "tetrazolium oxidases" similarly identified in the past by this test are in fact themselves superoxide dismutases (Lippitt and Fridovich, 1973).

The use of such assays has enabled quite highly purified preparations of superoxide dismutases to be obtained from bacterial, fungal, plant and animal sources (Fridovich, 1972). The superoxide dismutases of bovine erythrocytes, heart muscle, liver and brain (McCord and Fridovich, 1969) proved to be identical with proteins whose copper content and distinctive blue-green colour had previously brought them to the attention of several investigators who, being unsure of their functions, had given them the synonyms erythrocuprein, haemocuprein, hepatocuprein, cerebrocuprein or cytocuprein. The copper in bovine superoxide dismutase (erythrocuprein) is probably bound to a histidine residue at the catalytic site, and presumably undergoes valence changes in the course of the reaction (Forman *et al.*, 1973; Forman and Fridovich, 1973b; Stokes *et al.*, 1973; Fee and DiCorleto, 1973; Fielden *et al.*, 1974). Thus, even copper-containing chelates of histidine, lysine or tyrosine have appreciable superoxide dismutase activity (Joester *et al.*, 1972), though they have very low singlet oxygen scavenging ability (Paschen and Weser, 1973). The copper-containing enzyme D-galactose oxidase also has been reported to possess significant superoxide dismutase activity (Cleveland and Davis, 1974). The zinc that is also present in erythrocuprein has not been directly implicated in the catalytic process; it may play some essential role in maintaining the required three-dimensional structure of the enzyme molecule (Rotilio *et al.*, 1972a; Rotilio *et al.*, 1972c, 1973b; Fee, 1973; Forman and Fridovich, 1973b).

Similar superoxide dismutases containing copper or zinc have been isolated from such diverse, but invariably eukaryotic, sources as yeast (Goscin and Fridovich, 1972; Weser *et al.*, 1972), *Neurospora crassa* (Misra and Fridovich, 1972b), the luminous fungus *Pleurotus olearius* (Lavelle *et al.*, 1974), other fungi (Rapp *et al.*, 1973), wheat germ (Beauchamp and Fridovich, 1973) and spinach leaves (Asada *et al.*, 1973). In rat liver cells the CuZn-superoxide dismutase was present in the cytosol, and was therefore located in the same compartment as the soluble $O_2^{\cdot -}$-producing xanthine oxidase (Rotilio *et al.*, 1973a). Other oxidases, which do not generate superoxide anions, were confined either to the peroxisomes (D-amino-acid, glycollate and urate oxidases) or to the microsomes (diamine oxidase). Thus the superoxide anion-generating and scavenging enzymes were appropriately contiguous. With all of the catalase of these cells being contained within the peroxisomes, there would evidently be some danger of product (hydrogen peroxide) inhibition of the superoxide dismutase (Symonyan and Nalbandyan, 1972; Rotilio *et al.*, 1973b; Bray *et al.*, 1974). However, the presence in the cytosol of glutathione peroxidase could ensure the required protection. Chicken-liver and pig-heart cells also contained a CuZn-superoxide dismutase in their cytosols, but also possessed a second and quite distinct Mn-superoxide dismutase in their mitochondria (Weisiger and Fridovich, 1973a, b).

Manganese- or iron-containing superoxide dismutases have been found in all aerobic prokaryotes examined to date, and the cupro-zinc enzyme has not been discovered in any but eukaryotic cells (Fridovich, 1972)*. Two forms of superoxide dismutase have been isolated from *Streptococcus mutans* (Vance *et al.*, 1972) and an iron-containing enzyme has been reported in the marine luminescent bacteria *Photobacterium leiognathi* and *Photobacterium sepia* (*see* Lavelle *et al.*, 1973). Not unnaturally, though, it is the superoxide dismutase activity of *Escherichia coli* that has been most thoroughly studied. *Escherichia coli* B contains two enzymes with superoxide dismutase activity, (1) a cytoplasmic mangano-enzyme whose synthesis is enhanced by increased oxygenation of the growing culture (Keele *et al.*, 1970; Gregory and Fridovich, 1973a, b), and (2) a possibly periplasmic iron-enzyme produced under anaerobic, as well as under aerobic, growth conditions in amounts determined by the iron content of the growth medium (Gregory *et al.*, 1973; Yost and Fridovich, 1973). Discussion of the roles of these bacterial superoxide dismutases is

*But see *Note added in proof*.

best deferred until we consider their possible involvement in protection against oxygen toxicity (p. 217).

C. Mechanisms of Oxygen Toxicity: Current Hypotheses

The various hypotheses that have been proposed over the years to explain the oxygen sensitivity of strictly anaerobic bacteria were recently summarized by Morris and O'Brien (1971). They are: (1) oxygen itself is the toxic agent and is lethal to the cell; (2) normal growth and metabolism of the anaerobe is only possible within certain stringent limits of culture redox potential (E_h), and the presence of free oxygen in the medium is incompatible with the attainment and maintenance of the low E_h value required for growth. Reducing agents exhibit a "protective action" by virtue of their ability to poise the E_h value of a culture medium at a suitably low voltage or to otherwise assist the organism to sustain the favoured redox state; (3) the cell contains key components bearing free —SH groups (e.g. enzymes) whose oxidation by oxygen to the —S—S— form halts growth and metabolism. Added thiols such as cysteine or glutathione might be especially protective of such molecules; (4) preferential reduction of oxygen unproductively consumes the cell's "reducing power" leaving it insufficient to undertake necessary biosyntheses; (5) oxygen indirectly controls cellular activity by determining the intracellular concentration of a key metabolic regulator, presumably a participant in a redox couple liable to direct oxidation by oxygen or in in equilibrium with another such couple; and (6) it is not oxygen *per se* but products of the interaction of oxygen with the cells and/or components of their culture media that are the actual toxic agents. Hydrogen peroxide was initially favoured as the major culprit, though contemporary variants of this hypothesis would direct attention to free radicals (especially superoxide anions and hydroxyl radicals) and to singlet oxygen.

Since many of these theories have been re-iterated in discussions of the toxicity to aerobic organisms of high pressure oxygen (HPO or hyperbaric oxygen), in our appraisal of them we should be prepared to draw not only on findings made with obligate anaerobes exposed to low oxygen concentrations, but also on the very considerable body of information now available concerning the response of microbes of all kinds to hyperbaric oxygen (recently reviewed by Gottlieb, 1971).

(a) *Hypothesis 1: Oxygen itself is the toxic agent.* It was Pasteur (1861) who, finding that his butyric anaerobes lost both their motility and fermentative ability in the presence of air, first proposed that molecular oxygen was itself the lethal agent. Though this hypothesis is not immediately helpful in pinpointing the actual mechanism of oxygen's action, its later proponents have directed attention to the primary encounter between the organism and molecular oxygen, and thus to those enzymes and cellular components capable of directly interacting with molecular oxygen. By inference, whilst conceding that by-products of oxygenation (e.g. hydrogen peroxide or an elevated culture E_h value) can mimic certain of the toxic effects of molecular oxygen, they would claim that none could wholly reproduce its unique action. As to the invariable lethality of exposure to oxygen, so much of course depends on the conditions under which the organism actually encounters the molecular oxygen (e.g. the concentration of dissolved oxygen employed, duration of exposure and nature of the growth medium) that it is impossible to give any sensible ruling (but see Fredette *et al.*, 1967; Brown and Huggett, 1968). It is evident that amongst even obligate anaerobes there exists a complete spectrum of oxygen sensitivity, ranging from those bacteria for which oxygen is apparently bactericidal at very low concentrations, to those which can tolerate limited exposure to air and for which even "atmospheric oxygen" may, in the short term, be reversibly bacteriostatic rather than bactericidal (Smith, 1967; Loesche, 1969; O'Brien and Morris, 1971). This range of oxygen sensitivity is very often illustrated even by different members of the same genus. Thus, in the genus *Clostridium*, whilst *Cl. oedematiens* Type D would be considered to behave as a strict anaerobe, other species, such as *Cl. carnis*, *Cl. histolyticum* and *Cl. tertium*, would appear to be relatively aerotolerant (Willis, 1969). The situation is further complicated by the finding that some organisms, which on primary isolation grow only as strict anaerobes, may become more oxygen tolerant after two or three subcultures (Willis, 1969), whilst even individual organisms in a pure culture may vary enormously in their sensitivity to oxygen (e.g. *Corynebacterim acnes*, quoted by Smith, 1967). However, the thesis that it is oxygen *per se* that is directly damaging has at least proved not to be tenable in the case of a strain of the facultative anaerobe *Lactobacillus plantarum* recently studied by Gregory and Fridovich (1974). This organism, though it lacked catalase, peroxidase or superoxide dismutase, was unduly resistant to oxygenation, retaining its viability over quite long periods of exposure to hyperbaric oxygen. The reason for this oxygen-indifference apparently lay in its inability to utilize oxygen. It would seem therefore that damage ensues from oxygen consumption, and is not an inevitable sequel to the encounter between organism and molecular oxygen.

(b) *Hypothesis 2: Strict anaerobes demand a low culture E_h value which cannot be maintained in the presence of free oxygen.* It is surprisingly difficult to comment on the suggestion that the adverse action of oxygen on anaerobes is due to its elevation of the culture E_h value, since we cannot examine (or define) culture E_h values independently of the agencies that have conspired to produce them. This means that, even if we can poise identical cultures of the one organism at precisely the same E_h value, by using different "dominant" redox agents, we have no assurance that the implications for the organism are in each case identical (p. 186). Knaysi and Dutky (1936) found that a butyric clostridium, which normally in anaerobic culture would develop an E_h value of −265 mV, would not grow in the same medium under aerobic conditions when the E_h value was +335 mV. Yet good growth was possible at +335 mV when this elevated E_h value was maintained with ferricyanide. In a similar experiment, O'Brien and Morris (1971) found that *Clostridium acetobutylicum* would grow anaerobically at the high E_h value of +350 mV when this was maintained by ferricyanide, but, at this same E_h value, growth stopped as soon as the culture was made aerobic (dissolved oxygen concentration of 40 μM). It might reasonably be concluded that it was the free oxygen in the medium, and not the elevated culture E_h value, that was the crucial factor in causing growth inhibition in both of these cases, yet there remains doubt as to whether anything meaningful has been incontrovertibly proved by such experiments. Meanwhile it remains a demonstrable fact that many of the most oxygen-sensitive anaerobes are noted for the very low culture redox potentials which must be established before growth can occur.

(c) *Hypothesis 3: Oxygen oxidizes essential thiols.* Quastel and Stephenson (1926) were inclined to the view that a low culture E_h value was a prerequisite for the growth of strict anaerobes since key redox couples within their cells had to be maintained in the reduced state. Cell components bearing —SH groups were thought to be especially vulnerable since their oxidation by molecular oxygen is accelerated by metal ions (e.g. iron, copper and cobalt). When, thereafter, the catalytic activity of several enzymes (e.g. glyceraldehyde 3-phosphate dehydrogenase) was found to depend on their retaining certain strategic —SH groups in the reduced state, and when the important metabolic roles played by thiol coenzymes (CoA and lipoic acid) were also recognized, it was considered even more likely that some low molecular-weight thiols (e.g. glutathione) may have as their prime function the maintenance of favourable intracellular reducing conditions. The existence of specific enzymes effecting the reduction of these molecules, at the expense of reduced nicotinamide nucleotides, encouraged this view, and increasing knowledge of what may be required of an intracellular anti-oxidant substantiated their qualifications to play such a role. Thus, the sulphur-containing amino acids are only somewhat less effective than their seleno-analogues in breaking free radical chains, decomposing lipid peroxides and repairing damage to —SH proteins (Menzel, 1970).

The suggestion that tetramethylazoformamide (i.e. "diamide") could prove to be a specific reagent for intracellular oxidation of reduced glutathione, cysteine and homocysteine (Kosower and Kosower, 1969) suggested its use as a means of determining what role these —SH compounds might play in obligately anaerobic bacteria. Would their anaerobic oxidation mimic the effects of oxygen on these organisms? Accordingly, "diamide" was used to poise an anaerobic culture of *Cl. acetobutylicum* at an E_h value of +80 mV (an E_h value close to that established by bacteriostatic aerobic conditions). Gratifyingly, growth of the organism was almost completely inhibited, the rate of glucose consumption fell by 30% and, as under aerobic conditions, butyrate production was disproportionately restricted. On the other hand, excretion of pyruvate ceased, in contrast to the enhanced pyruvate production observed under aerobic conditions (O'Brien and Morris, 1971). Unfortunately, the initial hope of being able to deduce from these findings the consequences of specific oxidation of the "protective thiols" was frustrated by the discovery that "diamide" also oxidizes lipoate and coenzyme A, and even non-thiol electron donors such as reduced ferredoxin, FMNH, FADH and, more slowly, NAD(P)H (O'Brien *et al.*, 1970).

Inactivation of —SH (and other) enzymes by oxygen is well documented, and the procedures employed to extract and purify such oxygen-labile proteins resemble those used to culture strict anaerobes (e.g. rigorous exclusion of air, incorporation of reducing agents, such as cysteine and dithiothreitol, in the various buffer mixtures; Wood, 1966). Though more prevalent in anaerobes, such enzymes are also found in aerobic cells of all types (Haugaard, 1968) and, while it is doubtless true that oxidation of intracellular thiols (both protein and non-protein) may contribute to the toxicity of molecular oxygen, there is no good evidence that this is the primary or general cause of the peculiar oxygen-sensitivity of strict anaerobes, or of the toxicity of hyperbaric oxygen to aerobes. In a recent study of the susceptibility to oxidation, by hyperbaric oxygen, of intracellular and surface —SH groups of *Escherichia coli*, for example, it was calculated that the organism normally contained some $4{\cdot}71 \times 10^6$ —SH groups per cell, of which 84% were carried by proteins and about 5% ($2{\cdot}2 \times 10^5$/cell) were present at the bacterial surface. On exposure of the organism to a concentration of oxygen (6·2 atm) that was instantly growth inhibitory, oxidation of these thiol groups proceeded relatively slowly, even the surface —SH groups being oxidized at only 0·2% min^{-1} (Stees and Brown, 1973). It was concluded that, at least in this instance, toxicity to molecular oxygen was not due to generalized oxidation of —SH groups, though preferential oxidation of a relatively small number of crucial —SH groups could not be ruled out. For example, there is evidence that —SH in certain thiol enzymes exists in the more highly nucleophilic (mercaptide ion) form due to interaction with a neighbouring histidine residue (Polgar, 1974) so that it would be unwise to consider all biological —SH groups to be identical in their properties (see Jocelyn, 1972).

(d) *Hypothesis 4: Unproductive "drainage" of the cell's reducing power.* We have seen that the fermentations practised by some strict anaerobes involve "low potential" enzyme systems associated with ferredoxins, or with folate and cobamide coenzymes, which might be particularly vulnerable to oxygenation. What is at first sight much more surprising is that many of the obligate anaerobes contain electron-transport agents capable of utilizing molecular oxygen as a terminal electron acceptor. High concentrations of flavins are frequently present in these organisms, and potent soluble flavoproteins, capable of functioning as NAD(P)H oxidases, have been isolated from a variety of obligate anaerobes (Dolin, 1961a, b). The specific activity of the soluble NADH oxidase of *Cl. acetobutylicum*, already quite high in anaerobically grown cells, was increased some five- to six-fold by growth of the organism in the presence of oxygen (O'Brien and Morris, 1971). The present hypothesis would have us believe that, by acting as an unusually avid electron acceptor, oxygen can preferentially oxidize NAD(P)H at a rate which exceeds that at which it can continue to be generated by fermentation. This short-circuiting of the normally nicely balanced anaerobic cycle of NAD(P)H production and consumption could provoke a depletion of the intracellular pools of reduced electron donors which could be sufficient to halt biosynthesis, and thus growth. Aubel and Perdigon (1940, 1945) indeed found that oxygenation of washed suspensions of *Clostridium saccharobutyricum* provoked a marked change in the fermentation pattern, with proportionately much more acetate being produced at the expense of diminished formation of butyrate. O'Brien and Morris (1971) found precisely the same effect with batch cultures of *Cl. acetobutylicum* whose growth had been halted by the establishment of an aerobic state (viz. $E_h = +100$ mV; dissolved O_2 40 to 50 μM; Fig. 5). Though glucose continued to be consumed at some 30% to 40% of the anaerobic rate, butyrate was not produced, extra acetate (per mol of glucose used) being formed in its place, and a greater quantity of pyruvate being excreted. To maintain the culture in this aerobic state took a comparatively high, and sustained, rate of aeration, and it is reasonable to assume that the lack of butyrate excretion reflected a deprivation of intracellular NADH accountable for by its continuous consumption via the enhanced activity of the potent soluble NADH oxidase. As soon as the culture was returned to the anaerobic state, production of butyrate resumed immediately (Fig. 5). When Brunker and Brown (1971) examined the effects of hyperbaric oxygen (6·2 atm) on the levels of both oxidized and reduced NAD^+ and $NADP^+$, in *Escherichia coli* they found that, whereas the intracellular

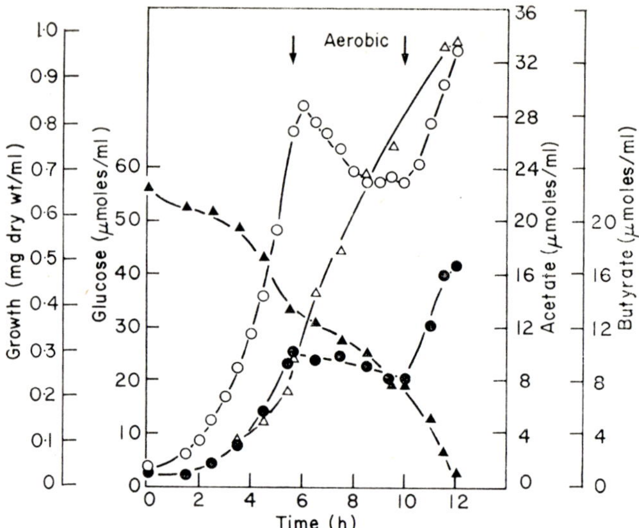

FIG. 5. Effects of aerobic conditions on growth and fermentation of *Clostridium acetobutylicum*. ○, Growth; ▲, glucose in medium; △, acetate in medium; ●, butyrate in medium. The culture initially growing anaerobically, was maintained under aerobic conditions [$E_h = +100$ mV; dissolved oxygen = 40–50 μM] for the period indicated. At the end of this time, anaerobiosis was re-established. From O'Brien and Morris (1971).

pools of the reduced nicotinamide nucleotides and of $NADP^+$ were unchanged after two hours exposure to this bacteriostatic concentration of oxygen, the intracellular NAD^+ concentration had decreased by some 60%. No explanation of this finding was forthcoming, but it stands as a salutary example of the danger of too facile extrapolation from oxygen effects found with any one bacterial species.

(e) *Hypothesis 5: Inhibition effected via oxidation of a key redox couple.* In one sense this hypothesis is a somewhat more sophisticated variant of hypotheses 3 and 4 in which, instead of conceiving of generalized debilitating but non-specific drainage of reducing power, it is supposed that the oxidation of a single redox agent, which regulates the operation of some crucial metabolic process, could be sufficient to prevent growth. There are indeed examples of the oxidation of intracellular redox agents producing potent metabolic inhibitors (e.g. glutathione disulphide as an inhibitor of protein synthesis; Kosower *et al.*, 1972) or of "normal" electron-transport factors acting as allosteric regulators of key enzymes in energy-generating pathways (Weitzman and Jones, 1968), but no precise mechanism has been proposed by those that see this hypothesis as a real possibility. Presumably its attraction lies in the fact that, by assuming ready reversibility of the oxidation/reduction of the metabolic regulator, this theory can account for the very rapid and total recovery on re-establishment of anaerobic conditions that are enjoyed by those anaerobes which have suffered merely a period of oxygen-provoked bacteriostasis. Here again, one must be careful not to generalize from individual cases. One of the most striking and immediate effects of establishing aerobic conditions in a culture of *Cl. acetobutylicum* was the precipitate decrease in the ATP content of the organisms (Fig. 6). Simultaneous assay of in-

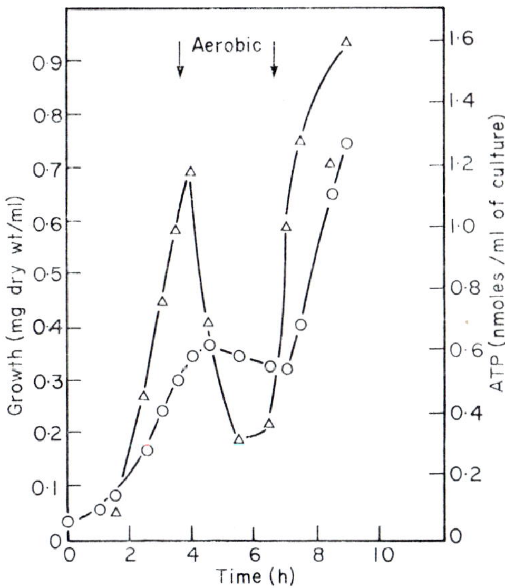

FIG. 6. Fall in intracellular concentration of ATP in *Clostridium acetobutylicum* exposed to aerobic conditions. A growing anaerobic culture was made aerobic for the period shown, and was then returned to its former anaerobic state. o, Growth; △, intracellular ATP. From O'Brien and Morris (1971).

tracellular ADP showed that the decrease in ATP was accompanied by an equal increase in the bacterial content of ADP. Re-establishment of anaerobic conditions immediately caused the ATP and ADP concentrations to revert to their pre-aerobic levels, and (or should it be *as*?) growth resumed (O'Brien and Morris, 1971). There was indeed some curtailment of glucose fermentation in the aerobic phase of bacteriostasis, but when growth was halted by a variety of techniques which similarly curtailed glucose utilization (e.g. deprivation of amino-acids, treatment with alternative oxidants such as "diamide" or metronidazole), the ATP content of the organisms did not fall in a similar fashion. With the indirect evidence also available that oxygen acted to drain away "reducing power" generated by the organism (Fig. 5), it would appear that organisms in the aerobic phase were maintained with an abnormally low NADH/NAD ratio (i.e. high internal E_h value) and an abnormally low adenylate energy charge. Could this be a sufficient explanation of the cessation of macromolecule synthesis and the halting of growth? If so, then the very rapid re-establishment of the usual higher levels of these reducing, and energy-charge, expressions could account for the remarkably rapid restoration of normal growth when the culture was again made anaerobic. If this is too simplistic a view of the situation in this fermentative *Clostridium* (and indeed there is no knowledge of the mechanism whereby molecular oxygen provokes the observed fall in intracellular ATP), there remains the possibility that molecular oxygen may act in some such fashion in other types of anaerobic organism. For example, in photosynthetic bacteria the "ATP energy pressure" plays an important role in the generation of reduced nicotinamide nucleotides at the expense of exogenous reduced substrates and, in turn, reduced nicotinamide nucleotides probably play an important regulatory role in influencing the conversion of light energy to ATP (Gest, 1972). It is by no means improbable that, besides its longer term effects on such organisms (e.g. inhibition of bacteriochlorophyll biosynthesis), molecular oxygen may act by disturbing this nicely balanced cyclical control system to the detriment of ATP synthesis. Or again, it has been suggested that all primitive anaerobes may resemble *Clostridium pasteurianum* and *Desulfovibrio desulfuricans* in possessing an inorganic pyrophosphatase that is reductant-activated (Ware and Postgate, 1970). Inorganic pyrophosphate is produced by many nucleotide triphosphate-consuming biosynthetic reactions (e.g. generation of adenylate amino-acids and nucleotide sugars), and it has been assumed that such reactions are in large measure "pulled" by the pyrophosphatase-catalysed removal of PP_i, with the regeneration of inorganic phosphate. Ware and Postgate (1970) argued that possession of a reductant-activated pyrophosphatase is likely to prove of advantage to the strict anaerobe confronted with oxygen. With the rise in intracellular E_h value causing decreased activity of the pyrophosphatase, several quite crucial biosynthetic processes could not proceed, and unprofitable expenditure of ATP would be prevented so long as conditions remained unpropitious for growth (i.e. a reductant-activated pyrophosphatase would, under conditions of oxygen stress, help to conserve ATP for more essential maintenance purposes). Yet what would thus appear to be an important part of the organism's armoury could prove to be its Achilles' heel, for it is possible to take the contrary view, viz. that its possession of a primitive reductant-dependent pyrophosphatase is a major cause of its sensitivity to oxygen.

One could multiply such examples of the manner in which oxygen might profoundly affect an anaerobic organism via its action on a single redox couple whose oxidation has growth-inhibitory consequences; it is indeed the diversity of such proposals that chiefly sustain this hypothesis.

(f) *Hypothesis 6: It is not molecular oxygen but products of its utilization that are toxic to obligate anaerobes.* The proposition that oxygen toxicity is a function not of molecular oxygen, but of secondary products of oxygen's interaction with the living cell, deserves special attention since it has been the source of the most durable unitary hypothesis and is currently very much in vogue. The theory that obligate anaerobes commit suicide on exposure to oxygen by building up lethal concentrations of hydrogen peroxide is generally attributed to McLeod and Gordon (1923). The fact that species of *Clostridium* were in general unable to synthesize haem, and hence cytochromes and catalase, had suggested to several investigators that such organisms must inevitably produce and accumulate hydrogen peroxide in toxic quantities when exposed to oxygen. The view that the common basis of obligate anaerobiosis could be deprivation of a single detoxifying agent (viz. catalase) was persuasive in its elegant simplicity. McLeod and Gordon themselves held to this "peroxide theory" of oxygen toxicity over a period of thirty years (Gordon *et al.*, 1953). Hydrogen peroxide can indeed be produced during oxygen consumption by the flavin-rich cell-free extracts, washed suspensions and (less often) growing cultures of many facultative and obligate anaerobes. It can be demonstrated to be both bacteriostatic and bactericidal, and to inhibit the activity of key enzymes in a variety of organisms (p. 201). From the chemistry of peroxides it is evident that they can cause devastating cellular damage (p. 195), and it comes as no surprise to find that catalase is a normal component of aerobic eukaryotic cells (p. 200). Yet even in the case of species of *Clostridium* known not to contain a conventional catalase, evidence of detrimental hydrogen peroxide production is far from satisfactory (see review by Morris and O'Brien, 1971). It is not unusual for hydrogen peroxide production to be demonstrable with washed suspensions of organisms, only to be absent from growing cultures exposed to molecular oxygen. Even in the absence of catalase and peroxidase activities, destruction of hydrogen peroxide is still possible by reducing compounds discharged by the organisms into the medium (e.g. SH compounds and/or pyruvate; Grunberg Manago *et al.*, 1952; Mickelson, 1966). Such "normal" extracellular destruction of intracellularly produced hydrogen peroxide need not necessarily be very effective but in many instances even addition of catalase to the culture affords no significant protection against the adverse effects of aeration (Mateles and Zuber, 1963). Some species of *Clostridium* may in fact contain peroxidase(s) (Mallin and Seeley, 1958), but none of those species in which some form of catalase has been detected in low activity was the better able to grow aerobically (Smith, 1967). In their study of six strains of *Bifidobacterium*, all of which were incapable of aerobic growth on agar plates, de Vries and Stouthamer (1969) found that they divided into three groups. Two strains were capable of growing in aerated liquid cultures, and no hydrogen peroxide was accumulated. Since both strains were capable of producing hydrogen peroxide it was assumed that the weak catalase activity of one strain, and chemical destruction of hydrogen peroxide in the medium, together conspired to prevent build-up of peroxide. The two strains of the second group, however, did accumulate hydrogen peroxide and this appeared to inhibit growth by inactivating the fructose 6-phosphate phosphoketolase that is a key enzyme in the glucose fermentation pathway employed by the organism. The remaining two strains were the most oxygen-sensitive; they did not grow in aerated media and did not accumulate peroxide, and it was concluded that oxygen prevented their growth by maintaining a too high culture E_h value. Even if we accept that the primary cause of oxygen toxicity in the case of two of these six strains was the accumulation of hydrogen peroxide, the intraspecies variation of response to oxygen revealed by this one study alone holds out little hope that the "peroxide theory" can be the unitary explanation of oxygen toxicity envisaged by McLeod and Gordon (1923). The universality of the hypothesis that oxygen toxicity results from its reduction to hydrogen peroxide which, in the absence of catalase, is accumulated in lethal quantity, was in any event called into question by the finding that the strictly anaerobic propionibacteria produce a potent catalase (Sherman, 1926); other strict anaerobes that similarly display catalase activity can also not tolerate exposure to air (Prevot and Thouvenot, 1952). Conversely, several organisms capable of aerobic growth do not contain conventional catalase activity, for example *Agromyces ramnosus* (Jones *et al.*, 1970). *Streptococcus faecalis* and *Lactobacillus plantarum* (McCord *et al.*, 1971), though at least in the case of some strains of these last two species non-haem-type catalatic activity has been reported (Jones *et al.*, 1964; Johnston and Delwiche, 1965). Such muted criticisms of the "peroxide theory" as were made in the past have not prevented its being cited in almost every textbook of bacteriology as *the* explanation of the susceptibility of strict anaerobes to oxygen; perhaps this is in large measure due to the fact that the experimental evidence brought forward to confound the theory has often been as suspect as some of the claims made in its defence.

A most significant alternative to the "peroxide" theory has recently been suggested by Fridovich and his colleagues (Fridovich, 1972). They concluded that the production of superoxide anions, which invariably accompanies oxygen utilization by living cells of all types, is potentially hazardous, and that anaerobes are particularly at risk because they do not possess the enzyme *superoxide dismutase*, which scavenges these free radicals as they are formed in both facultative and obligate aerobes (p. 201). In a survey of the distribution of catalase and superoxide dismutase in various microorganisms, McCord *et al.* (1971) discovered that strict anaerobes possessed no superoxide dismutase and generally no catalase activity. All aerobes which possessed cytochromes also contained both superoxide dismutase and catalase. Their findings (Table 2) were indeed impressively clear cut and would seem to support their contention that the prime physiological function of superoxide dismutase is protection of oxygen-utilizing organisms against the potentially harmful effects of $O_2^{\cdot-}$ generated by univalent reduction of molecular oxygen. Especially interesting was the finding that their strain of *Lactobacillus plantarum*, though aerotolerant, contained no superoxide dismutase. What at first

TABLE 2. Superoxide dismutase and catalase contents of a variety of micro-organisms (from McCord *et al.*, 1971)

	Superoxide dismutase (units/mg)	Catalase (units/mg)
Aerobes:		
Escherichia coli	1·8	6·1
Salmonella typhimurium	1·4	2·4
Halobacterium salinarium	2·1	3·4
Rhizobium japonicum	2·6	0·7
Micrococcus radiodurans	7·0	289
Saccharomyces cerevisiae	3·7	13·5
Mycobacterium sp.	2·9	2·7
Pseudomonas sp.	2·0	22·5
Strict Anaerobes:		
Veillonella alcalescens	0	0
Clostridium pasteurianum, sticklandii, lentoputrescens, cellobioparum, barkeri	0	0
Clostridium acetobutylicum	0	—
Clostridium sp. (strain M.C.)	0	0
Butyrivibrio fibrisolvens	0	0·1
N2C3*	0	< 0·1
Aerotolerant Anaerobes:		
Butyribacterium rettgeri	1·6	0
Streptococcus fecalis	0·8	0
Streptococcus mutans	0·5	0
Streptococcus bovis	0·3	0
Streptococcus mitis	0·2	0
Streptococcus lactis	1·4	0
Zymobacterium oroticum	0·6	0
Lactobacillus plantarum	0	0

* N2C3 is an unclassified celluloytic Gram-negative rod isolated from the rumen of an African zebu steer.

sight threatened the universality of their "superoxide theory" served to strengthen it when they further discovered that this strain of *L. plantarum* consumed no oxygen under their test aerobic conditions. Being oxygen-indifferent, it would not require to be protected against the hazardous by-products of oxygen reduction. In further work with this organism Gregory and Fridovich (1974) found it to be unusually resistant to hyperbaric oxygen, even though it lacked catalase and peroxidase as well as superoxide dismutase. It was, however, killed by exposure to photochemically generated $O_2^{\cdot-}$ whose lethality wasaugm ented by hydrogen peroxide.

Fridovich and his colleagues have been careful to point out that other factors besides lack of superoxide dismutase (e.g. presence of oxygen-labile enzymes or autoxidation of membrane lipids) could contribute to an organism's inability to survive in the presence of oxygen. They have also foreseen the main criticism that may be made of their unitary hypothesis, namely that were an organism obligately anaerobic due to some other cause, there is no good reason why it should "carry the genetic burden" of catalase and superoxide dismutase.

Gratifying confirmation of the invariant possession of superoxide dismutase by oxygen-utilizing bacteria has come from the examination of those aerobic bacteria which, by being destitute of any catalase-like activity, have long been irritating exceptions to the peroxide theory. Thus *Bacillus popilliae*, a cytochrome-containing aerobe without catalase or peroxidase activity (Pepper and Costilow, 1965) has been rehabilitated by the discovery that it possesses a potent superoxide dismutase (Costilow and Keele, 1972; Yousten *et al.*, 1973). Furthermore, there have so far been no reports of exceptions to the precept that a strict anaerobe will prove to be devoid of superoxide dismutase (e.g. *Bacteroides melaninogenicus* and *Bacteroides ruminocola* lack the enzyme; Fridovich, 1972). It would be interesting though to learn whether types of non-fermenting anaerobes not tested by McCord *et al.* (1971) will also obey this rule (e.g. the cytochrome-containing anaerobic sulphate reducers and photosynthetic anaerobes)*.

Synthesis of superoxide dismutase by both *Streptococcus faecalis* and *E. coli* B was induced by molecular oxygen. Thus *S. faecalis* grown under 20 atm oxygen had sixteen-times more superoxide dismutase than did anaerobically grown cells, but no conventional catalase. Exposure of a growing culture of *E. coli* B to 5 atm oxygen increased its superoxide dismutase activity some twenty-five-fold though its catalase activity was not increased (Gregory and Fridovich, 1973a). This ability to produce populations of both organisms, containing low and high activities of superoxide dismutase, was exploited to determine whether possession of

*See *Note added in proof.*

increased amounts of the enzyme conferred any protection to toxic concentrations of oxygen. In both cases, it was found that organisms with high levels of superoxide dismutase fared better on exposure to hyperbaric oxygen than did anaerobically grown cells with lesser activities of the enzyme. Superoxide dismutase activity in *Bacillus subtilis* was not induced by oxygen, whereas the catalase content could be increased twelve-fold by this means. It was therefore possible with this organism to determine whether cells with identical superoxide dismutase activity, but elevated levels of catalase, would prove more resistant to oxygen; in the event, the increase in catalase content conferred no such benefit (Gregory and Fridovich, 1973b). Of the two distinct superoxide dismutases produced by *E. coli* B, the iron-containing enzyme is possibly a periplasmic protein and its synthesis, though not induced by oxygen, could be curtailed by depriving the cells of iron during growth. It was the intracellularly located manganese-containing enzyme whose synthesis proved responsive to the oxygen concentration prevailing during growth. In a pleasing experiment, Gregory *et al.* (1973) independently manipulated the levels of these enzymes in *E. coli* B so as to produce two types of cells; (1) those grown "anaerobically with high iron", possessing maximum activity of the periplasmic iron-enzyme and low activity of the intracellular mangano-enzyme; and (2) those grown "oxygenated in low iron medium" and having the two enzymes in reversed ratio of activities. Type (1) cells, containing a preponderance of the periplasmic iron-protein superoxide dismutase, survived exposure to exogenous $O_2^{\cdot-}$ free radicals much better than did the second type of cells which were rich only in the intracellular mangano-enzyme. Addition to the medium of purified (bovine) superoxide dismutase protected the latter organisms, preventing their rapid death (Fig. 7). On the other hand, only those organisms with high levels of the mangano-enzyme demonstrated enhanced resistance to hyperbaric oxygen and to streptonigrin, an antibiotic whose effectiveness is markedly enhanced by molecular oxygen and whose action probably involves intracellular generation of $O_2^{\cdot-}$ (White and Dearman, 1965; White *et. al.*, 1971). It is reasonable to conclude that the intracellular location of the mangano-enzyme is particularly appropriate if its function is to scavenge endogenously produced superoxide anions while the iron enzyme's periplasmic location is similarly propitious if its function is to protect the organism against exogenous $O_2^{\cdot-}$. To test this proposition will require the selection of mutant strains of *E. coli* B specifically deficient in either or both of the superoxide dismutases; mutants with a temperature-sensitive defect in their ability to accumulate superoxide dismutase have already been isolated and have been found to display a comparable defect in their ability to grow in the presence of oxygen (McCord *et al.*, 1973). However, not all strains of *E. coli* respond

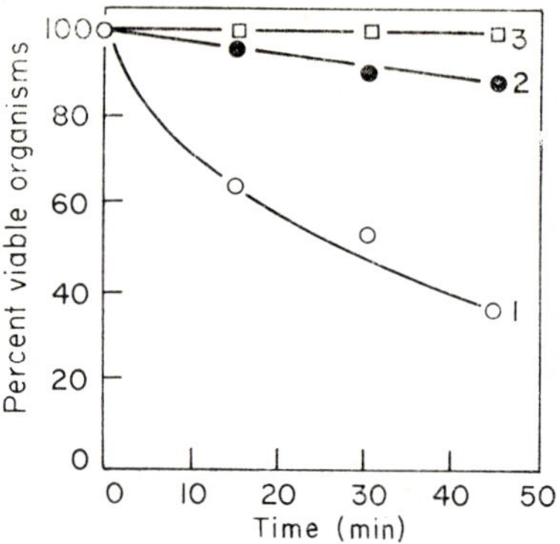

FIG. 7. Effects on viability of *Escherichia coli* B of exogenous, enzymically generated $O_2^{\cdot-}$. *Escherichia coli* B, grown either in an iron-deficient, aerated medium, or in iron-rich, anaerobic medium, was diluted to 10^7 cells per ml in 0·002 *M*-potassium phosphate, 6×10^{-5} *M*-EDTA, 0·025 *M*-sodium carbonate, 1·2% glucose, 5×10^{-4} *M*-xanthine and $3{\cdot}67 \times 10^{-8}$ *M*-xanthine oxidase at pH 9·0, and was incubated with stirring and aeration at 25°C under the following conditions: 1. o—o, iron-deficient organisms; 2. ●—●, iron-rich organisms; 3. □—□, iron-deficient or iron-rich organisms in the presence of 26·7 μg bovine superoxide dismutase per ml. At intervals, samples were withdrawn and a viable count of surviving organisms was obtained by normal dilution and plating procedures. From Gregory *et al.* (1973).

to elevated concentrations of oxygen by producing greater amounts of superoxide dismutase. Thus, Gregory *et al.* (1974) found that *E. coli* K-12 *his*⁻ formed no more superoxide dismutase when growing under one atmosphere of oxygen than when growing anaerobically, though its contents of catalase and peroxidase were substantially increased. Comfortingly, growth in the presence of one atmosphere of oxygen also did not confer upon this K-12 strain of *E. coli* any resistance towards 20 atmospheres of oxygen. When the yeast *Saccharomyces cerevisiae* var. *ellipsoideus* was grown under 100% oxygen, it contained 6·5-times more superoxide dismutase and 2·3-times more catalase than when it was grown anaerobically (Gregory *et al.*, 1974). The activities of its cyanide-sensitive (cupro-zinc, cytosol-located) and cyanide-insensitive (manganese, mitochondrial) superoxide dismutases were elevated to an equal degree by growth under oxygen, and the resulting organisms were more tolerant of hyperbaric oxygen. Similarly in the case of the cyanobacterium *Anacystis nidulans*, growth in the presence of air induced the synthesis of its superoxide dismutase and promoted resistance to photo-oxidative death (Abeliovich *et al.*, 1974).

The ubiquity and strategic intracellular distribution of the superoxide dismutases of oxygen-utilizing cells would seem to confirm that superoxide production poses a potentially lethal threat to all living organisms

(Fridovich, 1972; Rotilio *et al.*, 1973a; Zimmermann *et al.*, 1973). Yet it may not, in fact, be the accumulation of superoxide anions *per se*, but the singlet oxygen that would be formed by their non-enzymic dismutation (p. 192), or the OH· free radicals formed by their interaction with hydrogen peroxide (p. 194), that might be the most devastating penalty of not possessing an $O_2^{\cdot-}$-scavenging (and singlet oxygen-preventing) superoxide dismutase. Certainly there is evidence enough that both $^1O_2^*$ and OH· are hyper-reactive species that it would be most undesirable to harbour in any quantity within a living cell (p. 199), and evidence is mounting that superoxide dismutase may be of especial value because it yields no singlet oxygen in its catalysis of $O_2^{\cdot-}$ dismutation (Paschen and Weser, 1973).

Even if the greater sensitivity of strict anaerobes to oxygen is eventually proved to be not entirely due to their complete lack of superoxide dismutase, the emergence of $O_2^{\cdot-}$ as a villainously toxic by-product of reduction of molecular oxygen is a notable advance in that it draws attention to the possible damage wrought by oxygen-derived free radicals of all types. The possible importance of such free radicals in mediating oxygen's toxic action was first explored not so much by "anaerobiologists", who tended to be mesmerized by the peroxide theory, but by mammalian physiologists and those who discerned distinct resemblances between the effects on living cells of ionizing radiation, on the one hand, and hyperbaric oxygen on the other (Hedén and Malmborg, 1961; Gerschman, 1964; Haugaard, 1968; Menzel, 1970). I have already briefly discussed some of the means of initiation and propagation of these free radicals and have considered the sorts of cellular damage they may cause (e.g. lipid peroxidation in membranes, disruption of DNA; p. 196); some recent review articles more fully examine the bases of what might be termed free radical pathology (Demopoulos, 1973a, b; Pryor, 1973; Tappel, 1973). Suffice it to conclude that the potential threat posed by $O_2^{\cdot-}$ could well be amplified by its interaction with hydrogen peroxide to yield OH·, and it would doubtless be in the interest of the aerobe to prevent hydrogen peroxide as well as $O_2^{\cdot-}$ accumulation. Indeed, Gregory and Fridovich (1974), in their study of their oxygen super-tolerant strain of *Lactobacillus plantarum*, discovered that the lethality of $O_2^{\cdot-}$ was augmented by hydrogen peroxide and mitigated by mannitol (which scavenges OH· but has no action on $O_2^{\cdot-}$ or hydrogen peroxide). This suggested that the actual lethal agent was primarily OH· generated by the interaction of $O_2^{\cdot-}$ with hydrogen peroxide. Hydroxyl free radicals produced during flavin enzyme activity have been implicated in the peroxidation of lysosomal membranes in eukaryotic cells (Fong *et al.*, 1973). However, despite the mounting evidence which implicates free radicals in many aspects of oxygen-toxicity, we cannot assume that the lethal action of oxygen on bacteria is invariably, or solely, due to such agents. For example, in studies on oxygen-induced loss of viability of airborne or freeze-dried *E. coli*, it was evident that water loss from the bacteria accentuated their sensitivity to oxygen, which was attributed to reaction of molecular oxygen with resultant free radicals to produce damaging peroxyradicals (Benbough, 1969). Yet this conclusion has recently been challenged by Cox and Heckly (1973) who found no correlation between the production of oxygen-induced free radicals and loss of viability of freeze-dried *E. coli* B or *Serratia marcescens* 8UK.

D. Summary

Surveying the multiple mechanisms whereby exposure to molecular oxygen can cause cellular damage and metabolic malfunctions, it would appear that no one unitary hypothesis can adequately account for the oxygen sensitivity of all obligate anaerobes in all media. Yet by using elements abstracted from all of the main hypotheses it is possible to construct a reasonably comprehensive explanation of the major consequences of oxygenation. Preferential reduction of oxygen by an anaerobe can afford it some temporary degree of protection, but only at the expense of the diversion of "reducing power". Reversible metabolic disfunctions might result from the consequent elevation of intracellular E_h value (e.g. drop in adenylate energy charge and halting of biosyntheses), but in other respects the reduction of molecular oxygen is accomplished in a relatively innocuous manner. Full recovery may still be possible if return of the organisms to anaerobic conditions is not unduly delayed. This might be thought of as the bacteriostatic Phase 1 of oxygen action. Once the reductive defences of the organism are overwhelmed by continued (excessive) exposure to oxygen, the free ingress of oxygen into the cell may cause irreversible damage by a variety of means. Direct interaction of molecular oxygen with oxygen-labile cell components (including —SH compounds), univalent reduction to yield $O_2^{\overline{\cdot}}$ (in the course of reaction with flavins, flavoproteins and ferredoxins), consequential formation of singlet oxygen, concurrent production of hydrogen peroxide and other peroxides, generation of hydroxyl free radicals, all these (and possibly other agencies) may conspire to cause potentially lethal damage. This is the devastating Phase 2 of oxygen toxicity. Concentration of molecular oxygen within cell membranes (by virtue of its greater solubility in lipophilic media) could prove exceptionally hazardous. In such a location, singlet oxygen and oxygen-derived free radicals could wreak maximum havoc, causing such major structural damage as to seriously impair membrane functions; loss of K^+ ions and cellular metabolites, and diminished rates of substrate uptake, would follow (Young, 1968; Allen *et al.*, 1973).

The types of damage catalogued under Phase 2 of oxygen's action will be common to all living cells receiving an excessive dose of molecular oxygen. The merit of the "superoxide theory", like that of the "peroxide theory" before it, is that it provides a reasonable explanation for the super-sensitivity of the strict anaerobe. Whilst deprivation of catalase and peroxidase, with accumulation of hydrogen peroxide on exposure to oxygen need not, as was once thought, prove invariably fatal (possibly due to decomposition of hydrogen peroxide by non-conventional catalases, medium components or, in mixed cultures, by neighbouring organisms), it would doubtless prove additionally disadvantageous to an organism devoid of superoxide dismutase.

Even assuming that anaerobes and aerobes can be segregated on the basis of whether or not they contain a singlet-oxygen-preventing superoxide dismutase (and an even greater range of anaerobes must first be examined to prove this point), one would still have to explain the tremendous spectrum of sensitivity to oxygen that is found amongst the superoxide dismutase-less species. Presuming, for example, that both *Butyrivibrio* sp. and *Clostridium perfringens* are devoid of superoxide dismutase, why should exposure to air of a thin layer of a broth culture of *Butyrivibrio* result in the death of 99·9% of the organisms in six minutes, whilst *Cl. perfringens* in broth culture can stand similar exposure for as many hours with little decrease in viability (Smith, 1967)?* Does the difference between the strict and moderate anaerobe reside in the fact that the strict anaerobe possesses no means of diverting "reducing power" to mop up even small quantities of oxygen so that Phase 1 of oxygen action is virtually non-existent? Perhaps the strict anaerobe, by virtue of its content of the requisite enzymes and/or sensitizers, produces HO_2^-, $O_2^-\cdot$, ${}^1O_2^*$ and $OH\cdot$ at more rapid rates than does the moderate anaerobe. Alternatively, these damaging by-products of oxygen-utilization may be produced at similar rates by both extreme and moderate anaerobes, but the former contain components that are notably more susceptible to attack by these reagents. In the search for a unitary hypothesis, such questions have not yet attracted the attention that they warrant. Most "anaerobiologists" have sought solace in the statement that the strict anaerobe demands an extraordinarily low culture E_h value, whilst the moderate anaerobe can flourish at more positive oxidation-reduction potentials. Yet we have seen that the culture E_h value is not so much a determinant, but a mirror of events within the culture. Possibly those that place considerable emphasis on the low culture E_h values for the strictest anaerobes are inferring that the first of my suggestions is the most likely (i.e. that such organisms having no "spare" reducing power are immediately plunged into Phase 2 by exposure to even small concentrations of oxygen). Yet when an extreme anaerobe

*We have recently found a strain of *Cl. perfringens* to contain some superoxide dismutase (see *Note added in proof*).

discharges reducing power during growth (as hydrogen gas) its supersensitivity to oxygen remains something of a paradox.

McElroy and Seliger (1962) proposed that the bioluminescent systems now found in a wide variety of organisms have their origin in detoxifying processes for the removal of oxygen. These may have appeared in early anaerobic organisms when evolutionary experimentation with the use of various organic substances to remove molecular oxygen by direct reduction led to the formation of an excited phase that could emit light. In McElroy and Seliger's view, it was the struggle to maintain anaerobic conditions that led to the selection of organisms with specific oxidases (luciferases) catalysing the rapid removal of oxygen at very low oxygen tensions.

The thesis that oxygen-toxicity in anaerobes can be delayed, if not totally prevented, by diversion of reducing power to scavenge molecular oxygen, has its parallel in some aerobic bacteria that use this mechanism to shield key oxygen-labile components from direct contact with oxygen. For example, when *Azotobacter* species are fixing molecular nitrogen, a considerable proportion of the NADH generated during aerobic growth is sacrificed in order that free oxygen is denied access to the oxygen-sensitive nitrogenase (Drozd and Postgate, 1970; Jones *et al.*, 1973). The aerobic organism *Mycobacterium flavum* 301 similarly possesses an oxygen-labile nitrogenase. Its activity is inhibited at ordinary (atmospheric) oxygen tension unless fixed nitrogen is first supplied to enable the culture to achieve a sufficiently dense population whose respiratory activity is then adequate to remove the molecular oxygen (Hill *et al.*, 1972). This situation is analogous to that of the strict anaerobe whose growth is ensured only if a sufficiently large inoculum is employed.

The ability to provide strictly localized protection for key intracellular components is also a real possibility in anaerobes and aerobes alike. There are doubtless many examples of multiple enzyme complexes in which one or other component has a protective or activating reducing role (e.g. the pyruvate formate lyase complex of *E. coli* studied by Knappe *et al.*, 1969). There is also the example of conformational protection of the nitrogenase of *Azotobacter* whereby a reversible structural change in the nitrogenase complex renders the oxygen-sensitive sites less accessible to oxygen at the expense of temporary inactivity of the nitrogenase (Dalton and Postgate, 1969). In the light of all such possibilities, it would be a mistake to think of oxygen-toxicity as an "all or none" effect. As the diagnostic bacteriologist has oftentimes empirically discovered, even the strictest anaerobe can be aided in its attempts to moderate oxygen's potential toxicity by suitable modifications to the growth medium.

Finally, we have still to explain what is perhaps the greatest mystery of all, viz. why strict anaerobes can in nature inhabit locations which would seem to pose considerable dangers of oxygen-toxicity (p. 173). The nature of the chemical or physical factors that afford them protection in these natural environments has been little studied, and our ignorance of the behaviour of anaerobes, in mixed culture with facultative and obligate aerobes, is almost total (Smith, 1967).

VI. Obligate Anaerobes as Primitive Organisms

A. Phylogenetic Considerations

The oldest lithified microfossils unearthed to date have been bacteria-like structures located in those South African Fig Tree and Onverwacht cherts that are about $3{\cdot}1 \times 10^9$ and 3·4 to $3{\cdot}5 \times 10^9$ years old, respectively. This allows us to suppose that prokaryotic life forms could have originated some 4 to $3{\cdot}5 \times 10^9$ years ago (Schopf, 1970; Ponnamperuma, 1972). By computerized analysis of the divergences of composition of contemporary molecules of cytochrome *c* and tRNAs from various organisms, with careful assessment of the number of genetic changes involved, it has been estimated that the genetic code evolved approximately $3{\cdot}4 \times 10^9$ years ago (Dayhoff, 1971). Some of the microstructures found in the black Fig Tree cherts ($3{\cdot}1 \times 10^9$ years) appear to be similar to present day blue-green algae, and certainly the younger microfossils found in Canadian Gunflint iron formation (1·6 to 2×10^9 years), Australian Bungle Bungle dolomite ($1{\cdot}5 \times 10^9$ years) and the Australian Bitter Springs deposits (9×10^8 years) are composed predominantly of cyanobacteria (blue-green algae), with traces of eukaryotic algae appearing as early as $1{\cdot}3 \times 10^9$ years ago (Schopf, 1970). Yet in the Early Precambrian era ($>2{\cdot}75 \times 10^9$ years ago) the primitive atmosphere was still wholly anaerobic. Its initial major components had been probably a highly reducing mixture of methane and hydrogen, with lesser amounts of water, nitrogen, hydrogen sulphide, ammonia, argon and helium (Miller and Urey, 1959). Escape of the light gas, hydrogen, into interplanetary space, and consequential quite rapid decline in the proportions of methane and ammonia, would very likely mean that, at about the time when life first made its appearance on earth, the major atmospheric gases were nitrogen with carbon dioxide, argon and water vapour (Holland, 1962; Cloud, 1968). The pre-biotic course of chemical evolution in these anaerobic conditions was powered by a diversity of energy sources, including electrical discharges in the atmosphere, volcanic and hot spring activities, solar heat and, of course, ultraviolet irradiation which would have acted on the atmosphere and the oceans to generate reactive free radicals and ions. The quantity, and enormous variety, of complex organic chemicals that could have been synthesized in this way over a period of hundreds of millions of years have been well described in recent review articles (Calvin, 1969; Kenyon and Steinman, 1969; Buvet and Ponnamperuma, 1971; Kimball and Oro, 1971; Rutten, 1971; Ponnamperuma, 1972; Ponnamperuma and Gabel, 1974). Under these circumstances, it seems more than likely that the earliest life form would have been a strictly anaerobic, fermentative prokaryote. Possibly the diversity of fermentative capacities that we find represented in a contemporary genus such as *Clostridium* reflects the ready evolution of variants of this primitive anaerobe, each feasting on different components of the primaeval "soup". Such little oxygen as was formed in this period was probably chiefly the result of photolysis of water, and the element was not formed at a sufficient rate to accumulate in any significant concentration in the atmosphere. Cloud (1972) concluded that, prior to about $1{\cdot}9 \times 10^9$ years ago, the earth's atmosphere contained only trivial and transient amounts of free oxygen. However, due to the action of intense ultraviolet-irradiation, some ozone was produced which, rising in the atmosphere, gradually accumulated in an ozone belt and, by acting as an ultraviolet filter, lowered the intensity of low-wavelength irradiation arriving at the surface of the earth (Gaffron, 1962). It was only when the potential mutagenic hazard presented by such ultraviolet-irradiation had been mitigated by the establishment of the ozone shield, that life could emerge from the waters and develop on land (Wald, 1964). But presumably long before this happened, the ozone "shield" had made its presence known in another way. As the build-up of fermentable organic substrates in the pre-ozone era had largely been due to the action of high-intensity solar ultraviolet-irradiation, when this was diminished by the action of the ozone shield, the regeneration of these materials by chemical means would have been sorely curtailed. Competition for fermentable substrates would have increased, and this period might have seen the first "famine" to have swept this planet. The resultant evolutionary stress encouraged the emergence of anaerobic bacteria utilizing an alternative source of energy, and employing inorganic reductants as sources of reducing power (viz. the strictly anaerobic photosynthetic bacteria). These in turn were the progenitors of the cyanobacteria, and of the oxygen-generating photosynthesis which, in various forms, was to be retained thereafter in the green algae and higher plants (Stanier, 1974). Although the Pre-Cambrian microfossil record suggests the early appearance of cyanobacteria, free oxygen might have been only slowly released to the atmosphere because of the large quantities of autoxidizable inorganic materials (e.g. ferrous ions) present in solution in the oceans. Activities of the photosynthetic sulphur bacteria could have contributed to the creation of the "sulphuretum" wherein sulphate again was reduced to elemental sulphur and hydrogen sulphide by the action of obligate anaerobes such as

Desulfovibrio (Peck, 1974). As the concentration of molecular oxygen in the atmosphere mounted (albeit slowly), oxygen-tolerant, and eventually oxygen-respiring, organisms made their appearance, thus laying the foundation for a burgeoning aerobic biosphere. According to Cloud (1972), the geological evidence suggests a marked increase in atmospheric oxygen about 680 million years ago which would account for the apparently abrupt appearance of metazoa at about this period (see also Berkner and Marshall, 1964). In his delightful review, Smith (1967) describes with biblical imagery both the origin of anaerobes, and the later genesis of aerobic, photosynthetic and respiring organisms: "In the warm, dark, shallow sea, some anaerobic form of life developed and thrived, generation following generation, for hundreds of thousands of years", and, "Once the primitive green plants had started to develop, once the first bubble of oxygen rose through the water of the sea, the surface of the earth began to change and to become suitable for the aerobic forms of life, for mice and men and molds, for all those things that must have gaseous oxygen or die."

The succession so envisaged would be as summarized in Fig. 8, and is the more persuasive because it largely accords with phylogenetic schemes based on other essential attributes (Klein and Cronquist, 1967) and because, in any present day aerobe, all of the peculiarly "aerobic processes" appear to be grafted onto the basic stock of anaerobic cellular metabolism (Wald, 1964). Not all of the present day anaerobes need of course be filial descendants of the original "patriarchal" prokaryotes;

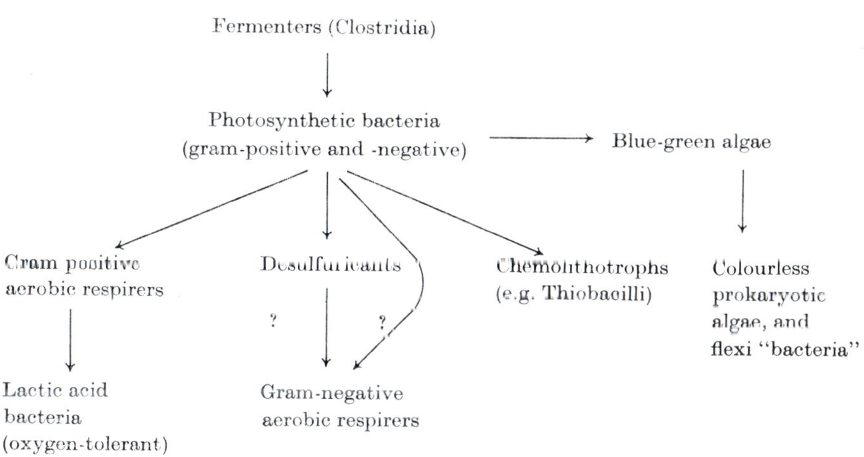

FIG. 8. A proposed scheme for the evolution of bacterial respiration. From Broda (1971).

they might have acquired their strictly anaerobic character by selective loss of capacities during successive generations of sheltered existence in a specialized anaerobic environment. There is some evidence, for example, that the oxygen-tolerant, but micro-aerophilic, lactic acid bacteria may have degenerated from aerobes (Bryan-Jones and Whittenbury, 1969; Broda, 1971).

It has been proposed that one major evolutionary advance that prepared the way for the origin of eukaryotic cells was the development of a cell membrane of increased "flexibility", capable of participating in pinocytosis. This possibly coincided with the acquisition of the aerobic ability to synthesize sterols and polyunsaturated fatty acids whose incorporation into cell membranes paradoxically carried the penalty of enhanced susceptibility to oxygen free radical damage. The progenitor(s) of eukaryotic cells might then have arisen as the product of sequential specific symbioses in which oxygen-evolving photosynthetic prokaryotes (protochloroplasts) and, later in the course of evolution, aerobically-respiring prokaryotes (protomitochondria) could have been ingested by a suitable host and taken up permanent symbiotic residence therein. According to this view, hereditary endosymbiosis has been a significantly evolutionary mechanism in the origin of the eukaryotic cell and is the reason for the otherwise unbridgeable discontinuity between prokaryotic and eukaryotic lines (Margulis, 1970, 1971; Stanier, 1970). de Duve (1973) has suggested that the primitive phagocyte, in which the ancestors of mitochondria allegedly settled, relied on peroxisomes for its respiratory metabolism. Distinct, and perhaps mutually incompatible, differentiations of the cell membrane of a common ancestor (the aerobic bacterium with peroxisomal respiration) could have given rise, on the one hand, to proliferation in intracellular membranes with an increase in cell size and acquisition of phagocytic tendencies (leading to the host) and, on the other, to the development of a membrane-integrated respiratory chain with coupled phosphorylating systems (the protomitochondrion). The fact that chicken liver and pig heart cells contain two types of superoxide dismutase: (1) the "eukaryotic" cupro-zinc enzyme in the cytosol; and (2) a mangano-enzyme located in the mitochondrion, has been interpreted as lending support to the endosymbiotic, prokaryotic origin of the mitochondrion (Weisiger and Fridovich, 1973a; Fridovich, 1974). The resemblance between the bacterial and mitochondrial superoxide dismutases is indeed striking, for while both of the enzymes from *E. coli* have a molecular weight of 40,000 daltons, and are composed of two subunits of equal size, the mitochondrial mangano-enzyme from chicken liver has a molecular weight of 80,000 daltons and is made up of four equal subunits (Forman and Fridovich, 1973b). Whilst there is approximately 80% sequence homology among the superoxide dismutases from *E. coli* and chicken liver mitochondria, there is no significant homology between any of these enzymes and the cupro-zinc enzyme from bovine erythrocytes (Steinman and Hill, 1973). If this line of reasoning is followed, one would either have to assume that the original phagocytic host to the protomitochondrion (de Duve, 1973) had not acquired a superoxide dismutase, or, if it possessed the mangano-enzyme appropriate to its prokaryotic ancestry, this was lost with the subsequent evolution of a new cupro-zinc enzyme to serve the same functions in the cytosol. The need for such an enzyme was possibly aggravated by the progressive increase in the oxygen content of the atmosphere which carried on until green-plant photosynthesis and aerobic respiration achieved the nice balance which now maintains a constant proportion of molecular oxygen in our atmosphere (Wald, 1964). In any event, the possibly polyphyletic origin of superoxide dismutase is evidence itself of the considerable service that this enzyme must render the aerobic cell. It is possible that a prokaryotic superoxide dismutase was first evolved in oxygen generating photosynthetic organisms. It is for this reason that it would be particularly interesting to learn whether present-day species of strictly anaerobic photosynthetic bacteria are devoid of the enzyme (p. 216).

Though very many questions relevant to the course of prokaryotic and eukaryotic evolution remain to be answered, it does seem to be well established that all living organisms may have had their ultimate origin in a strictly anaerobic prokaryote. This makes it particularly intriguing to consider what vestiges of their primitive ancestry may remain in those strict anaerobes which may now be the aborigines of the contemporary microbial world.

B. "Primitive" Metabolic Features of Obligate Anaerobes

After some two thousand million years of aerobiosis there may be no truly primitive organisms left on this earth (Gaffron, 1962), though there is always the possibility that some "living fossil" might be exhumed from an ecological niche in which the primal conditions in which it once flourished have fortuitously been sustained or recreated. Such, at least, was the suggestion in the case of an organism, superficially resembling the Pre-Cambrian microfossil *Kakabekia umbellata*, which Siegel and Giumarro (1966) unearthed in Wales from a site adjacent to the walls of Harlech Castle. Even so, if present day strict anaerobes are the filial descendants of the most primaeval living organisms, then the anaerobiologist has the pleasurable hope that while exploring the physiology of contemporary species he may turn up "memories of earlier evolutionary days" (Gaffron, 1962). But how may these relics of the ancient past be recognized? Though several features of anaerobic metabolism have been interpreted as "primitive", the criteria applied have frequently not been made clear. Generally, however, simplicity of construction, lack of sophistication in a control mechanism, inefficiency in performance, even oxygen-lability, have all been deemed sufficient reason to designate as "primitive" a particular cellular component or metabolic process.

In energy metabolism, the employment of diverse non-glycolytic routes of fermentation, reliance on "low potential" reactions involving ferredoxins, pteroylglutamates and cobamides, possession of the pyruvate phosphoroclastic system, synthesis and utilization of inorganic polyphosphates as energy-coupling agents, have all been cited as primitive characteristics (Decker *et al.*, 1970; Kulaev, 1971; Baltscheffsky, 1971; Pantskhava, 1971). Since it is now known that ferredoxins are to be found in non-photosynthetic aerobes (e.g. *Azotobacter* spp.), as well as in photosynthetic and fermentative organisms (Yoch and Arnon, 1972; Yoch and Valentine, 1972), it would seem that possession of these non-haem iron proteins can no longer be considered to betoken a peculiarly ancient ancestry. However, Lipmann (1971) suggested that in its very simplicity of construction the ferredoxin of *Clostridium butyricum* might be considered as a model of an early enzyme. Sequence analysis of the ferredoxins from the fermentative anaerobes has clearly demonstrated their homology, and has revealed a symmetry between the two halves of the molecule which might betoken early gene duplication (Eck and Dayhoff, 1966; Tanaka *et al.*, 1971). A phylogenetic tree drawn on the basis of the amino-acid sequences of various ferredoxins helpfully suggests possible evolutionary relationships between the fermentative anaerobes, photosynthetic bacteria and sulphate reducers (Fig. 9).

The fermentative production of hydrogen gas has been singled out as a device of ancient origins since, by serving to dispose of electrons released in metabolic oxidations, it can be considered to represent a primitive alternative to an aerobe's cytochrome oxidase (Gray and Gest, 1965). The discovery that *Methanobacillus omelianskii* was in fact a mixed culture consisting of a hydrogen producer (S organism) and a hydrogen-consuming, methane producer (H organism), suggested that interspecies hydrogen transfer could form the basis for more widespread symbiotic associations amongst anaerobic bacteria (Bryant, 1969). The periplasmic location of hydrogenase in *Desulfovibrio gigas* similarly suggests a role in the utilization of exogenous hydrogen rather than in the generation of molecular hydrogen (Peck, 1974). It is not too fanciful, therefore, to imagine that such co-operativity between hydrogen-producing and hydrogen-consuming organisms mirrors a onetime much more general phenomenon.

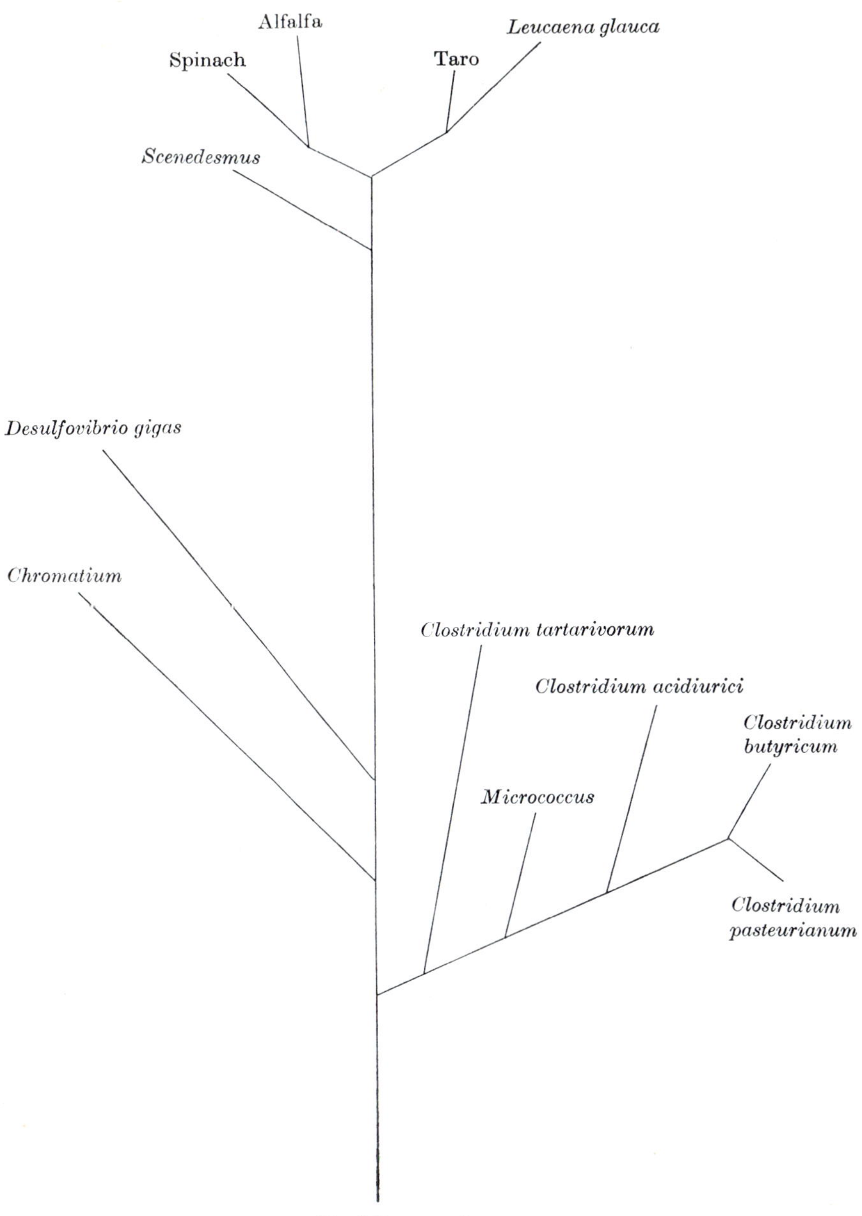

FIG. 9. Phylogenetic tree for ferredoxins. From Peck (1974), based on work by R. V. Eck, J. Travis and H. D. Peck, Jr.

Since the ability to fix molecular nitrogen is restricted to prokaryotic micro-organisms, is associated with an oxygen-labile enzyme complex operating at "low potential" and is particularly widespread amongst organisms having other characteristics (e.g. ferredoxin and hydrogenase) regarded as primitive, dinitrogen-fixation, or at least the possession of a nitrogenase, has been viewed as being of ancient origin (de Ley and Park, 1966). Silver and Postgate (1973) postulated that since nitrogenase may have evolved in strict anaerobes supplied (due to the current ammonia-containing environment) with ample fixed nitrogen, it could have had as its primary function the reductive detoxification of potentially hazardous constituents of the primaeval environment (e.g. cyanide or cyanogen). More recently, however, Postgate (1974) has declared himself more attracted to the view that the ability to fix nitrogen has appeared relatively recently on this planet, possibly when the eukaryotes were so far evolved as to be unable to accommodate the *nif* gene cluster. The restriction of *nif* to prokaryotes would be explained by their providing suitably anaerobic host cells still susceptible to genetic modification by incorporation of "foreign" DNA.

Properties of the sulphate-reducing bacteria, which might legitimately be regarded as primitive, were discussed by Peck (1966) and more recently by Le Gall and Postgate (1973). They included possession of hydrogenase and ferredoxin, ability to conduct both the reductive carboxylation of acetate and the pyruvate carboxylation reaction, possession of a citrate synthase of variable stereospecificity (Gottschalk, 1968), production of traces of methane originating in the methyl group of pyruvate (Wolfe, 1971), and content of "pristine" hydrocarbon components (Han and Calvin, 1969). From the fact that sulphur enriched in ^{32}S has been found in ancient rocks (2 to $3{\cdot}5 \times 10^9$ years old), it might appear that dissimilatory sulphate reduction was a very early evolutionary advance (Peck, 1974), though other evidence (e.g. Fig. 8, p. 225) would suggest that the sulphate reducers succeeded rather than preceded the photosynthetic sulphur bacteria. The finding that *Chloropseudomonas ethylica* is a mixed culture of *Chlorobium* with an unidentified sulphate-reducing bacterium, which together carry out a partial sulphur cycle (Gray, 1972), has led Peck (1974) to speculate that the establishment of a sulphur cycle was possibly a vital stage in bacterial evolution.

Though not a great deal is as yet known of the mechanism of proton flux in anaerobes, it is likely that a transmembrane pH gradient can be established at the expense of fermentatively generated ATP, and that proton coupling can account for the accumulation of substrates (Harold *et al.*, 1970; Kashket and Wilson, 1973; Niven *et al.*, 1973). The membrane adenosine triphosphatases of strict anaerobes would doubtless repay study, as they might be of greater "antiquity" than those involved in oxidative phosphorylation. Indeed, Mitchell (1968) has suggested that a reversible proton-translocating adenosine triphosphatase, and a proton-translocating oxido-reduction loop system, may have separately arisen during evolution in primitive prokaryotic organisms as alternatives for generating the pH difference and membrane potential required for nutrient uptake and ionic regulation via porter systems. The accidental occurrence of both systems in the same cell may then have provided the means of storing the free energy of oxido-reduction either in ATP synthesized by the reversal of the adenosine triphosphatase, or in some other anhydride such as pyrophosphate.

There are numerous examples of enzymes obtained from anaerobic bacteria being adjudged primitive because they are not subject to the more sophisticated (usually allosteric) fine controls which regulate the activities of their counterparts in "more advanced" organisms. Thus the phosphofructokinase of *Clostridium pasteurianum* proved to be exceptional in that it required ammonium ions and that, although subject to regulation by the prevailing fructose 6-phosphate and ADP concentrations, it was not inhibited by ATP, phosphoenolpyruvate or citrate. Uyeda and Kurooka (1970) thus concluded that the enzyme in *Cl. pasteurianum* exhibits perhaps the simplest regulatory mechanism of all allosteric phosphofructokinases. The ADP-glucose pyrophosphorylase of the same organism, which is involved in its biosynthesis of granulose, similarly appeared relatively unsophisticated, in comparison with analogous enzymes from glycogen-synthesizing bacteria, in that it was not subject to allosteric inhibition by intermediates of the energy-generating fermentation pathway (Robson *et al.*, 1972). An atypical (R)-citrate synthase present in *Clostridium kluyveri* (Gottschalk and Barker, 1966, 1967) has to date only been found in some strict anaerobes, and it has been proposed that it antedated the now usual (S)-citrate synthase which possibly appeared when prokaryotes became prototrophic for glutamate.

In the search for clues of an organism's primitive ancestry one may also take account of evidence afforded by the absence of biochemical characters known to have developed later in the progress of evolution (e.g. ability to synthesize haem, or to produce certain types of carotenoid or sterols). Such evidence (of the "dog that did not bark in the night-time" variety) is often quite as valuable as that supplied by an organism's possession of an "aboriginal" trait, though the possibility of loss of some acquired character (by retrogressive evolution) can prove misleading.

The few examples here instanced of properties of contemporary strict anaerobes, that might point to their primitive ancestry, are only a selection of many that have been reported in the extensive literature on the physiology of these organisms. Yet however diligently we winnow out vestigial characters from current species, and attempt therefrom to create a composite picture of the primitive prokaryote, we will never know with certainty how successful we have been. It follows that we must not invest our phylogenetic speculations with too great significance. Stanier (1970), with his usual felicity, put it thus: "Evolutionary speculation constitutes a kind of metascience which has the same intellectual fascination for some biologists that metaphysical speculation possessed for some mediaeval scholastics. It can be considered a relatively harmless habit, like eating peanuts, unless it assumes the form of an obsession; then it becomes a vice."

VII. Conclusion

The oxygen sensitivity of the strictly anaerobic bacteria has paradoxically been both the feature that has guaranteed their continuing fascination for microbiologists and the major deterrent to their study. The popularity of *Escherichia coli* amongst molecular biologists and microbial geneticists is no mystery to those who have worked with obligate anaerobes. Yet those who have pioneered studies on the energy-yielding metabolism of such obligately anaerobic organisms have found in them a wealth of novelty and metabolic ingenuity that has more than repaid the extra effort involved in handling some of the more fastidious species (e.g. Barker, 1956, 1961; Wood, 1961; Wolfe, 1971; Stadtman, 1973). Equal rewards await those who choose to study other aspects of their biochemistry, for surprisingly little is known of the general physiology of even species of clinical or industrial importance. Knowledge of their genetics is virtually non-existent, and ecological studies are in their infancy (Hungate, 1966; Hobson, 1971). After many years in which no progress was made beyond the assumptions of the peroxide theory, the recent confluence of ideas from the fields of free radical chemistry, radiation biology and membrane biochemistry has re-invigorated the quest for rational explanations of the strict anaerobe's super-sensitivity to oxygen. The emergence of the superoxide anion, singlet oxygen and hydroxyl free radical as prime suspects, far from marking the end of this quest, has given new impetus to enquiries into mechanisms of oxygen-toxicity to anaerobes and aerobes alike, for the broad spectrum of oxygen-sensitivity displayed even by obligate anaerobes has yet to be explained. Perhaps the single most valuable lesson learned to date is that it is very unlikely that any one unitary hypothesis can account for all of the many manifestations of this oxygen sensitivity.

One need be no prophet to foresee a burgeoning interest in obligately anaerobic organisms in the immediate future. With our current awareness of energy and petroleum shortages, the methanogenic anaerobic digestion of domestic sewage and agricultural effluents is certain to be exploited on a far larger scale than hitherto. There might even be a revival of interest in industrial-scale solvent-yielding fermentations. The use of anaerobic bacteria in biological fuel cells is not beyond the bounds of possibility (Lewis, 1966), whilst present pilot-scale use of *Desulfovibrio* to produce sulphur from industrial wastes and sewage, although not currently economically viable, might easily become so as the world's resources of native sulphur are progessively depleted (Le Gall and Postgate, 1973). The elaborate precautions taken in the early days of lunar exploration to decontaminate returning astronauts and their space capsules, dramatized the possible existence on our neighbouring planets of a novel, necessarily anaerobic, microbial flora. Yet while the discovery of an extraterrestrial obligate anaerobe would doubtless prove sensational, the physiological problems still posed by telluric species are so intriguing that one is as much amazed by what is not known, as impressed by what has already been discovered. It is possibly more true of the properties of obligately anaerobic bacteria than of any others, that

> "We dance round in a ring and suppose
> But the secret sits in the middle and knows".
> (ROBERT FROST, "*The Secret Sits*")

VIII. Acknowledgements

Work in our laboratory was supported by the Science Research Council and I am particularly indebted to my former colleague Dr. R. W. O'Brien for his many contributions both "at the bench" and in discussion.

REFERENCES

Abeliovich, A., Kellenberg, D. and Shilo, M. (1974). *Photochemistry and Photobiology* **19**, 379.

Adams, B. G. and Parks, L. W. (1969). *Journal of Bacteriology* **100**, 370.

Allen, J. E., Goodman, D. B. P., Besarab, A. and Rasmussen, H. (1973). *Biochimica et Biophysica Acta* **320**, 708.

Allen, B. G., Stjernholm, R. L. and Steele, R. H. (1972). *Biochemical and Biophysical Research Communications* **47**, 679.

Anbar, M. and Pecht, I. (1964). *Journal of Physical Chemistry* **68**, 352.

Ardon, M. (1965). "Oxygen. Elementary Forms and Hydrogen Peroxide." W. A. Benjamin, New York.

Arneson, R. M. (1970). *Archives of Biochemistry and Biophysics* **136**, 352.

Asada, K., Urano, M. and Takahashi, M. (1973). *European Journal of Biochemistry*, **36**, 257.

Aubel, E. and Perdigon, E. (1940). *Compte rendu hebdomadaire des séances de l'Académie des sciences, Paris* **211**, 439.

Aubel, E. and Perdigon, E. (1945). *Revue Canadienne de Biologie* **4**, 498.

Autor, A. P. (1974). *Life Sciences* **14**, 1309.

Balny, C. and Douzou, P. (1974). *Biochemical and Biophysical Research Communications* **56**, 386.

Baltscheffsky, H. (1971). *In* "Chemical Evolution and the Origin of Life", (R. Buvet and C. Ponnamperuma, eds.), pp. 466–474. North-Holland, Amsterdam and London.

Barker, H. A. (1956). "Bacterial Fermentations." John Wiley & Sons, New York.

Barker, H. A. (1961). *In* "The Bacteria", (I. C. Gunsalus and R. Y. Stanier, eds.), Vol. 2, pp. 151–207. Academic Press, New York.

Barnes, E. M. (1969). *In* "Methods in Microbiology", (J. R. Norris and D. W. Ribbons, eds.), Vol. 3B, pp. 151–160. Academic Press, London.

Barton, L. L., Le Gall, J. and Peck, H. D., Jr. (1970). *Biochemical and Biophysical Research Communications* **41**, 1036.

Barton, L. L. and Peck, H. D., Jr. (1971). *Bacteriological Proceedings* 156.

Beauchamp, C. and Fridovich, I. (1971). *Analytical Biochemistry* **44**, 276.

Beauchamp, C. O. and Fridovich, I. (1973). *Biochimica et Biophysica Acta* **317**, 50.

Beechey, R. B. and Ribbons, D. W. (1972). *In* "Methods in Microbiology", (J. R. Norris and D. W. Ribbons, eds.), Vol. 6B, pp. 25–53, Academic Press, London.

Behar, D., Czapski, G., Rabani, J., Dorfman, L. M. and Schwarz, H. A. (1970). *Journal of Physical Chemistry* **74**, 3209.

Benbough, J. E. (1969). *Journal of General Microbiology* **56**, 241.

Berkner, L. V. and Marshall, L. C. (1964). *Discussions of the Faraday Society* **37**, 122.

Bielski, B. H. J. and Chan, P. C. (1973). *Archives of Biochemistry and Biophysics* **159**, 873.

Bloch, K., Baronowsky, P., Goldfine, H., Lennarz, W. J., Light, R., Norris, A. T. and Scheuerbrandt, G. (1961). *Federation Proceedings. Federation of American Societies for Experimental Biology* **20**, 921.

Bray, R. C., Cockle, S. A., Fielden, E. M., Roberts, P. B., Rotilio, G. and Calabrese, L. (1974). *Biochemical Journal* **139**, 43.

Broda, E. (1971). *In* "Chemical Evolution and the Origin of Life", (R. Buvet and C. Ponnamperuma, eds.), pp. 446–452. North-Holland, Amsterdam and London.

Brown, D. E. (1970). *In* "Methods in Microbiology", (J. R. Norris and D. W. Ribbons, eds.), Vol. 2, pp. 125–174. Academic Press, London.

Brown, O. R. and Huggett, D. O. (1968). *Applied Microbiology* **16**, 476.

Brunker, R. L. and Brown, O. R. (1971). *Microbios* **4**, 193.

Bryan-Jones, D. G. and Whittenbury, R. (1969). *Journal of General Microbiology* **58**, 247.

Bryant, M. P. (1969). *Abstracts of the 158th Meeting of the American Chemical Society; Microbiology Section* p. 18.

Buvet, R. and Ponnamperuma, C. (1971). "Chemical Evolution and the Origin of Life. Molecular Evolution." Vol. 1. North-Holland, Amsterdam and London.

Calvin, M. (1969). "Chemical Evolution." Oxford University Press, Oxford and London.

Campbell, J. E. and Dimmick, R. L. (1966). *Journal of Bacteriology* **91**, 925.

Cannan, R. K., Cohen, B. and Clark, W. M. (1926). *U.S. Public Health Reports. Supplement 55*, p. 1.

Cardon, B. P. and Barker, H. A. (1946). *Journal of Bacteriology* **52**, 629.

Cato, E. P., Cummins, C. S., Holdeman, L. V., Johnson, J. L., Moore, W. E. C., Smibert, R. M. and Smith, L. D. S. (1970). Outline of Clinical Methods in Anaerobic Bacteriology. Virginia Polytechnic Institute and State University Anaerobe Laboratory, Blacksburg, Virginia.

Chan, H. W. S. (1971). *Journal of the American Chemical Society* **93**, 2357.
Chandler, J. L. R. and Gholson, R. K. (1972). *Biochimica et Biophysica Acta* **264**, 311.
Chandler, J. L. R., Gholson, R. K. and Scott, T. A. (1970). *Biochimica et Biophysica Acta* **222**, 523.
Clark, L. C. and Gollan, F. (1966). *Science, New York* **152**, 1755.
Clark, W. M. (1924). *Journal of the Washington Academy of Sciences* **14**, 123.
Cleveland, L. R. (1925). *Biological Bulletin* **48**, 455.
Cleveland, L. and Davis, L. (1974). *Biochimica et Biophysica Acta* **341**, 517.
Cloud, P. E., Jr. (1968). *Science, New York* **160**, 729.
Cloud, P. E., Jr. (1972). *American Journal of Science* **272**, 537.
Cohen, G. and Somerson, N. L. (1969). *Journal of Bacteriology* **98**, 547.
Cohen-Bazire, G., Sistrom, W. R. and Stanier, R. Y. (1957). *Journal of Cellular and Comparative Physiology* **49**, 25.
Connell, W. E. and Patrick, W. H. (1968). *Science, New York* **159**, 86.
Coon, M. J., Stroebel, H. W., Autor, A. P., Heidema, J. and Duppel, W. (1972). *In* "Biological Hydroxylation Mechanisms". Biochemical Society Symposia No. 34, pp. 45–54.
Costilow, R. N. and Keele, B. B., Jr. (1972). *Journal of Bacteriology* **111**, 628.
Cowan, S. T. (1968). A Dictionary of Microbial Taxonomic Usage. Oliver & Boyd, Edinburgh.
Cox, C. S. and Heckly, R. J. (1973). *Canadian Journal of Microbiology* **19**, 189.
Cusachs, L. C. and Steele, R. H. (1967). *Journal of Quantum Chemistry Symposium* **1**, 175.
Czapski, G. (1972). *Israel Journal of Chemistry* **10**, 987.
Dagley, S. (1971). *Advances in Microbial Physiology* **6**, 1.
Dalton, H. (1974). *Critical Reviews in Microbiology* **3**, 183.
Dalton, H. and Mortenson, L. E. (1972). *Bacteriological Reviews* **36**, 231.
Dalton, H. and Postgate, J. R. (1969). *Journal of General Microbiology* **54**, 463.
Davies, R. C., Gorchein, A., Neuberger, A., Sandy, J. D. and Tait, G. H. (1973). *Nature, London* **245**, 15.
Dayhoff, M. O. (1971). *In* "Chemical Evolution and the Origin of Life", (R. Buvet and C. Ponnamperuma, eds.), Vol. 1. pp. 392–419. North-Holland, Amsterdam and London.
Decker, K., Jungermann, K. and Thauer, R. K. (1970). *Angew. Chemie. International Edition* **9**, 138.
Demopoulos, H. B. (1973a). *Federation Proceedings. Federation of American Societies of Experimental Biology* **32**, 1859.
Demopoulos, H. B. (1973b). *Federation Proceedings. Federation of American Societies of Experimental Biology* **32**, 1903.
Dolin, M. I. (1961a). *In* "The Bacteria", (I. C. Gunsalus and R. Y. Stanier, eds.), vol. 2, pp. 319–363. Academic Press, New York.
Dolin, M. I. (1961b). *In* "The Bacteria", (I. C. Gunsalus and R. Y. Stanier, eds.), vol. 2, pp. 425–460. Academic Press, New York.
Dowell, V. R., Jr. (1972). *American Journal of Clinical Nutrition* **25**, 1335.
Dowell, V. R., Jr. and Hawkins, T. M. (1968). Laboratory Methods in Anaerobic Bacteriology. Public Health Service Publication No. 1803. U.S. Government Printing Office, Washington D.C.
Dowty, B., Laseter, J. L., Griffin, G. W., Politzer, I. R. and Walkinshaw, C. H. (1973). *Science, New York* **181**, 669.
Drozd, J. and Postgate, J. R. (1970). *Journal of General Microbiology* **63**, 63.
Dürner, G., Roomer, R. and Schwartz, W. (1965). *Zeitschrift für Allgemeine Mikrobiologie* **5**, 395.

Dutton, P. L. and Evans, W. C. (1969). *Biochemical Journal* **113**, 525.
de Duve, C. (1973). *Science, New York* **182**, 85.
de Duve, C. and Baudhuin, P. (1966). *Physiological Reviews* **46**, 323.
Eck, R. V. and Dayhoff, M. O. (1966). *Science, New York* **152**, 363.
Emerson, R. and Held, A. A. (1969). *American Journal of Botany* **56**, 1103.
Epel, B. L. and Neumann, J. (1973). *Biochimica et Biophysica Acta* **325**, 520.
Evans, M. C. W. and Whatley, F. R. (1970). *Symposium of The Society for General Microbiology* **20**, 203–220.
Faria Oliveira, O. M. M., Sanioto, D. L. and Cilento, G. (1974). *Biochemical and Biophysical Research Communications* **58**, 391.
Farmilio, A. and Wilkinson, F. (1973). *Photochemistry and Photobiology* **18**, 447.
Farrington, J. A., Ebert, M., Land, E. J. and Fletcher, K. (1973). *Biochimica et Biophysica Acta* **314**, 372.
Fee, J. A. (1973). *Biochimica et Biophysica Acta* **295**, 87.
Fee, J. A. and DiCorleto, P. E. (1973). *Biochemistry, New York* **12**, 4893.
Fee, J. A. and Hildenbrand, P. G. (1974). *Federation of European Biochemical Societies Letters* **39**, 79.
Fenchel, T. M. and Riedl, R. J. (1970). *Marine Biology* **7**, 255.
Fielden, E. M., Roberts, P. B., Bray, R. C., Lowe, D. J., Mautner, G. N., Rotilio, G. and Calabrese, L. (1974). *Biochemical Journal* **139**, 49.
Finazzi-Agro, A., Giovagnoli, C., De Sole, P., Calabrese, L., Rotilio, G. and Mondovi, B. (1972). *Federation of European Biochemical Societies Letters* **21**, 183.
Flohe, L., Günzler, W. A. and Shock, H. H. (1973). *Federation of European Biochemical Societies Letters* **32**, 132.
Fong, K.-L., McCay, P. B., Poyer, J. L., Keele, B. B. and Misra, H. (1973). *Journal of Biological Chemistry* **248**, 7792.
Foote, C. S. (1968). *Science, New York* **162**, 963.
Foote, C. S., Chang, Y. C. and Denny, D. W. (1970a). *Journal of the American Chemical Society* **92**, 5216.
Foote, C. S., Denny, R. W., Weaver, L., Chang, Y. and Peters, J. (1970b). *Annals of the New York Academy of Sciences* **171**, 139.
Forman, H. J., Evans, H. J., Hill, R. L. and Fridovich, I. (1973). *Biochemistry, New York* **12**, 823.
Forman, H. J. and Fridovich, I. (1973a). *Archives of Biochemistry and Biophysics* **158**, 396.
Forman, H. J. and Fridovich, I. (1973b). *Journal of Biological Chemistry* **248**, 2645.
Fredette, V., Planté, C. and Roy, A. (1967). *Journal of Bacteriology* **94**, 2012.
Fridovich, I. (1970). *Journal of Biological Chemistry* **245**, 4053.
Fridovich, I. (1972). *Accounts of Chemical Research* **5**, 321.
Fridovich, I. (1974). *Life Sciences* **14**, 819.
Fried, R., Fried, L. W. and Babin, D. R. (1973). *European Journal of Biochemistry* **33**, 439.
Gaffron, H. (1962). *In* "Horizons in Biochemistry", (M. Kasha and B. Pullman, eds.), pp. 59–89. Academic Press, New York.
Gerschman, R. (1964). *In* "Oxygen and the Animal Organism". Proceedings of a Symposium, London (1963), pp. 475–494.
Gest, H. (1972). *Advances in Microbial Physiology* **7**, 243.
Gibson, F. and Pittard, J. (1968). *Bacteriological Reviews* **32**, 465.
Gillespie, L. J. (1920). *Soil Science* **9**, 199.
Goda, K., Chu, J.-W., Kimura, T. and Schaap, A. P. (1973). *Biochemical and Biophysical Research Communications* **52**, 1300.

Goldfine, H. (1965). *In* "Oxygen". Proceedings of a Symposium sponsored by the New York Heart Association, pp. 253–268. Little Brown, Boston.
Gordon, J., Holman, R. A. and McLeod, J. W. (1953). *Journal of Pathology and Bacteriology* **66**, 527.
Goscin, S. A. and Fridovich, I. (1972). *Biochimica et Biophysica Acta* **289**, 276.
Gottlieb, S. F. (1971). *Annual Review of Microbiology* **25**, 111.
Gottschalk, G. (1968). *European Journal of Biochemistry* **5**, 346.
Gottschalk, G. and Barker, H. A. (1966). *Biochemistry, New York* **5**, 1125.
Gottschalk, G. and Barker, H. A. (1967). *Biochemistry, New York* **6**, 1027.
Gray, B. H. (1972). Ph.D. Thesis, University of Tennessee.
Gray, C. T. and Gest, H. (1965). *Science, New York* **148**, 186.
Gray, C. T., Wimpenny, J. W. T., Hughes, D. A. and Mossman, M. R. (1966a). *Biochimica et Biophysica Acta* **117**, 22.
Gray, C. T., Wimpenny, J. W. T. and Mossman, M. R. (1966b). *Biochimica et Biophysica Acta* **117**, 33.
Green, A. R. and Cruzov, G. (1968). *Nature, London* **220**, 1095.
Greenlee, L., Fridovich, I. and Handler, P. (1962). *Biochemistry, New York* **1**, 779.
Gregory, E. M. and Fridovich, I. (1973a). *Journal of Bacteriology* **114**, 543.
Gregory, E. M. and Fridovich, I. (1973b). *Journal of Bacteriology* **114**, 1193.
Gregory, E. M. and Fridovich, I. (1974). *Journal of Bacteriology* **117**, 166.
Gregory, E. M., Yost, F. J. and Fridovich, I. (1973). *Journal of Bacteriology* **115**, 987.
Gregory, E. M., Goscin, S. A. and Fridovich, I. (1974). *Journal of Bacteriology* **117**, 456.
Gross, D., Feige, A. and Schütte, H. R. (1967). *Zeitschrift für Naturforschung B* **22**, 835.
Gross, D., Feige, A., Stecher, R., Zurech, A. and Schütte, H. R. (1965). *Zeitschrift für Naturforschung B* **20**, 116.
Grunberg, M. (1948). *Annales de l'Institut Pasteur, Paris* **74**, 216.
Grunberg, M. (1949). *Année Biologique* **53**, 36.
Grunberg-Manago, M., Szulmajster, J. and Delavier, C. (1952). *Annales de l'Institut Pasteur, Paris* **83**, 102.
Gunsalus, I. C., Lipscomb, J. D., Marshall, V., Frauenfelder, H. Greenbaum, E. and Münck, E. (1972). *In* "Biological Hydroxylation Mechanisms". Biochemical Society Symposia No. 34, pp. 135–157.
Gunsalus, I. C. and Shuster, C. W. (1961). *In* "The Bacteria", (I. C. Gunsalus and R. Y. Stanier, eds.), Vol. 2, pp. 1–58. Academic Press, New York.
Guyer, M. and Hegeman, G. D. (1969). *Journal of Bacteriology* **99**, 906.
Haber, F. and Weiss, J. (1934). *Proceedings of the Royal Society Series A* **147**, 332.
Hamilton, G. A., Libby, R. D. and Hartzell, C. R. (1973). *Biochemical and Biophysical Research Communications* **55**, 333.
Han, J. and Calvin, M. (1969). *Proceedings of the National Academy of Sciences of the United States of America* **64**, 436.
Harold, F. M., Pavlasova, E. and Baarda, J. R. (1970). *Biochimica et Biophysica Acta* **196**, 235.
Harrison, D. E. F. (1972). *Journal of Applied Chemistry and Biotechnology* **22**, 417.
Harrison, D. E. F. (1973). *Critical Reviews in Microbiology* **2**, 185.
Hastings, J. W., Balny, C., Le Peuch, C. and Douzou, P. (1973). *Proceedings of the National Academy of Sciences of the United States of America* **70**, 3468.
Haugaard, N. (1968). *Physiological Reviews* **48**, 311.
Hayaishi, O. and Nozaki, M. (1969). *Science, New York* **164**, 389.

Heden, C. G. and Malmborg, A. S. (1961). *Scientific Reports of the Istitutio di Superiore de Sanita, Rome* **1**, 213.
Hegeman, G. D. (1967). *Archiv für Mikrobiologie* **59**, 143.
Hewitt, L. F. (1950). "Oxidation-Reduction Potentials in Bacteriology and Biochemistry", 6th edn. Livingstone, Edinburgh.
Hill, S., Drozd, J. W. and Postgate, J. R. (1972). *Journal of Applied Chemistry and Biotechnology* **22**, 541.
Hobson, P. N. (1971). *Progress in Industrial Microbiology* **9**, 41.
Hodgson, E. K. and Fridovich, I. (1973a). *Biochemical and Biophysical Research Communications* **54**, 270.
Hodgson, E. K. and Fridovich, I. (1973b). *Photochemistry and Photobiology* **18**, 451.
Holdeman, L. V. and Moore, W. E. C. (1972). Editors of "Anaerobe Laboratory Manual". Virginia Polytechnic Institute and State University Anaerobe Laboratory, Blacksburg, Virginia.
Holland, H. D. (1962). *In* "Petrologic Studies: A Volume in Honour of A. F. Buddington" (A. E. J. Engel, H. L. James and B. F. Leonard, eds.), p. 447. Geological Society of America.
Holmberg, K. and Hollander, H. O. (1973). *Archives of Oral Biology* **18**, 423.
Howes, R. M. and Steele, R. H. (1972). *Research Communications in Chemical Pathology and Pharmacology* **3**, 349.
Hughes, D. E. and Wimpenny, J. W. T. (1969). *Advances in Microbial Physiology* **3**, 197.
Hungate, R. E. (1966). "The Rumen and its Microbes." Academic Press, New York.
Hungate, R. E. (1969). *In* "Methods in Microbiology", (J. R. Norris and D. W. Ribbons, eds), Vol. 3B, pp. 117–132. Academic Press, London.
Hurlbert, R. E. (1967). *Journal of Bacteriology* **93**, 1346.
Ingold, K. V. (1969). *Accounts of Chemical Research* **2**, 1.
Isquith, A. J. and Moat, A. G. (1965). *Bacteriological Proceedings* 74.
Isquith, A. J. and Moat, A. G. (1966). *Biochemical and Biophysical Research Communications* **22**, 565.
Jacob, H.-E. (1970). *In* "Methods in Microbiology", (J. R. Norris and D. W. Ribbons, eds.), Vol. 2, pp. 91–123. Academic Press, London.
Jacobs, N. J., Jacobs, J. M. and Sheng, G. S. (1969). *Journal of Bacteriology* **99**, 37.
Jocelyn, P. C. (1970). *Biochemical Journal* **117**, 951.
Jocelyn, P. C. (1972). "Biochemistry of the SH Group." Academic Press, London.
Joester, K.-E., Jung, G., Weber, U. and Weser, U. (1972). *Federation of European Biochemical Societies Letters* **25**, 25.
Johnston, M. A. and Delwiche, E. A. (1965). *Journal of Bacteriology* **90**, 352.
Jones, C. W., Brice, J. M., Wright, V. and Ackrell, B. A. C. (1973). *Federation of European Biochemical Societies Letters* **29**, 77.
Jones, D., Deibel, R. H. and Niven, C. F., Jr. (1964). *Journal of Bacteriology* **88**, 602.
Jones, D., Watkins, J. and Meyer, D. J. (1970). *Nature, London* **226**, 1249.
Kasha, M. and Khan, A. U. (1970). *Annals of the New York Academy of Sciences* **171**, 5.
Kashket, E. R. and Wilson, T. H. (1973). *Proceedings of the National Academy of Sciences of the United States of America* **70**, 2866.
Katagiri, M. and Takemori, S. (1973). *Molecular and Cellular Biochemistry* **2**, 137.
Kearns, D. R. (1971). *Chemical Reviews* **71**, 395.
Keele, B. B., McCord, J. M. and Fridovich, I. (1970). *Journal of Biological Chemistry* **245**, 6176.

Kenyon, D. H. and Steinman, G. (1969). "Biochemical Predestination." McGraw-Hill, New York.

Khan, A. U. (1970). *Science, New York* **168**, 476.

Khan, A. U. and Kasha, M. (1970). *Annals of the New York Academy of Sciences* **171**, 24.

Kimball, A. P. and Orò, J. (1971). "Prebiotic and Biochemical Evolution." North-Holland & American Elsevier, New York.

Kirsch, E. J. and Sykes, R. M. (1971). *Progress in Industrial Microbiology* **9**, 155.

Klebanoff, S. J., Clem, W. H. and Luebke, R. G. (1966). *Biochimica et Biophysica Acta* **117**, 63.

Klein, H. P. (1955). *Journal of Bacteriology* **69**, 620.

Klein, R. W. and Cronquist, A. (1967). *Quarterly Review of Biology* **42**, 105

Klug, D., Rabani, J. and Fridovich, I. (1972). *Journal of Biological Chemistry* **247**, 4839.

Knappe, J., Schacht, J., Mockel, W., Hopner, T., Vetter, H. Jr. and Eden-Harder, R. (1969). *European Journal of Biochemistry* **11**, 316.

Knaysi, G. and Dutky, S. R. (1936). *Journal of Bacteriology* **31**, 137.

Knowles, P. F., Gibson, J. F., Pick, F. M. and Bray, R. C. (1969). *Biochemical Journal* **111**, 53.

Kosower, E. M. and Kosower, N. S. (1969). *Nature, London* **224**, 117.

Kosower, N. S., Vanderhoff, G. A. and Kosower, E. M. (1972). *Biochimica et Biophysica Acta* **272**, 623.

Kulaev, I. S. (1971). *In* "Chemical Evolution and the Origin of Life", (R. Buvet and C. Ponnamperuma, eds.), Vol. 1, pp. 458–465. North Holland, Amsterdam and London.

Kumar, R. P., Ravindranath, S. D., Vaidyanathan, C. S. and Rao, N. A. (1972). *Biochemical and Biophysical Research Communications* **49**, 1422.

Kwiatkowski, L. D. and Kosman, D. J. (1973). *Biochemical and Biophysical Research Communications* **53**, 715.

Lagnado, J. R. and Sourks, T. L. (1958). *Canadian Journal of Biochemistry and Physiology* **36**, 587.

Lagnowski, J. J. (1973). "Modern Inorganic Chemistry." Marcel Dekker, New York.

Lamolo, A. A., Yamane, T. and Trozzolo, A. M. (1973). *Science, New York* **179**, 1131.

Lascelles, J. and Altschuler, T. (1969). *Journal of Bacteriology* **98**, 721.

Lavelle, F., Durosay, P. and Michelson, A. M. (1974). *Biochemie* **56**, 451.

Lavelle, F., Michelson, A. M. and Dimitrijevic, L. (1973). *Biochemical and Biophysical Research Communications* **55**, 350.

Le Gall, J. and Postgate, J. R. (1973). *Advances in Microbial Physiology* **10**, 82.

Leman, A. (1965). *Archiv für Mikrobiologie* **50**, 194.

Lengyel, Z. L. and Nyiri, L. (1965). *Biotechnology and Bioengineering* **7**, 91.

Lewis, K. (1966). *Bacteriological Reviews* **30**, 101.

de Ley, J. and Park, T. W. (1966). *Antonie van Leeuwenhoek* **32**, 6.

Lipmann, F. (1971). *In* "Chemical Evolution and the Origin of Life", (R. Buvet and C. Ponnamperuma, eds.), Vol. 1, pp. 381–391. North-Holland, Amsterdam and London.

Lippitt, B., McCord, J. M. and Fridovich, I. (1972). *Journal of Biological Chemistry* **247**, 4688.

Lippitt, B. and Fridovich, I. (1973). *Archives of Biochemistry and Biophysics* **159**, 738.

Lloyd, D. (1969). *Symposium of The Society for General Microbiology* **19**, 299.

Loesche, W. J. (1969). *Applied Microbiology* **18**, 723.

Loschen, G., Azzi, A. and Flohé, L. (1973). *Federation of European Biochemical Societies Letters* **33**, 84.

Low, I. E., Eaton, M. D. and Proctor, P. (1968). *Journal of Bacteriology* **95**, 1425.

Low, I. E. and Zimkus, S. M. (1973). *Journal of Bacteriology* **116**, 346.

Mallin, M. L. and Seeley, H. W. (1958). *Archives of Biochemistry and Biophysics* **73**, 306.

Margulis, L. (1970). "Origin of Eukaryotic Cells." Yale University Press, New Haven.

Margulis, L. (1971). *In* "Chemical Evolution and the Origin of Life", (R. Buvet and C. Ponnamperuma, eds.), Vol. 1, pp. 480–484. North-Holland, Amsterdam and London.

Massey, V., Strickland, S., Mayhew, S. G., Howell, L. G., Engel, P. C., Mathews, R. G., Schumann, M. and Sullivan, P. A. (1969). *Biochemical and Biophysical Research Communications* **36**, 891.

Mateles, R. I. and Zuber, B. L. (1963). *Antonie van Leeuwenhoek* **29**, 249.

Maugh, T. H. (1973). *Science, New York* **182**, 44.

May, S. W. and Abbott, B. J. (1973). *Journal of Biological Chemistry* **248**, 1725.

May, S. W., Abbott, B. J. and Felix, A. (1973). *Biochemical and Biophysical Research Communications* **54**, 1540.

Mazia, D. and Ruby, A. (1967). *Experimental Cell Research* **45**, 505.

Menzel, D. B. (1970). *Annual Reviews of Pharmacology* **10**, 379.

Merkel, P. B. and Kearns, D. R. (1972). *Journal of the American Chemical Society* **94**, 7244.

Meyer, D. J. and Jones, C. W. (1973). *International Journal of Systematic Bacteriology* **23**, 459.

Michelson, A. M. (1973). *Biochemie* **55**, 465.

Mickelson, M. N. (1966). *Journal of General Microbiology* **43**, 31.

Miller, S. L. and Urey, H. C. (1959). *Science, New York* **130**, 245.

Miraglia, G. J. (1974). *Critical Reviews in Microbiology* **3**, 161.

Misra, H. P. (1974). *Journal of Biological Chemistry* **249**, 2151.

Misra, H. P. and Fridovich, I. (1971). *Journal of Biological Chemistry* **246**, 6886.

Misra, H. P. and Fridovich, I. (1972a). *Journal of Biological Chemistry* **247**, 188.

Misra, H. P. and Fridovich, I. (1972b). *Journal of Biological Chemistry* **247**, 3170.

Misra, H. P. and Fridovich, I. (1972c). *Journal of Biological Chemistry* **247**, 3410.

Mitchell, P. (1968). "Chemiosmotic Coupling and Energy Transduction." Glynn Research, Bodmin, Cornwall, U. K.

Montgomery, H. A. C., Thom, N. S. and Cockburn, A. (1964). *Journal of Applied Chemistry, London* **14**, 280.

Moore, W. E. C. (1966). *International Journal of Systematic Bacteriology* **16**, 173.

Moore, W. E. C. and Holdeman, L. V. (1972). *American Journal of Clinical Nutrition* **25**, 1306.

Morris, J. C. and Stumm, W. (1967). *In* "Equilibrium Concepts in Natural Water Systems", (R. F. Gould, ed.), p. 270. American Chemical Society, Washington, D.C.

Morris, J. G. (1974). "A Biologist's Physical Chemistry," 2nd edn. Edward Arnold, London.

Morris, J. G. and O'Brien, R. W. (1971). *In* "Spore Research 1971", (A. N. Barker, G. W. Gould and J. Wolf, eds.), pp. 1–37. Academic Press, London.

Murphy, M. J., Siegel, L. M., Kamin, H., Dervartanian, D. V., Lee, J. P. Le Gall, J. and Peck, H. D., Jr. (1973). *Biochemical and Biophysical Research Communications* **54**, 82.

Murphy, M. J., Siegel, L. M., Tove, S. R. and Kamin, H. (1974). *Proceedings of the National Academy of Sciences of the United States of America* **71**, 612.
Murray, R. W. and Lin, J. W.-P. (1970). *Annals of the New York Academy of Sciences* **171**, 121.
Myers, L. S. (1973). *Federation Proceedings. Federation of American Societies for Experimental Biology* **32**, 1882.
McCord, J. M. and Fridovich, I. (1968). *Journal of Biological Chemistry* **243**, 5753.
McCord, J. M. and Fridovich, I. (1969). *Journal of Biological Chemistry* **244**, 6049.
McCord, J. M., Keele, B. B. and Fridovich, I. (1971). *Proceedings of the National Academy of Sciences of the United States of America* **68**, 1024.
McCord, J. M., Beauchamp, C. O., Goscin, S., Misra, H. P. and Fridovich, I. (1973). *In* "Oxidases and Related Redox Systems", (T. E. King, H. S. Mason and M. Morrison, eds.), Proceedings of the 2nd International Symposium, Memphis, Tennessee, June, 1971. University Park Press, Baltimore.
McElroy, W. D. and Seliger, H. H. (1962). *In* "Horizons in Biochemistry", (M. Kasha and B. Pullman, eds.), pp. 91–101. Academic Press, New York.
McLeod, J. W. and Gordon, J. (1923). *Journal of Pathology and Bacteriology* **26**, 326.
McRipley, R. J. and Sbarra, A. J. (1967a). *Journal of Bacteriology* **94**, 1417.
McRipley, R. J. and Sbarra, A. J. (1967b). *Journal of Bacteriology* **94**, 1425.
Nilsson, R. and Kearns, D. R. (1974). *Photochemistry and Photobiology* **19**, 181.
Niven, D. F., Jeacocke, R. E. and Hamilton, W. A. (1973). *Federation of European Biochemical Societies Letters*, **29**, 248.
Noland, L. E. and Gojdics, M. (1967). *In* "Research in Protozoology", (T. Tuan Chen, ed.), Vol. 2, pp. 217–266. Pergamon Press, Oxford.
O'Brien, R. W. and Morris, J. G. (1971). *Journal of General Microbiology* **68**, 307.
O'Brien, R. W., Weitzman, P. D. J. and Morris, J. G. (1970). *Federation of European Biochemical Societies Letters* **10**, 343.
O'Donnell, J. H. and Sangster, D. F. (1970). "Principles of Radiation Chemistry." American Elsevier, New York.
Orme-Johnson, W. H. and Beinert, H. (1969). *Biochemical and Biophysical Research Communications* **36**, 891.
Pantskhava, E. (1971). *In* "Chemical Evolution and the Origin of Life", (R. Buvet and C. Ponnamperuma, eds.), Vol. 1, pp. 475–479. North-Holland, Amsterdam and London.
Paschen, W. and Weser, U. (1973). *Biochimica et Biophysica Acta* **327**, 217.
Pasteur, L. (1861). *Compte rendu hebdomadaire des séances de l'Académie des sciences* **52**, 344.
Paul, B. B., Strauss, R. R., Selvaraj, R. J. and Sbarra, A. J. (1973). *Science, New York* **181**, 849.
Pauling, L. (1931). *Journal of the American Chemical Society* **53**, 3225.
Paynter, M. J. B. and Hungate, R. E. (1968). *Journal of Bacteriology* **95**, 1943.
Peck, H. D., Jr. (1966). *Biochemical and Biophysical Research Communications* **22**, 112.
Peck, H. D., Jr. (1974). *Symposium of the Society for General Microbiology* **24**, 241.
Pederson, T. C. and Aust, S. D. (1972). *Biochemical and Biophysical Research Communications* **48**, 789.
Pederson, T. C. and Aust, S. D. (1973). *Biochemical and Biophysical Research Communications* **52**, 1071.
Pepper, R. E. and Costilow, R. N. (1965). *Journal of Bacteriology* **89**, 271.
Peters, J. W., Pitts, J. N. Jr., Rosenthal, I. and Fuhr, H. (1972). *Journal of the American Chemical Society* **94**, 4348.

Pfennig, N. (1967). *Annual Reviews of Microbiology* **21**, 286.
Polgár, L. (1974). *Federation of European Biochemical Societies Letters* **38**, 187.
Politzer, I. R., Griffin, G. W. and Laseter, J. L. (1971). *Chemico-Biological Interactions* **3**, 73.
Ponnamperuma, C. (1972). "The Origins of Life." Thames and Hudson, London.
Ponnamperuma, C. and Gabel, N. W. (1974). *Symposium of the Society for General Microbiology* **24**, 393.
Porter, D. J. T. and Ingraham, L. L. (1974). *Biochimica et Biophysica Acta* **334**, 97.
Postgate, J. R. (1968). *Proceedings of the Royal Society, Series B* **171**, 67.
Postgate, J. R. (1974). *Symposium of the Society for General Microbiology* **24**, 263.
Potter, M. C. (1910). *Durham University Philosophical Society Proceedings* **3**.
Potter, M. C. (1911). *Proceedings of the Royal Society, Series B* **84**, 260.
Powers, E. L. (1972). *Israel Journal of Chemistry* **10**, 1199.
Prevot, A. R. and Thouvenot, H. (1952). *Annales de l'Institut Pasteur, Paris* **83**, 443.
Pryor, W. A. (1973). *Federation Proceedings. Federation of American Societies for Experimental Biology* **32**, 1862.
Quastel, J. H. and Stephenson, M. (1926). *Biochemical Journal* **20**, 1125.
Rabani, J. and Nielsen, S. O. (1969). *Journal of Physical Chemistry* **73**, 3736.
Rabinowitz, M., Getz, G. S., Casey, J. and Swift, H. (1969). *Journal of Molecular Biology* **41**, 381.
Rabotnowa, I. L. (1963). "Die Bedentung physikalisch-chemischer Faktoren (pH und rH) für die Lebenstätigkeit der Mikroorganismen." Fischer-Verlag, Jena.
Rao, P. S. and Hayon, E. M. (1973). *Biochemical and Biophysical Research Communications* **51**, 468.
Rapp, U., Adams, W. C. and Miller, R. W. (1973). *Canadian Journal of Biochemistry* **51**, 158.
Rawls, H. R. and van Santen, R. J. (1970). *Annals of the New York Academy of Sciences* **171**, 135.
Reiter, B., Pickering, A. and Oram, J. D. (1964). 4th International Symposium on Food Microbiology, p. 297. SIK, Goteburg, Sweden.
Roberton, A. M. and Wolfe, R. S. (1970). *Journal of Bacteriology* **102**, 43.
Robson, R. L., Robson, R. M. and Morris, J. G. (1972). *Biochemical Journal* **130**, 4P.
Rosebury, R. (1962). "Microorganisms Indigenous to Man." McGraw-Hill, New York.
Rotilio, G., Bray, R. C. and Fielden, E. M. (1972a). *Biochimica et Biophysica Acta* **268**, 605.
Rotilio, G., Calabrese, L., Bossa, F., Barra, D., Agro, A. F. and Mondovi, B. (1972b). *Biochemistry, New York* **11**, 2182.
Rotilio, G., Morpugo, L., Giovagnoli, C., Calabrese, L. and Mondovi, B. (1972c). *Biochemistry, New York* **11**, 2187.
Rotilio, G., Calabrese, L., Finazzi-Agrò, A., Argento-Cerù, M. P. (1973a) *Biochimica et Biophysica Acta* **321**, 98.
Rotilio, G., Morpugo, L., Calabrese, L. and Mondovi, B. (1973b). *Biochimica et Biophysica Acta* **302**, 229.
Rutten, M. G. (1971). "The Origin of Life by Natural Causes." Elsevier, Amsterdam.
Saunders, B. C., Holmes-Siedle, A. G. and Stark, B. P. (1964). "Peroxidase. The Properties and Uses of a Versatile Enzyme and of some related Catalysts." Butterworths, London.
Schilpkoter, H.-W. and Bruch, J. (1973). *Zentralblatt für Bakteriologie, Parasitenkunde Infektionskrankheiten und Hygiene, Abteilung A* **156**, 486.

Schopf, J. W. (1970). *Biological Reviews* **45**, 319.

Scott, T. A., Bellion, E. and Mattey, M. (1969). *European Journal of Biochemistry* **10**, 318.

Scott, T. A. and Hussey, H. (1965). *Biochemical Journal* **96**, 9C.

Seeley, H. W. and Vandemark, P. J. (1951). *Journal of Bacteriology* **61**, 27.

Shapton, D. A. and Board, R. G. (eds.) (1971). "Isolation of Anaerobes." Society for Applied Bacteriology Technical Series No. 5. Academic Press, London.

Sherman, J. M. (1926). *Journal of Bacteriology* **11**, 417.

Shuster, C. W. and Gunsalus, I. C. (1958). *Federation Proceedings. Federation of American Societies for Experimental Biology* **17**, 310.

Siegel, S. M. and Giumarro, C. (1966). *Proceedings of the National Academy of Sciences of the United States of America* **55**, 349.

Silver, W. S. and Postgate, J. R. (1973). *Journal of Theoretical Biology* **40**, 1.

Sisler, F. D. (1971). *Progress in Industrial Microbiology* **9**, 1.

Sistrom, W. R. (1965). *Journal of Bacteriology* **89**, 403.

Smith, L. (1961). *In* "The Bacteria", (I. C. Gunsalus and R. Y. Stanier, eds.), Vol. 2, pp. 365–396. Academic Press, New York.

Smith, L. DS. (1967). *In* "The Anaerobic Bacteria", (V. Fredette, ed.), pp. 13–24. Institute of Microbiology and Hygiene of Montreal University, Montreal, Canada.

Somerson, N. L., Walls, B. E. and Chanock, R. M. (1965). *Science, New York* **150**, 226.

Sorokin, Yu. I. (1966). *Nature, London* **210**, 551.

Squires, R. W. and Hosler, P. (1958). *Industrial and Engineering Chemistry* **50**, 1263.

Stadtman, T. C. (1967). *Annual Reviews of Microbiology* **21**, 139.

Stadtman, T. C. (1973). *Advances in Enzymology* **38**, 413.

Stanier, R. Y. (1970). *Symposium of the Society for General Microbiology* **20**, 1.

Stanier, R. Y. (1974). *Symposium of the Society for General Microbiology* **24**, 219.

Stanier, R. Y., Doudoroff, M. and Adelberg, E. A. (1971). "General Microbiology", 3rd ed. Macmillan, London.

Stees, J. L. and Brown, O. R. (1973). *Microbios* **7**, 257.

Steinman, H. M. and Hill, R. L. (1973). *Proceedings of the National Academy of Sciences of the United States of America* **70**, 3725.

Stephenson, L. M., McClure, D. E. and Sysak, P. K. (1973). *Journal of the American Chemical Society* **95**, 7888.

Stokes, A. M., Hill. H. A. O., Bannister, W. H. and Bannister, J. V. (1973). *Federation of European Biochemical Societies Letters* **32**, 119.

Stouthamer, A. H. (1969). *In* "Methods in Microbiology", (J. R. Norris and D. W. Ribbons, eds.), Vol. 1, pp. 629–663. Academic Press, London.

Stouthamer, A. H. (1973). *Antonie van Leeuwenhoek* **39**, 545.

Stouthamer, A. H. and Bettenhaussen, C. (1973). *Biochimica et Biophysica Acta* **301**, 53.

Strickland, S. and Massey, V. (1973). *In* "Oxidases and Related Redox Systems", (T. E. King, H. S. Mason and M. Morrison, eds.) 2nd Symposium, Memphis, Tennessee, June 1971. University Park Press, Baltimore.

Symonyan, M. A. and Nalbandyan, R. M. (1972). *Federation of European Biochemical Societies Letters* **28**, 22.

Tanaka, M., Haniu, M., Matsueda, G., Yasunobu, K., Himes, R. H., Akagi, J. M., Barnes, E. M. and Devanathan, T. (1971). *Journal of Biological Chemistry* **246**, 3953.

Tappel, A. L. (1973). *Federation Proceedings. Federation of American Societies for Experimental Biology* **32**, 1870.
Taube, H. (1965). "Oxygen: Chemistry, Structure and Excited States." Little Brown, New York.
Taylor, B. F., Campbell, W. L. and Chinoy, I. (1970). *Journal of Bacteriology* **102**, 430.
Tengerdy, R. P. (1961). *Biotechnology and Bioengineering* **3**, 241.
Thauer, R. K., Jungermann, K., Henninger, H., Wenning, J. and Decker, K. (1968). *European Journal of Biochemistry* **4**, 173.
Tolbert, N. E. (1971). *Annual Review of Plant Physiology* **22**, 45.
Tollin, G. and Fox, J. L. (1967). *In* "Magnetic Resonance in Biological Systems", (A. Ehrenberg, B. G. Malmstrom and T. Vanngard, eds.), pp. 289–297. Pergamon, Oxford.
Uyeda, K. and Kurooka, S. (1970). *Journal of Biological Chemistry* **245**, 3315.
Vallee, B. L. and Williams, R. J. P. (1968). *Proceedings of the National Academy of Sciences of the United States of America* **59**, 498.
Vallery-Radot, R. (1901). "The Life of Pasteur." Constable, London.
Vance, P. G., Keele, B. B. and Rajagopalan, K. (1972). *Journal of Biological Chemistry* **247**, 4782.
Vosjan, J. H. (1970). *Antonie van Leeuwenhoek* **36**, 584.
de Vries, W., van Wijck-Kapteyn, W. M. C. and Oosterhuis, S. K. H. (1974). *Journal of General Microbiology* **81**, 69.
de Vries, W., van Wijck-Kapteijn, W. M. C. and Stouthamer, A. H. (1972). *Journal of General Microbiology* **71**, 515.
de Vries, W. and Stouthamer, A. H. (1969). *Archiv für Mikrobiologie* **65**, 275.
Wald, G. (1964). *Proceedings of the National Academy of Sciences of the United States of America* **52**, 595.
Ware, D. and Postgate, J. R. (1970). *Nature, London* **226**, 1250.
Watson, J. A. and Schubert, J. (1969). *Journal of General Microbiology* **57**, 25.
Weisiger, R. and Fridovich, I. (1973a). *Journal of Biological Chemistry* **248**, 3582.
Weisiger, R. A. and Fridovich, I. (1973b). *Journal of Biological Chemistry* **248**, 4793.
Weitzman, P. D. J. and Jones, D. (1968). *Nature, London* **219**, 270.
Weser, U., Fretzdorff, A. and Prinz, R. (1972). *Federation of European Biochemical Societies Letters* **27**, 267.
Weser, U. and Paschen, W. (1972). *Federation of European Biochemical Societies Letters* **27**, 248.
Wever, R., Oudega, B. and van Gelder, B. F. (1973). *Biochimica et Biophysica Acta* **302**, 475.
Wheland, G. W. (1937). *Transactions of the Faraday Society* **33**, 1499.
White, J. R. and Dearman, H. H. (1965). *Proceedings of the National Academy of Sciences of the United States of America* **54**, 887.
White, J. L., Vaughn, T. O. and Yeh, W. S. (1971). *Abstracts of the Proceedings of the Federation of American Societies for Experimental Biology* **30**, 1145.
Willis, A. T. (1964). "Anaerobic Bacteriology in Clinical Medicine." Butterworths, London.
Willis, A. T. (1969). *In* "Methods in Microbiology", (J. R. Norris and D. W. Ribbons, eds.), Vol. 3B, pp. 79–115. Academic Press, London.
Wimpenny, J. W. T. (1969). *Symposium of the Society for General Microbiology* **19**, 161.
Wimpenny, J. W. T. (1970). *Biotechnology and Bioengineering* **11**, 623.

Wimpenny, J. W. T. and Necklen, D. K. (1971). *Biochimica et Biophysica Acta* **253**, 352.
Wolfe, R. S. (1971). *Advances in Microbial Physiology* **6**, 107.
Wood, N. P. (1966). *Methods in Enzymology* **9**, 3.
Wood, W. A. (1961). *In* "The Bacteria", (I. C. Gunsalus and R. Y. Stanier, eds.), Vol. 2, pp. 59–149. Academic Press, New York.
Yang, K. S. and Waller, G. R. (1965). *Phytochemistry* **4**, 881.
Yoch, D. C. and Arnon, D. I. (1972). *Journal of Biological Chemistry* **247**, 4514.
Yoch, D. C. and Valentine, R. C. (1972). *Annual Review of Microbiology* **26**, 139.
Yost, F. J. and Fridovich, I. (1973). *Journal of Biological Chemistry* **248**, 4905.
Young, H. L. (1968). *Nature, London* **219**, 1068.
Yousten, A. A., Bulla, L. A. Jr. and McCord, J. M. (1973). *Journal of Bacteriology* **113**, 524.
Zimmermann, R., Flohé, L., Weser, U. and Hartmann, H. J. (1973). *Federation of European Biochemical Societies Letters*, **29**, 117.

Note added in proof

Since this essay was written (February 1974), reports on superoxide radical, superoxide dismutase and singlet oxygen have appeared in the literature at an accelerating rate. A textbook on "Molecular Mechanisms of Oxygen Activation" (Hayaishi, 1974) includes one of the two recent reviews of Superoxide Dismutase written by Fridovich (1974 a, b) and a collection of articles on the properties and biological functions of a large number of oxidases, oxygenases and peroxidases. There have been several reports that superoxide dismutase does not quench singlet oxygen as previously suggested (pp. 200 & 219), but rather catalyses the dismutation of superoxide anion radicals by a mechanism which leads to the formation of triplet oxygen, unlike the non-enzymatic dismutation which yields singlet oxygen (Mayeda and Bard, 1974; Schapa *et al.*, 1974; Goda *et al.*, 1974; Michelson, 1974). The discovery by Puget and Michelson (1974) that the marine bacterium *Photobacterium leiognathi* possesses a copper and zinc-containing superoxide dismutase (which they have called bacteriocuprein), demonstrates that possession of a cupro-zinc enzyme is not the exclusive prerogative of the eukaryote as was formerly supposed (p. 227). Furthermore, it suggests that the original hypothesis of functional convergence from polyphyletic origins (p. 227) should be discarded in favour of a monophyletic evolutionary process from prokaryote to eukaryote. Though in addition to its copper-zinc enzyme, *Photobacterium leiognathi* produces an iron-containing superoxide dismutase, this bacterium contains no trace of the manganese enzyme which is the more common prokaryotic superoxide dismutase (Puget and Michelson, 1974). The facultatively anaerobic flagellates *Tritrichomonas foetus* and *Monocercomonas* sp. have been found to contain cyanide-insensitive superoxide dismutases, which in *T. foetus* are located both in the cytosol and the hydrogenosomes (Lindmark and Müller, 1974). This would appear to be the first report of a cyanide-insensitive superoxide dismutase in the cytoplasm of a eukaryote. In this laboratory we have found superoxide dismutase(s) (albeit in low activity) in a number of bacteria usually classified as obligate anaerobes (Hewitt and Morris, 1975). These included the photosynthetic anaerobes *Chlorobium thiosulfatophilum* NCIB 8346 and *Chromatium* sp. NCIB 8348, the sulphate reducers *Desulfotomaculum nigrificans* NCIB 8395 and *Desulfovibrio desulfuricans* NCIB 8307, and several species of *Clostridium* known to be moderately oxygen tolerant e.g. *Cl. pasteurianum* ATCC 6013 and *Cl. perfringens* type A NCIB 11105.

The cyanide-insensitive superoxide dismutase of *Cl. perfringens* was partially purified in the course of this work. Evidently, an organism can contain *some* superoxide dismutase activity and yet be incapable of growth in air. But this does not mean that its limited content of the enzyme renders this organism no significant service, for the spectrum of aero-tolerance extends into the region of obligate anaerobiosis (see p. 206). Further studies should now be undertaken with other species of *Clostridium* in order to determine whether there is any correlation between their differing degrees of oxygen-tolerance and their contents of superoxide dismutase; and attempts should be made to manipulate the concentration of this enzyme within such organisms, as has already been done with facultatively aerobic species (Fridovich, 1974a; see pp. 216–218). Looking further ahead, it seems likely that the coming years will see much more effort expended on identification of specific targets of oxygen radical and singlet oxygen damage, with attention predictably being focused on effects of oxygen on bacterial membranes.

Additional references

Fridovich, I. (1974a). *In* "Molecular Mechanisms of Oxygen Activation", (O. Hayaishi, ed.), pp. 453–477. Academic Press, New York and London.

Fridovich, I. (1974b). *Advances in Enzymology* **41**, 35.

Goda, K., Kimura, T., Thayer, A. L., Kees, K. and Schaap, A. P. (1974). *Biochemical and Biophysical Research Communications* **58**, 660.

Hayaishi, O. (1974). Ed. "Molecular Mechanisms of Oxygen Activation", pp. 678. Academic Press, New York and London.

Hewitt, J. and Morris, J. G. (1975). *Federation of European Biochemical Societies Letters*—in Press.

Lindmark, D. G. and Müller, M. (1974). *Journal of Biological Chemistry* **249**, 4634.

Mayeda, E. A. and Bard, A. J. (1974). *Journal of the American Chemical Society* **96**, 4023.

Michelson, A. M. (1974). *Federation of European Biochemical Societies Letters* **44**, **97**.

Puget, K. and Michelson, A. M. (1974). *Biochemical and Biophysical Research Communications* **58**, 830.

Schaap, A. P., Thayer, A. L., Faler G. R., Goda, K. and Kimura, T. (1974). *Journal of the American Chemical Society* **96**, 4025.

DNA Replication in Bacteria

Tatsuo Matsushita and Herbert E. Kubitschek

*Division of Biological and Medical Research,
Argonne National Laboratory,
Argonne, Illinois* 60439 *U.S.A.*

I. Introduction

The discovery of the structure of deoxyribonucleic acid (DNA) by Watson and Crick (1953) sparked a most impressive period of progress in biology in which DNA was firmly established as the central molecule of molecular biology. Deoxyribonucleic acid is an essential component of chromosomes and comprises the genetic material of all cells. The basic features of how the informational content of DNA (the genetic code) defines the structure of proteins, through the processes of RNA synthesis (transcription) and protein biosynthesis (translation), have now been elucidated; however, some of the basic features of DNA replication are still unsolved.

A. Purpose of the Review

This review has been written for students and research workers who may not necessarily be studying DNA replication as such, but who may find the information of value as a teaching aid or as a guide to understanding physiological phenomena associated with this complex process. We will try to make it clear to readers why such a basic biological phenomena has not been elucidated, in spite of the strenuous efforts of many laboratories, and will show that the complexities of the DNA replication process preclude a simple description such as the mere addition of deoxyribonucleoside triphosphates by DNA polymerase. We shall attempt to describe the many physical, biochemical and genetic requirements for cellular DNA replication; in addition, we hope to help workers in this field by including findings published after the reviews of Smith (1973) and Klein and Bonhoeffer (1972).

We will mainly restrict our review to DNA replication in bacteria, emphasizing work done with *Escherichia coli* and *Bacillus subtilis*, the two most widely used organisms. In general, the complexity of DNA replication is expected to increase with phylogenetic order, with bacteriophage DNA replication being less complex than mammalian chromosome synthesis. Consequently, more is known about phage DNA replication than bacterial or eukaryotic chromosome synthesis. However, bacteria are among the simplest of free-living organisms and, as such, should be among the most amenable systems for attacking the problem of chromosome replication. Certainly, much of the large amount of information available on replication has come from bacterial studies, although other levels of organization (phage and eukaryotic organisms) will be discussed where they help to clarify the replication process in bacteria.

Unavoidably, a large number of abbreviations are used in the text, and these are listed below.

DNA forms, genes, bases, analogues, and precursors. A, C, G, and T: nucleic acid bases, adenine, cytosine, guanine, and thymine; A:T and G:C: hydrogen-bonded base pairs; araCTP, 1-β-D-arabinofuranosylcytosine triphosphate; dBrUTP, deoxybromouridine 5′-triphosphate; dHMCTP, hydroxymethyldeoxycytidine 5′-triphosphate; dN, deoxyribonucleosides of any base N; dNTP, deoxyribonucleoside 5′-triphosphate containing any base N; H–H, H–L, and L–L: duplex DNA containing BU or other heavy isotope in both strands, one strand, or neither, respectively; NTP, ribonucleoside 5′-triphosphate containing any base N; p, phosphodiester linkage in a nucleic acid strand; Pol A,

Pol B, and Pol C: genes required for the synthesis of DNA polymerase I, II, and III, respectively; Py, either pyrimidine base C or T (U); SS, single-stranded DNA or chromosome; RF, double-stranded replicative form of DNA or chromosome; SS-RF system, one in which a double-stranded RF is made from a single-stranded template; rN, ribonucleoside containing any base N.

Other abbreviations. ATP, adenosine triphosphate; BU, 5-bromouracil; CM, chloramphenicol; DEAE, diethylaminoethyl; DNAse, deoxyribonuclease; EDTA, ethylenediaminetetraacetate; HPUra, 6-(*p*-hydroxyphenylazo)-uracil; NAD, nicotinamide adenine dinucleotide; NMN, nicotinamide mononucleotide; *p*CMB, *para*-chloromercuribenzoate; *p*HMB, *para*-hydroxymercuribenzoate; pol I, pol II, and pol III: DNA polymerases I, II, and III, respectively; RNAse, ribonuclease; ts, temperature sensitivity; UV, ultraviolet radiation.

B. Early Observations on DNA Synthesis

1. *The Structure of DNA and its Relationship to Replication*

The structure of DNA was first elucidated by Watson and Crick (1953). Duplex DNA consists of a pair of polymers, each composed of chains of deoxyribonucleosides joined together by phosphodiester linkages between the C-3 atom (or 3′ position) of the deoxyribose of one nucleoside and the C-5 (or 5′ position) of the next (Watson, 1965). The two chains form a double helix with the nucleosides of one chain hydrogen-bonded to the nucleotides of the other chain through complementary bases, adenine to thymine (A:T) and guanine to cytosine (G:C). The chains have opposite polarity; thus, reading from one end of the duplex, the nucleosides on one strand are all arranged 3′ to 5′, and 5′ to 3′ on the other, complementary strand.

The duplex structure of DNA led Watson and Crick to propose a simple model of replication, in which parental strands separated and provided templates for the sequential synthesis of complementary macromolecular progeny strands. Although the detailed manner by which this process occurs is presently under intensive investigation in many laboratories, several properties of DNA replication were already proposed or established a decade ago. At that time, Kornberg (1961) had carried out the *in vitro* synthesis of DNA, and shown that the requirements for the reaction, in the presence of Mg^{2+}, were: (1) the purified enzyme, DNA polymerase (now called DNA polymerase I); (2) macromolecular DNA to act as a template; and (3) the deoxyribonucleoside 5′-triphosphates of the four bases: adenine, thymine, guanine and cytosine. Most early models of replication were based on these *in vitro* studies and utilized DNA polymerase I as the replicating enzyme.

DNA polymerase I has since been shown to be involved in excision repair, and its role in DNA replication is uncertain.

TABLE 1. DNA Synthesis in Replication and Repair

Property	DNA Replication	DNA Repair
Quantity of DNA synthesis	Whole chromosome	Short regions
Mode of synthesis	Semiconservative	Dispersive
Essential polymerizing activity	DNA polymerase III ?DNA polymerase I	DNA polymerase I ?DNA polymerase II ?DNA polymerase III

Repair of DNA represents the synthesis of *short* regions of the chromosome which have been damaged from UV, X-rays or other agents (Table 1). However, when we speak of replication, we mean the formation of the *whole* chromosome, not simply the filling in of short regions of the chromosome, as in DNA repair. And, although the polymerization of DNA in both repair and replication appears to involve the four deoxynucleoside triphosphates, the essential polymerizing activities differ, with DNA polymerase I being mainly responsible for repair, and DNA polymerase III (perhaps along with DNA polymerase I) being essential for replication (see Section III, C, p. 287).

2. Characteristics of Replication

(a) *Semiconservative replication.* The Watson and Crick model of DNA structure suggests a semiconservative mode of replication (Fig. 1). The first implication from the model is that the genetic code is conserved in the sequence of nucleotides which are *covalently* linked in each strand. This structure argues against a dispersive mode of replication in which the parental strands are not kept intact and where the covalent phosphodiester bonds are broken during replication. Instead it seems more reasonable to suppose that each parental DNA strand is conserved in its base sequence and the covalent bonds, if they are broken during replication, are reformed into their original sequence before the next round of replication starts. Furthermore, the two parental strands are *complementary* to each other in base sequence and are held together by hydrogen bonding. So the parental strands are capable of being separated when these weaker (than covalent) chemical bonds are broken. A completely conservative mode of replication would have no separation of the parent strands, whereas a semiconservative mode requires a separation of the two strands (Fig. 1).

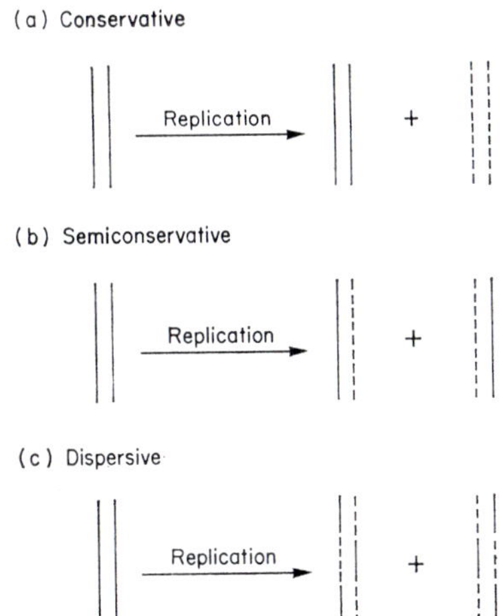

FIG. 1. Possible modes of DNA synthesis: (a) conservative, (b) semiconservative, and (c) dispersive. The solid lines represent parental (template) DNA strands and the broken lines newly synthesized DNA strands.

The semiconservative mode of replication has been shown repeatedly since the first classical demonstration in *E. coli* by Meselson and Stahl (1958). This repetition is necessary since the semiconservative mode is a keystone for replication studies; thus, only when one demonstrates that DNA synthesis is semiconservative can one be sure that replication, and not repair, is being studied. The original Meselson and Stahl experiment showed that semiconservative replication could be demonstrated biophysically. They recognized that if one grew organisms in media that contained DNA precursors with a different density from those present in the parental culture, then the new daughter DNA would have a different density from that of the original parental DNA. By using CsCl gradients, they showed that, one generation after the density shift, all of the DNA was composed of one peak of intermediate density to that produced either before the shift (where both the parental strands were heavy) or much later (where both the daughter strands were light). This intermediate band was shown to be a hybrid of one heavy and one light DNA strand (Fig. 1b). The result is incompatible with a conservative replication mode, since after one generation a completely conservative mode would have yielded two density peaks, one composed of heavy-heavy parental DNA and one composed of light-light daughter DNA (Fig. 1a). The experiment demonstrated clearly the conservation of covalent bonding in the parental strands and the separation of the parental strands from each other by hydrogen-bond breaking, a result completely compatible with the semiconservative mode of replication.

Further studies showed that, after *two* generations of density shift, there were two equal peaks of DNA, one having the hybrid density and the other the daughter density, again clearly showing semiconservative replication. This also eliminated the dispersive mode of replication (Fig. 1c) since this would have yielded many intermediate density peaks (since all daughter strands would include both parental and daughter DNA).

(b) *Replication is sequential and follows gene order*. Meselson and Stahl's early confirmation of semiconservative replication, which was predicted by the Watson and Crick model, led to serious consideration of the plausibility of sequential replication. Semiconservative replication could be non-sequential in several ways: genes could be synthesized in random order at multiple origins on the chromosome, or one strand could be synthesized in its entirety in one direction on template strand A, followed by synthesis of the other daughter strand on template strand B. However, if the chromosome replicates sequentially from one end (origin) to the other (terminus) with both strands replicating simultaneously, a "replication fork" should be formed (Fig. 2). Cairns (1963a) observed such replication forks in autoradiographic studies of *E. coli* chromosomes at different stages of replication. At the level of the gene, the strongest support was provided by the demonstration (Yoshikawa and Sueoka, 1963a, b) that gene frequencies during exponential growth depended upon their mapped position on the chromosome. Thus the frequency of a genetic marker near the origin should be nearly twice as high as that of a marker near the terminus, if sequential replication is occurring (see Fig. 2); this they clearly demonstrated. Another kind of early evidence for sequential replication was the observation that enzyme

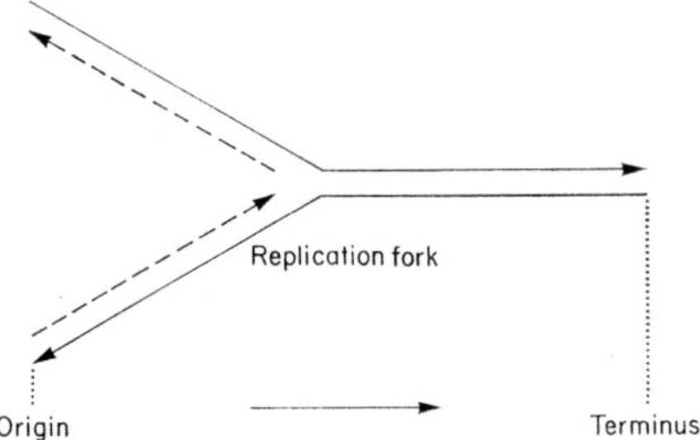

FIG. 2. A model of the chromosome undergoing replication. The solid lines represent the parental (template) strand and the broken lines the newly synthesized strands with the arrows representing strand polarity. Replication is from left to right. Adapted from Sueoka and Yoshikawa (1963).

synthesis in synchronized cultures increased during the cycle in the same sequential order as that appearing in the genetic map (Masters and Pardee, 1965).

(c) *A fixed origin for replication, generation after generation.* The formation of the replication fork has been called chromosome initiation since it appears to be genetically distinct (Section III, C-8, p. 304) from the movement of the replication fork (elongation of the chromosome). If genes are synthesized sequentially in their mapped order, then chromosome initiation should occur from a common origin, and not randomly at points throughout the chromosome. The concept of a fixed origin was first implied by the results of Yoshikawa and Sueoka (1963a, b) and Lark *et al.* (1963). A common origin was also assumed in the replicon model of Jacob *et al.* (1963). The accurate mapping of the origin has only recently been accomplished in *E. coli* and appears to be very near the *ilv* locus (see Section II, B1, p. 261). In *B. subtilis* it is seemingly located near *ade*16 (Yoshikawa and Sueoka, 1963a; O'Sullivan and Sueoka, 1967), but the detailed genetic map at the origin is still not worked out (Hara and Yoshikawa, 1973; Yamaguchi and Yoshikawa, 1973). Nagata and Meselson (1968) extended the evidence for a fixed origin when they pulse-labelled chromosomes in exponential-phase cultures of *E. coli* with ^3H-thymidine and then pulse-labelled with heavy isotopes at various times after the tritium pulse. This resulted in incorporation of a varying proportion of tritium label into the hybrid-density DNA peaks isolated by CsCl centrifugation. A striking periodicity, corresponding to the generation time of the bacteria, was demonstrated over some five generations for the hybrid-peak tritium label. This periodicity is clearly incompatible with an origin randomly located with respect to the origin from the previous replication. Thus the origin of DNA replication appears to be at the same locus, generation after generation.

(d) *Simultaneous replication of both strands of DNA at the replication fork.* When Meselson and Stahl (1958) showed the simultaneous appearance of two hybrid daughters after one generation of semiconservative replication, it immediately suggested simultaneous synthesis of both DNA strands. This was confirmed by Cairns (1963a) who prepared autoradiographs of pulse-labelled DNA and found the label density, and range, to be the same for both strands at the replication fork.

(e) *Constant rate of DNA synthesis at a replication fork.* Although the observations of Cairns (1963a) were in agreement with rate constancy, the strongest early evidence was provided by the results of Yoshikawa *et al.* (1964) who showed that *B. subtilis* cells double their growth rate by doubling chromosome initiation frequency (dichotomous replication) not by accelerating the rate of replication-fork movement. This early evidence led Maaløe and Kjeldgaard (1966) to suggest that the rate of replication of DNA at a fork is always constant, irrespective of the growth rate of the cell, and that the replication fork is always saturated with substrates. The best evidence currently available for rate constancy is provided by the experimental results of Helmstetter and his co-workers (Helmstetter, 1967; Helmstetter and Cooper, 1968; Helmstetter *et al.*, 1968) and, for slowly growing cells, by Kubitschek and Freedman (1971). Whether there are small undetectable differences in the rate of replication under different conditions, or temporary hesitations during replication, is still unknown.

With some modifications, the above general features of DNA replication have held up extremely well and form the basis upon which current research into the intricate biochemical mechanisms must rest.

II. Chromosome Structure and Growth

The structure of the bacterial genome is conceptually simple, consisting of a circular helical duplex deoxyribonucleotide polymer, first described by Watson and Crick (1953). But difficulties arise when one considers the size of the chromosome which is approximately 1000-times as long as the cell in which it is confined. The need for such a huge molecule is understandable, since all of the information for the production and control of the many different cell products required for the organization and function of the cell, including DNA replication, must be encoded in the chromosome. Thus, the resultant great length of these DNA duplexes requires extensive folding of the DNA, folding that cannot be relieved as the cell replicates its DNA, or retrieves information from the chromosome, except for short regions approximately the size of a cell. Studies on chromosome organization are relatively new and have just begun to reveal the shape and structure of the replicating chromosome in living cells. Less is known of the manner in which this structure is maintained during replication and transcription.

The organizational complexity of the chromosome requires DNA replication to be examined at several levels. At one extreme, chromosomes can be observed to replicate and segregate into progeny cells. At the other extreme, one must determine the detailed biochemical processes occurring during the initiation of DNA synthesis and in chain elongation. In addition, DNA synthesis is characterized by intermediate macromolecular reactions regulating the kinetics of formation and stitching together of macromolecular DNA segments during unwinding, synthesis, and repair of DNA.

During the past several years, substantial advances have been made in understanding the structure of the whole bacterial chromosome. We know it is circular in both *E. coli.* and *B. subtilis* (Cairns 1963a, b; Wake, 1973) and is packaged in the cell by many folds which are stabilized by membrane, RNA and proteins. The membrane attachment of the chromosome has been recognized for almost a decade, and the relationship of this attachment to chromosome segregation and replication is under active study. The first sub-section (IIA) discusses DNA replication at the level of the whole chromosome, and the implications of circularity, stabilization and membrane attachment for DNA replication. The second sub-section (IIB) covers the intermediate macromolecular level of research, and discusses the mechanics of the different modes of chromosome replication within the cell. The biochemical level is discussed in Section III (p. 270) and includes the molecular mechanisms involved in DNA replication and a description of the enzymes involved in this process. In the final Section (IV, p. 308) we return to the level of the whole chromosome and discuss the relationship between chromosome replication and cell division. By defining these levels of discussion, we hope to present a clearer picture of how DNA replication is understood at present, and where important gaps in our knowledge exist.

A. Structure

1. *Circularity*

Although the genetic studies of Jacob and Wollman (1961) showed that the markers in the genome of *E. coli* could be arranged on a single, circular linkage map, this genetic information was insufficient to establish the circular structure of the chromosome. Physical evidence for circularity was first provided by Cairns (1963a, b) with autoradiographs of whole, replicating chromosomes. These autoradiographs also indicated that the chromosome is composed of one double-stranded DNA molecule.

These results were supported by observations in other bacteria. Genetic maps compatible with circularity could be drawn for *Salmonella typhimurium* (Sanderson and Demerec, 1964; Sanderson, 1967), *Streptomyces coelicolor* (Hopwood, 1967), and *Bacillus subtilis* (Dubnau, 1970; Young and Wilson, 1972). Using a detergent lysis method, Bode and Morowitz (1967) obtained electron micrographs of disentangled circular chromosomes of *Mycoplasma hominis*, and observed unambiguous circular double-stranded chromosomes. Later, replicating circular chromosomes of the Cairns-type configuration also were demonstrated in *B. subtilis* (Wake, 1973).

Circular forms are also common for chromosomes of most other prokaryotes, including bacteriophages and plasmids (fertility, colicinogenic, and resistance-transfer factors), and in most mitochondrial DNA species so far examined (see review by Smith, 1973). For further information on circular DNA and its properties, the reader is referred to the extensive review by Helinski and Clewell (1971).

At present there is no solid evidence for the advantage of a circular configuration during chromosome replication. Unwinding of the strands in the duplex during replication is at least as complex for circular as for linear chromosomes. Furthermore, during bidirectional replication (Section II, B-1, p. 261), the circular "track" of the parental strands does not conserve attachment of the replication complexes, which presumably meet at a terminus located at some distance from the origin of replication. It may be that the circular form prevents exonucleolytic attack of "loose ends" during critical periods in the cell cycle. Also loose ends, with single-stranded redundant tails, might lead to concatemers (chromosomes linked in tandem) larger than unit length chromosomes (Watson, 1972). Other speculations deal with initiation. Certain kinds of replication, as in conjugation (Section II, B 3, p. 269), might be initiated when the circle is nicked. Alternatively, replication may be initiated at a specific sequence without the need for a nick or chromosomal end. Whatever the advantages of circularity, its relationship to the replication process remains a question for the future.

2. *Size and Stabilization*

Autoradiographs of the DNA from chromosomes of *E. coli* and *B. subtilis* show these to be roughly of the same length, ranging from 700 to 1300 μm for *E. coli* (Cairns, 1963 a, b; Bleecken *et al.*, 1966) and from 900 to 1100 μm for *B. subtilis* (Wake, 1973). Molecular weights also are similar; thus, from measurements of viscosity and sedimentation coefficients, Massie and Zimm (1965) obtained preliminary values of $2{\cdot}4 \times 10^9$ and $2{\cdot}3 \times 10^9$ daltons, respectively, for the two species. Later estimates indicated that the molecular weight of *B. subtilis* DNA was slightly lower; that is, $2{\cdot}6$–$2{\cdot}8 \times 10^9$ as compared with $2{\cdot}8 \times 10^9$ daltons for *E. coli* (determined by renaturation kinetics; Bak *et al.*, 1970), and $2{\cdot}0 \times 10^9$ as compared with $2{\cdot}7 \times 10^9$ daltons (determined by viscoelastometry; Klotz and Zimm, 1972). These values accord well with Cairns' (1963 a, b) determination of the length of the DNA molecule; thus, assuming the mass per unit length of DNA to be 196 daltons per 0·1 nm (Lange *et al.*, 1967), a chromosome of 1100–1300 μm would have a molecular weight of $2{\cdot}2$–$2{\cdot}7 \times 10^9$ daltons. These estimated values are also in good agreement with chemical determinations of the molecular weight of the single equivalent genome, for *E. coli*, of $2{\cdot}5 \times 10^9$ daltons (Cooper and Helmstetter, 1968; Kubitschek and Freedman, 1971).

The linear dimensions (1000 μm) of the DNA in these bacterial chromosomes are far greater than the dimensions of the cells (one to several μm) in which the DNA resides, and this fact alone requires that chromosomal DNA must be highly folded. Stonington and Pettijohn (1971) developed a procedure of cell lysis with non-ionic detergents that allowed the folded DNA genome complex from *E. coli* to be isolated. They showed that this complex contained both protein and RNA. The protein was predominantly core RNA polymerase, and contained very little if any of the DNA-binding proteins, discussed later (p. 289). The RNA of the complex was mainly messenger RNA and ribosomal RNA chains. That the RNA component is somehow involved in stabilization was shown by unfolding of the DNA from these complexes upon exposure to ribonuclease (RNAse) or heat.

By altering DNA coiling with the intercalating compound ethidium bromide, Worcel and Burgi (1972) showed that these folded chromosomes are supercoiled (i.e., have twists superimposed on the DNA double helix). This was later confirmed directly from electron micrographs (Delius and Worcel, 1974). In addition to relaxing the degree of supercoiling with ethidium bromide, Worcel and Burgi (1972) obtained similar results by nicking the folded chromosome with deoxyribonuclease (DNAse), obtaining more open structures that sedimented more slowly. They observed that roughly 6 to 40 nicks per single DNA strand provided a completely "relaxed" complex, in agreement with an earlier estimate of about 30 attachment points obtained by Rosenberg and Cavalieri (1968). On the basis of these results, Worcel and Burgi (1972) suggested that the folded chromosome may consist of DNA looped around an RNA core that partitions the chromosome into 12 to 80 loops. Also, because unfolding of the DNA complexes by RNAse takes place as an all-or-none effect, they suggested that the stabilization may be provided by a single species of RNA molecule.

In addition, when lysis was carried out at temperatures that were 10–15°C below ambient, a second, more rapidly sedimenting, type of folded chromosome was observed in which the DNA was additionally attached to membrane fragments (Pettijohn *et al.*, 1973). These were shown to contain membrane proteins and phospholipids, as well as folded DNA and nascent RNA (Worcel and Burgi, 1974). When folded DNA was isolated from amino acid-starved cells that had been allowed to complete their rounds of DNA replication, no membrane-attached chromosomes were found. However, when amino-acid synthesis was again permitted, all of the chromosomes became attached to membranes before DNA synthesis was detectable. Worcel and Burgi (1974) suggested that both attachment to, and release of folded chromosomes from, membrane are directly associated with the act of DNA replication.

It appears, then, that the whole chromosome is organized in folds and is packaged in relationship to RNA, protein and membrane. The spatial requirement for this tight organization is obvious. But how does the chromosome function during replication and transcription? These questions are not entirely answered by *in vitro* experiments because the original chromosomal material is expanded into volumes far greater than the confines of a cell. In the bacterial cell, the DNA is packaged into a dense, tortuous network, and the rates of some biochemical reactions might be decreased by this packing. Nevertheless, this tight organization is maintained during replication.

3. *Membrane Attachment*

The replicon hypothesis of Jacob *et al.* (1963) assumed that the ordered segregation of chromosomes to daughter cells is brought about by attachment of chromosomes to the growing cell membrane. The first evidence for chromosome attachment to membrane invaginations, called mesosomes, was obtained by Ryter and Jacob (1963) from electron micrographs of *B. subtilis*, and their observations were supported by other electron-microscope observations (see Smith, 1973).

When gently lysed cell preparations of *B. subtilis* (Ganesan and Lederberg, 1965) or *E. coli* (Smith and Hanawalt, 1967) were sedimented in sucrose gradients, nascent DNA was found attached to a large structure of low density (presumably a membrane-bound complex) that separated rapidly from the bulk of the DNA. Later studies also showed that this newly synthesized DNA became soluble when these structures were exposed to other detergents, as Smith and Hanawalt (1967) had observed, or to lipase (see Lark, 1969). Further evidence that these structures contained membrane fragments was that they attached to hydrophobic surfaces of magnesium-sodium lauryl sarcosinate crystals in what was termed the M-complex (Earhart *et al.*, 1968). These membrane structures contained nascent RNA as well.

The appearance of hybrid density DNA in this material, after pulse-labelling the DNA with bromouracil (Smith and Hanawalt, 1967), indicated the presence of the growing point region of the chromosome. However, whether the growing point complexes (GP-complexes) were directly attached to the membrane was not determined. When GP-complexes were isolated from cells treated with a non-ionic detergent (Brij) they contained protein, but no evidence for membranous material was found (Fuchs and Hanawalt, 1970). Considering the evidence discussed above for membrane attachment points to folded chromosomes (Rosenberg and Cavalieri, 1968; Worcel and Burgi, 1972), entrainment of some non-membranous GP-complexes in membrane material might be expected because of density heterogeneity of the complexes, or because of the procedures employed. Thus, the evidence for membrane attachment of the replication fork in bacteria is, at present, inconclusive. With bacteriophages, however, there is evidence for membrane attachment of the replication fork (see review by Siegel and Schaechter, 1973). On the other hand, Huberman *et al.* (1973) found that DNA replication in mammalian cells was not restricted to the nuclear membrane but occurred throughout the nucleus, suggesting that, in these cells, the replication fork does not have to be membrane-attached for DNA elongation. Nevertheless membrane-DNA complexes have been isolated from mammalian cells (Pearson and Hanawalt, 1971) and the membrane's role, if any, at the replication fork will undoubtedly receive further investigation in both bacteria and mammalian cells.

In contrast, there is evidence for membrane attachment at the origin of replication. When membrane-DNA, as defined by the procedure of Smith and Hanawalt (1967), was used for transformation studies (Sueoka and Quinn, 1968), the *ade*16 marker was found at enriched frequency over other markers. Since the *ade*16 marker lies very close to the replication origin (see Fig. 7, p. 274), this higher frequency of the early marker suggested an origin attachment to the membrane. The frequency of the *met* marker (a very late gene) was also increased in the membrane-DNA preparation, suggesting a corresponding attachment of the terminus. Also, when synchronously replicating DNA from germinating spores was pulse-labelled with ^{3}H-thymidine at the origin, the label was retained in the membrane fraction during further cell growth in unlabelled thymidine. These results support the suggestion of Sueoka and Quinn (1968) that the chromosome origin, and possibly the terminus, are permanently attached to the membrane. Similar evidence for membrane attachment of origin and terminus in *B. subtilis* was obtained by Yamaguchi and Yoshikawa (1973). However the evidence for terminus attachment (as defined by the *met* marker) is now somewhat puzzling in the light of the demonstration by Wake (1973) of 50% bidirectional replication in *B. subtilis* (see Section II, B1, p. 261). If the *ade*16 to *met* part of the chromosome is linked and sequentially replicated in one direction (Yoshikawa and Sueoka, 1963a, b; Section I, B.2(b), p. 252), then the *met* marker cannot be closely linked to *ade*16 in a bidirectionally replicating circular chromosome. This suggests that there may be at least two distinct membrane attachment sites for the *B. subtilis* chromosome.

The membrane attachment at the origin immediately suggests a role for the membrane in initiation of DNA replication. As mentioned in the previous section, Worcel and Burgi (1974) obtained evidence suggesting that the release and attachment of the folded chromosome is involved in some replication process. Marvin (1968) proposed a model for control of initiation based on the assumption that formation of a new membrane site is inhibited within a critical distance of the old site. Thus, the old existing site controls initiation by repressing new site formation until new membrane is made to relieve the repression. Helmstetter (1974a, b) also supports this concept of the timing of initiation being determined by rate of cell-envelope growth.

In summary, the bacterial chromosome is attached to the cell membrane at one or more places. Although the specificity and permanency of these attachments are still under investigation, the origin seems to be one specific place of attachment. Very possibly the membrane-DNA interaction at this site could be linked to initiation of DNA replication. Membrane attachment of the replication fork, in bacteria, is not yet firmly established experimentally. Models for the segregation of replicated chromosomes to daughter cells have assumed membrane attachment (see Lark, 1969; Earhart, 1970), but the membrane linkages for segregation of chromosomes could arise from the origin-terminus attachments, or from transient attachments along the chromosome. Future studies on the attachment of membrane to specific functional chromosome locations will undoubtedly give some insight into how the membrane interaction with the chromosome contributes to the DNA replication process.

B. Replication Mechanics

In the previous section (II, A, p. 254) we discussed DNA replication at the level of the whole chromosome. Now we will consider some of the mechanics involved in DNA replication at an intermediate level of investigation between that on the whole chromosome and that on the biochemical events underlying replication. Perhaps another description of this "intermediate level" is that it is a study of the different ways a cell can increase the number of replication forks on a chromosome. Two replication forks may form at a common origin but travel in opposite directions, resulting in a chromosome which replicates bidirectionally (see p. 261). Also a second replication fork may form before the first fork has finished replicating in the same direction, resulting in multifork "dichotomous" replication (see p. 265). Sometimes the second fork only occurs on one template strand (asymmetrical replication), but more frequently it occurs on both template strands simultaneously (symmetrical replication; p. 266). By increasing the number of forks during one replication cycle, the cell achieves an increase in rate of DNA formation without increasing the fork travel speed. Therefore, the control of rate of chromosome formation is not through the rate of travel of the replication fork during elongation, but by fork formation (chromosome initiation). Viewed in this way, replication mechanics discusses the different ways in which the chromosome can initiate DNA replication.

1. *Bidirectional Replication*

Before 1968 it was commonly assumed from the prevailing evidence that DNA replication was unidirectional, with a single replication complex traversing the entire chromosome. The possibility that, in some biological systems, DNA replication may be bidirectional first became apparent from the observations of Huberman and Riggs (1968). They found that, in autoradiograph tracks of pulse-labelled mammalian (Chinese hamster) chromosomes, the labelled stretches terminated in decreased grain density at both ends of the tracks (when the label was diluted before removal), in agreement with replication in both directions. If replication were unidirectional, tracks should fade at only one end. Replication of DNA is not invariably bidirectional in all mammalian systems, however, since replication of mouse mitochondrial DNA is unidirectional (Kasamatsu and Vinograd, 1973).

In prokaryotic systems, the idea of bidirectional replication was discounted until Schnös and Inman (1970) found evidence for this in lambda-phage. Using electron microscopy, they observed replicating lambda-phage contained a high proportion of double-branched circular structures. By denaturation mapping, they located the position of the branch points and showed these represented two oppositely travelling replication forks. With this clear demonstration of bidirectionality in a prokaryotic system, the direction of bacterial DNA replication was examined more closely; this resulted in the demonstration of bidirectionality in a number of bacteria. Evidence for bidirectional replication in *E. coli* was obtained by Masters and Broda (1971), who compared marker frequencies of bacteria transduced with lysates of phage obtained from exponentially growing cultures of *E. coli* B and K12. Because gene frequencies are higher for genes replicated early in the cycle, as discussed previously (p. 252), marker frequencies indicate the time of gene replication during the cell cycle and follow a gradient of decreasing values from the origin to the terminus of replication. Masters and Broda observed two gradients of transduction, with marker frequencies decreasing in both directions from the putative origin located at about 65 minutes on the standard genetic map (Taylor and Trotter, 1972).

Their interpretation was soon supported by results from a variety of experiments with exponential cultures. Bird *et al.* (1972) constructed isogenic strains of *E. coli* K12 lysogenic for phage λ which integrates at a single position on the chromosome, and for phage Mu-1 which integrates at a number of different chromosomal locations. Phage Mu-1 inactivates the gene into which it is incorporated, thereby indicating the location of this phage. DNA from exponential cultures of each strain was extracted, denatured, immobilized on filters, and hybridized against a mixture of radioactive DNA from the two kinds of phages. The amount of hybridization with phage Mu-1 DNA provided the assay for Mu-gene dose, while that for phage λ DNA provided normalization for the amount of DNA present. Bird *et al.* (1972) also observed two gradients of marker frequency, and since the two decreased in the same manner their results support replication at the same velocity in opposite directions from the point of initiation of DNA synthesis. Similarly, Jonasson (1973) also found bidirectional replication by the method of the DNA/DNA hybridization in *E. coli* strain C, using phages P2 and λ. Evidence for bidirectional replication in *Salmonella typhimurium* has also been obtained (Nishioka and Eisenstark, 1970; Fujisawa and Eisenstark, 1973).

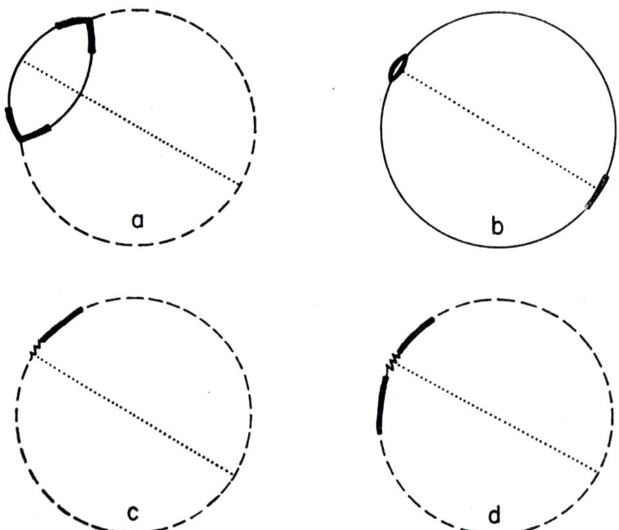

FIG. 3. The *Escherichia coli* chromosome in either a bidirectional mode (a, b, d) or a unidirectional mode (c). The dotted line divides the chromosome in half with the upper left end representing the origin of replication and the lower right end the terminus of replication. Solid lines indicate visualization of DNA; dashed lines indicate unsighted parts of the chromosome. Extra heavy lines are tritium label; the region ∿∿ contains bromouracil. See text for explanation of a, b, c, and d.

Autoradiographic evidence provided more direct support for bidirectional replication, both in *B. subtilis* and in *E. coli*. Cultures of *E. coli* strain TAU-bar were synchronized by amino-acid starvation, and DNA synthesis was initiated in the presence of ^{3}H-thymidine at low specific activity (Prescott and Kuempel, 1972). Then the cultures were shifted to higher levels of specific activity before stopping DNA synthesis. Autoradiographic tracks were observed to have increased grain densities at both ends of labelled stretches rather than at one end only as predicted for unidirectional replication. Similarly, with cultures of *B. subtilis*, synchronous initiation of DNA replication (achieved by germination of spores) followed by shifting briefly to a medium containing a higher specific activity of ^{3}H-thymidine showed the newly synthesized DNA to be sometimes present in loops that usually had dense regions at opposite position (Fig. 3a) (Gyurasits and Wake, 1973). Linear tracks also were seen, and are believed to arise by loop fragmentation. Later, Wake (1973) found labelled loops so large that they could account for bidirectional synthesis of at least 50% of the chromosome. It is interesting that some of these larger loops clearly showed a failure of both newly replicated end regions to separate.

Rodriguez *et al.* (1973) also obtained very clear evidence for bidirectional replication using a mutant of *E. coli* that is temperature-sensitive for initiation of DNA synthesis. By synchronizing the chromosomes, they incorporated labelled thymidine at high specific activity into the origins and termini of chromosomes uniformly labelled at lower specific activity. The terminus and origin should be contiguous if replication is unidirectional, but separated into regions at opposite locations on the circular chromosome if replication is bidirectional (Fig. 3b). They found chromosomes with two regions of high grain density.

The third kind of evidence supporting bidirectional replication is that provided by studies of marker order in synchronized cultures. Bird *et al.* (1972) starved *E. coli* K12 for its required amino acids and then allowed re-initiation of replication in the presence of bromouracil (BU) as a high-density label. They measured the amount of prophage Mu-1 marker-specific DNA in hybrid chromosomal DNA (half heavy and half light), by hybridization with phage Mu-1 DNA, to determine the sequence of marker replication for a number of markers. They found two sequences that fitted to the mapping order, in agreement with their gradient-frequency data. Hohlfeld and Vielmetter (1973) synchronized *E. coli* B/r, both by filter elution and by a transient block of DNA synthesis with nalidixic acid, and determined marker order by sequential mutagenesis with N-methyl-N′-nitro-N-nitrosoguanidine. This mutagen acts predominantly at or near the replication fork, so mutant frequencies therefore reflect the position of the replicating fork in these organisms. Again, mutant frequencies of known markers revealed two directions of replication from a common origin.

In a rather different approach, used earlier by Weintraub (1972) in studies of mammalian DNA, McKenna and Masters (1972) distinguished between uni-and bidirectional replication by fragmentation of newly synthesized DNA. After amino-acid starvation, replication was initiated with a pulse of BU in a thymine-requiring strain of *E. coli* B/r, and synthesis was then permitted for a longer period of time in the presence of thymine. The DNA was extracted and irradiated with UV, which leads to specific breakage of the DNA phosphodiester backbone adjacent to incorporated BU residues. If replication is unidirectional, then newly-synthesized DNA, labelled with ^{3}H-thymidine to distinguish it from unlabelled parental DNA, would have essentially the same length whether exposed to UV, and broken, or not, since the incorporated BU would be at one end of the strand (Fig. 3c). If replication is bidirectional, however, the incorporated BU would be located in the middle of the new strand (Fig. 3d) and breakage after exposure to UV would produce two tritium-labelled strands of half the unirradiated length, and these were found. A further result of this simple yet powerful technique was the observation that there was no linkage between parental and daughter DNA, corroborating the finding by Stein and Hanawalt (1972). Thus no replication mechanism which requires either linkage of parent and daughter DNA or unidirectional replication, such as the rolling circle model (see Smith, 1973), describes DNA replication of *E. coli* grown under these conditions.

Bidirectional replication of the *B. subtilis* chromosome (Wake, 1973) has been supported by genetic evidence by Hara and Yoshikawa (1973), who found the replication order of origin markers would only correspond to the mapped order if a bidirectional mode was assumed. Again, if the replication from *ade*16 to *met* is sequential, then the *B. subtilis* chromosome might replicate bidirectionally in an asymmetric manner (Sueoka *et al.*, 1973). By checking the replication order of markers at the terminus, A. O'Sullivan, K. Howard and N. Sueoka (personal communication) observed that a later marker (*glt*A14) replicated ahead of an earlier marker (*cit*K). Thus the terminus seems to occur between these two markers but with one replicon terminating before the other. However, these genetic data disagree with autoradiographic studies which suggest complete bidirectionality of *B. subtilis* replication (R. G. Wake, personal communication). To make the picture even more complex, J. C. Copeland (personal communication) has suggested that there may be another origin in the middle of the *ade*16 to *met* replicon, since there is a difference in replication order from the mapped order for the *rec*A and *aro*A markers. It is clear that more work is needed to resolve the degree of bidirectionality in *B. subtilis* DNA replication.

The situation in *E. coli* appears to be more straightforward. With increases in experimental accuracy, the origin of replication is observed to be in the region of the *ilv* marker (Bird *et al.*, 1972; Hohlfeld and Vielmetter, 1973). These data indicate that the origin of replication is located at 75 ± 3 minutes on the standard genetic map, a later position than earlier determinations had suggested. Assuming that replication velocities are the same in both directions, the calculated location of the terminus at 30 minutes on the standard 90-minute map is in excellent agreement with determinations of Hohlfeld and Vielmetter (1973). Their result, and the symmetry of marker gradients in the studies of Bird *et al.* (1972), clearly show that velocities and distances of replication are the same, or nearly so, in both directions of replication from the origin in *E. coli*. Thus bidirectional replication in *E. coli* appears to be equal in both directions.

2. *Dichotomous or Multifork Replication*

In bacteria, the DNA content per cell increases with growth rate, as first observed by Schaechter *et al.* (1958) in steady-state cultures of *S. typhimurium* grown at a variety of rates in different media. Similar results were obtained for *E. coli* 15T$^-$ by Lark and his coworkers (see review by K. G. Lark, 1966). Such increases in cellular DNA content are a result of dichotomous replication, as first demonstrated by Sueoka and his coworkers (Yoshikawa *et al.*, 1964; Oishi *et al.*, 1964) who showed that frequencies of genes near the origin of replication were four times greater than those near the terminus in rapidly growing synchronized cultures of *B. subtilis*. This dichotomous replication occurs when new rounds of chromosome replication are initiated before the original round is terminated, thus giving rise to multiple forks as shown in Fig. 4.

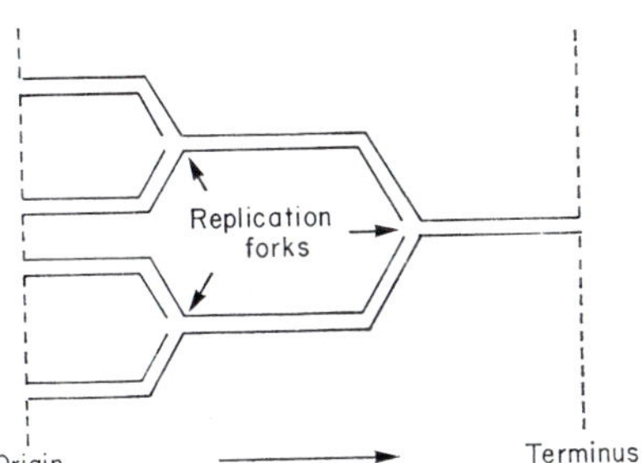

FIG. 4. Dichotomous or multifork replication. The direction of replication is from left to right on this single, multifork chromosome. A linear chromosome is represented for simplicity although multifork chromosomes probably exist in the circular bidirectional form (Wake, 1972).

Although this figure shows only three replication forks, later studies of the same system gave evidence for as many as 15 forks (Yoshikawa and Haas, 1968).

Other observations also are consistent with dichotomous replication. In synchronized cultures of *E. coli*, with generation times of about an hour, Clark and Maaløe (1967) and Helmstetter (1967) observed that the rate of DNA synthesis doubled about halfway through the cell cycle, in agreement with the introduction of extra replication forks. A similar abrupt increase in the rate of nitrosoguanidine-induced mutagenesis was observed (Ward and Glaser, 1969). A visual demonstration of multifork chromosomes was achieved, in *B. subtilis*, by Wake (1972); using autoradiography, he was able to show a bidirectionally replicating multifork loop, representing the part of the chromosome that had undergone multiple initiations (see Fig. 6a).

As mentioned previously (p. 253), the rate of DNA chain elongation appears to be constant and independent of generation time. Constant rates of DNA elongation were observed in rapidly-growing synchronized cells of *E. coli* by Helmstetter and Cooper (1968) and Helmstetter *et al.* (1968), and also were suggested for slow-growing steady-state cultures by Kubitschek and Freedman (1971). As proposed by Maaløe and Kjeldgaard (1966) such constancy requires that the rates of DNA synthesis be regulated by the frequency of initiation of new rounds of DNA regulation. Because DNA is replicated at a constant rate, cells with but a single replication fork would be limited to a generation time of about 40 minutes, the time for the fork to traverse the entire genome. The presence of multiple forks, however, allows different segments of the genome to be copied simultaneously at different forks, and thereby allows the entire genome to be copied in a shorter time. In this way, generation times of 20 minutes or less can be achieved in *E. coli*. Detailed verification of multifork replication in rapidly growing cultures was provided by Helmstetter and Cooper (1968), and their model for DNA synthesis during the cell cycle will be discussed later (p. 308).

3. *Symmetry*

Bacterial DNA replication appears always to duplicate both strands of the duplex parental DNA in the region of the replication fork. Symmetry also occurs in dichotomous chromosomes with the simultaneous initiation of replication forks in daughter duplexes (Fig. 5a). This mode of dichotomous replication is called *symmetric*. If only one of the daughter duplexes initiates a second round of DNA synthesis, the replication mode is termed *asymmetric* (Fig. 5b). Since the bacterial chromosome is circular and replicates bidirectionally, we have also included more representative diagrams for symmetric and asymmetric modes of replication (Fig. 6).

In the cell, dichotomous replication is symmetric, and appears to be the natural mode during balanced exponential growth in bacterial

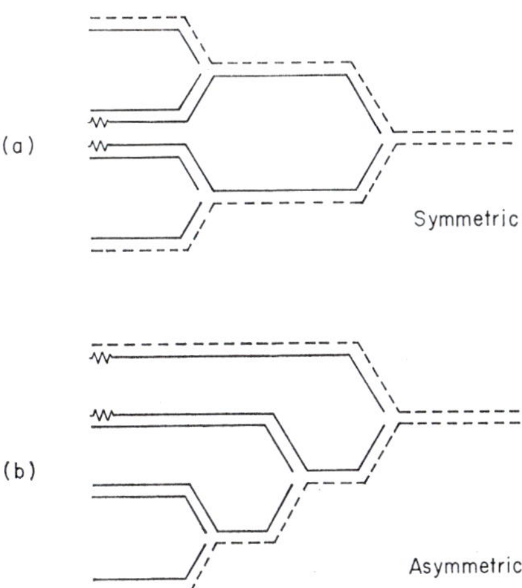

FIG. 5. Different modes of dichotomous (multifork) replication (linear representation). The dashed lines represent template strands, and the solid lines represent newly replicating strands. Replication is from left to right. (a) Symmetric mode of dichotomous replication, and (b) asymmetric mode of dichotomous replication in a tritiated thymidine pulse (see text). Adapted from Quinn and Sueoka (1970).

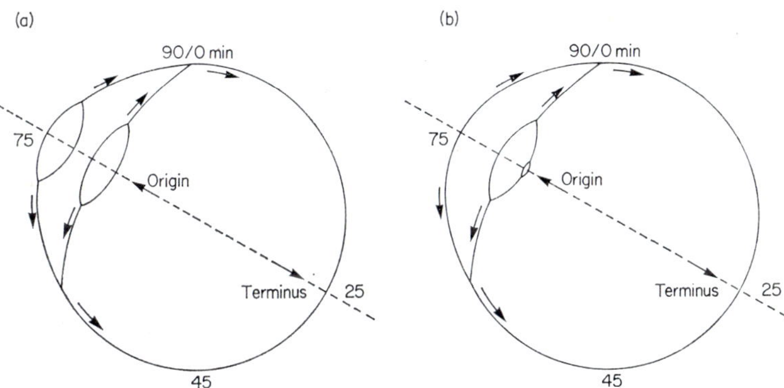

FIG. 6. Different modes of dichotomous (multifork) replication (circular representation). The *Escherichia coli* chromosome is shown as a bidirectionally replicating circle. (a) Symmetric mode of dichotomous replication, and (b) asymmetric mode of dichotomous replication.

batch cultures. As described earlier, a symmetric replication mode was suggested by the marker-frequency analysis in synchronized cultures of *B. subtilis* (p. 65). Replication symmetry was confirmed in synchronized dichotomous cultures in density-transfer experiments by Quinn and

Sueoka (1970). After pulse-labelling newly initiated daughter strands with ^{3}H-thymidine, and then shifting to medium containing heavy isotope, they observed that, following the next initiation, this label was associated only with completely heavy DNA (Fig. 5a). As may be seen from the figure, if replication is asymmetric the DNA at the origin of one of the strands of the original duplex will replicate twice before the entire chromosome is replicated once and the label would be found in both H–H and H–L DNA. This was not observed. Fritsch and Worcel (1971) demonstrated symmetric replication in rapidly growing exponential cultures of *E. coli* by an extension of the same method; no matter how branched was the chromosome, all of the duplex DNA was in the H–L form one generation after density transfer. Symmetric replication also has been observed in autoradiographs of re-initiated chromosomes of *B. subtilis* (Wake, 1972).

Symmetric replication is assumed in the Cooper and Helmstetter model for multifork replication of *E. coli*, and experimental data from rapidly growing synchronous cultures are in good agreement (Cooper and Helmstetter, 1968; Helmstetter and Cooper, 1968; Helmstetter *et al.*, 1968), as are measurements of average cellular DNA content at different growth rates (Kubitschek and Freedman, 1971). This symmetrical model is also supported by data from rapidly growing synchronous cultures of *S. typhimurium* (Cooper and Ruettinger, 1973). Symmetric replication was also observed under some conditions where growth and replication are known to be interrupted. Bird and Lark (1968) aligned chromosomes of *E. coli* by amino-acid starvation, and labelled origins and termini with ^{3}H-thymidine and ^{14}C-thymidine, respectively. During subsequent replication, initiation occurred at both of the available origins. A general mathematical solution for the age-distribution function of symmetric replication of bacterial chromosomes was given by Sueoka (1971).

Asymmetric replication is an alternative mode that appears to be available to many bacteria when grown under some non-steady state, or deleterious, conditions. In cultures of *E. coli* in which chromosomes were aligned by amino-acid starvation and then starved by thymine, it was found that, after re-initiation by the addition of thymine, both the rate of DNA synthesis and the rate of transfer of DNA to hybrid duplex suggested that only one of the two existing daughter duplexes had re-initiated replication (Pritchard and Lark, 1964). Later autoradiographic studies by Bird and Lark (1965) confirmed this asymmetric mode of replication after thymine starvation in *E. coli*. In the symmetric mode with three replication forks, addition of tritiated thymine would label six newly forming strands while, in the asymmetric mode in which only two forks are present, only four strands would be labelled. During subsequent generations these strands segregate to individual progeny cells and are identifiable by autoradiography as individual grain clusters within the colony. Bird and Lark (1965) observed a unimodal distribution of grain clusters per colony with the peak of the distribution at four grain clusters, as expected for asymmetric replication. The unimodal distribution also eliminated the possibility that half of the chromosomes had three replication forks and the other half had a single replication fork, which would have resulted in a bimodal distribution.

In thermosensitive strains, shifts to the non-permissive temperature can induce chromosome re-initiation. By using successive temperature shifts, Worcel (1970) introduced multiple forks in each chromosome in cells of *E. coli* and used density transfer to show that only a single fork was introduced at each shift. These premature asymmetric initiations always ocurred at the same point on the chromosome, that is at the same origin defined by amino-acid starvation (Stein and Hanawalt, 1969). Worcel (1970) also presented evidence that asymmetric replication repeatedly involved the same strand of the duplex, identified by density labelling. Schwartz and Worcel (1971) later showed that the induced re-initiation led to a synchronized round of chromosome replication, as was evident from phased increases in levels of inducible enzymes and from the sequence of mutations induced with nitrosoguanidine.

Replication of DNA also is highly asymmetric during bacterial conjugation. A single strand of donor DNA is displaced into the recipient cell while a new copy is synthesized and remains in the donor (Vapnek and Rupp, 1970; see also Curtiss, 1969). As suggested by Rupp and Ihler (1968) and by Ohki and Tomizawa (1968), transfer of DNA during conjugation is consistent with the asymmetric, rolling circle model of DNA replication proposed by Gilbert and Dressler (1968). In this model, one strand of the circular DNA duplex is cut and synthesis begins at the newly exposed 3′-OH end of the molecule using the unbroken circle as a template. In agreement with this model, more than an entire chromosome length can be transferred to the recipient (Fulton, 1965) and the 5′-end is transferred first (Rupp and Ihler, 1968).

As mentioned above, thymine starvation appears to induce premature initiations in *E. coli*, with the resulting chromosomes replicating in an asymmetric mode. Thymine starvation in *B. subtilis* also induces premature initiations (Kallenbach and Ma, 1968, Quinn and Sueoka, 1970). In exponentially growing cells, Kallenbach and Ma (1968) found the ratio of origin (*ade*16) to terminal (*met*) chromosome markers was three to one after a 30-minute period of thymine starvation. They also saw a doubling (rather than a tripling) in the rate of the DNA synthesis, also in keeping with an asymmetric mode of replication. These data would appear to be in conflict with the observations of Quinn and Sueoka (1970) and Wake (1972) (described above, p. 268) that synchronized chromosomes, which are prematurely initiated by thymine starvation, are symmetrically replicated in *B. subtilis*. The difference might arise from strain differences, different periods of thymine starvation, or the fact that Quinn and Sueoka (1970) used germinating spores for synchronization of growth in their experiments. In addition, the Kallenbach and Ma (1968) experiments were done in a non-synchronized system where there could be a mixture of chromosomes at different stages of dichotomy or non-dichotomy.

Other conditions also induce premature initiations in cells. Bromouracil induces premature initiation as shown by replication of labelled origins (Abe and Tomizawa, 1967; Wolf *et al.*, 1968). Abe and Tomizawa (1967) suggested that this BU-induced dichotomy resulted from premature initiations of both origins of a replicating chromosome. Again premature initiation occurred after exposure of cells to nalidixic acid, as indicated by rates of DNA synthesis (Boyle *et al.*, 1967). However, it was not determined whether these initiations were asymmetric or symmetric. In conclusion, symmetric replication appears to be the normal mode of DNA replication in vegetative cells, and in rapidly growing cultures this mode takes full advantage of dichotomous replication. Asymmetric replication occurs during conjugation, or under some conditions of stress that lead to premature initiations of some, but not all, of the available chromosome origins. Further research must be done to determine the specific conditions which lead to asymmetry, and whether this mode of synthesis is related to some specific control of the DNA replication rate.

III. The Biochemistry of DNA Replication

In this section, we examine DNA replication at the basic biochemical level and look at the specific events occurring at the replication fork. First, as background, it is necessary to review the principal *in vitro* DNA replication systems which have aided in our understanding of the functions of replication enzymes (see p. 271), and emphasize some of the special experimental difficulties encountered in studying the biochemistry of replication. Since the events at the replication fork are complex and probably involve many proteins, in sub-section B we simplify the discussion of replication proteins in the following way. Of several contemporary or recent models, we present the discontinuous mode of DNA synthesis and use it to anchor our description of the biochemical events at the replication fork. By inserting the possible replication proteins into the various stages of the discontinuous mode of synthesis, this biochemical model provides a framework within which to categorize the many replication proteins (Table 3). Finally, on page 287, we discuss each of the possible replication proteins in detail, including known enzymic activities and the known *dna* gene products.

A. *In vitro* DNA Replication Systems

The study of DNA replication has long been hampered by the lack of a good *in vitro* system. The major problem is that, once the replication complex is dismantled, the complete replication system cannot be re-assembled in a functional form. Since the structure of the intact chromosome probably is related to DNA replication (Section IIA, p. 255), the re-assembly of such a huge complex organization of DNA, RNA, membrane and protein has proved to be technically difficult. Isolated DNA polymerases I, II and III cannot by themselves perform an adequate amount of semiconservative replication. Besides, they cannot initiate new chains since they require a 3′-OH primer strand opposite the longer template strand. The development of the present *in vitro* replication systems was facilitated by the isolation of the DNA polymerase I mutant (DeLucia and Cairns, 1969). In this mutant, the incorporation of precursors into DNA represented DNA replication with little background repair incorporation.

Because of inactivation by dismantling, all bacterial systems were developed on the principle of minimal disturbance of the replication

Table 2. *In vitro* DNA Synthesis Systems in Bacteria (Adapted from Smith (1973)

System	Reference
A. *Permeable Cell Systems*	
1. Tris–EDTA or Tris-Mn^{2+}	Buttin and Kornberg (1966)
a. mostly repair	Buttin and Wright (1968)
2. Ether-treated	Vosberg and Hoffman-Berling (1971)
a. replication	Dürwald and Hoffman-Berling (1971)
3. Toluene treatment	Moses and Richardson (1970a)
a. replication	Matsushita *et al.* (1971)
4. Toluene treatment + Triton X-100	Moses (1972)
5. High sucrose	Wickner and Hurwitz (1972)
a. replication	
6. Freezing and thawing	Billen *et al.* (1971a, b)
a. repair and replication	
7. Freezing and thawing, sucrose-Mg^{2+}-sodium azide, Brij 58	Ganesan (1971)
a. replication	
B. *Lysed Cell Systems—All Replication*	
1. The cellophane system	Schaller *et al.* (1972)
2. The agar system	Smith *et al.* (1970)
3. The membrane systems	Ganesan and Lederberg (1965)
	Knippers and Strätling (1970)
	Okazaki *et al.* (1970)

complex. The replication complex was isolated intact but was accessible to precursors and replication proteins. This led to two major experimental approaches: (1) alteration of cell permeability by chemical treatment without destroying replication (permeable cell system); and (2) lysis of the cells, very gently, to isolate the replication complex in a functional state (lysed cell system). Table 2 lists many of the bacterial *in vitro* DNA synthesis systems in existence. However, we will focus on three systems: the toluenized cell system (Moses and Richardson, 1970a); the cellophane-disc system (Schaller *et al.*, 1972); and the single-stranded DNA phage system (Wickner, W. *et al.*, 1972). The first two are the most useful of each of the two major classes of bacterial *in vitro* systems. Although the third system is a phage system, it has been extremely helpful for studying bacterial DNA replication because of its dependency on host replication proteins. This system is an excellent example of how a phage system can help elucidate the functions of bacterial replication proteins.

1. Toluenized Cell System

(a) *The* Escherichia coli *toluenized cell system.* Moses and Richardson (1970a) first established that *E. coli polA*$^-$ cells treated with 1% toluene would incorporate deoxyribonucleotide triphosphates (dNTPs) into DNA in the presence of Mg^{2+} and high concentrations of ATP. Since these workers utilized a *polA*$^-$ mutant, they were relatively certain that this incorporation represented replication and not repair, and confirmed this by density-transfer studies (incorporation in the presence of heavy label dBrUTP indicated semiconservative replication over a period of 60 minutes). The ATP-stimulated synthesis was inhibited by sulphhydryl inhibitors such as N-ethylmaleimide (NEM) and *p*-hydroxymercuribenzoate (*p*HMB) but was not inhibited by antipolymerase I, cyanide or azide. Genetic evidence that this synthesis represented replication was provided by their demonstration that *dna* mutants were also unable to replicate their DNA at the non-permissive temperature in toluenized cells. Although electron microscopy showed that the cell wall and membrane were still intact after toluene treatment (Jackson and DeMoss, 1965), Moses and Richardson (1970a) were able to stimulate DNA repair with exogenous DNAse I in *polA*$^+$ strains, but not in *polA*$^-$ strains, an indication of penetration of this protein of molecular weight 31,000 daltons. Recently Moses (1972) obtained evidence that the addition of Triton X-100 increases permeability in this system but maintains replication. Although there is a diffusion of DNA polymerase I (molecular weight 109,000 daltons) and lactate dehydrogenase out of the toluenized cells, there is no indication that large molecular-weight proteins can penetrate and complement DNA replication in this system. This system does provide a distinct advantage over the lysed cell systems, however, in that it permits a greater amount of DNA synthesis. On the basis of amount of parental (template strand) labelled DNA appearing in the hybrid peak during density-transfer studies, toluenized cell synthesis represents around 5–15% of the chromosome, whereas lysed systems are generally less that 1% (except for the cellophane system (6%); Matsushita *et al.*, 1971; Burger, 1971; Knippers and Strätling, 1970; Smith *et al.*, 1970; Schaller *et al.*, 1972). This relatively greater synthesis in toluenized cells enables the product DNA from CsCl gradients to be analysed genetically by transformation assays, an advantage lacking in the other systems at present (see discussion below, p. 274).

(b) *The* Bacillus subtilis *toluenized cell system.* Toluene treatment was adapted to *B. subtilis* by Matsushita *et al.* (1971) and the *in vitro* synthesis again is semiconservative. In addition to demonstrating a

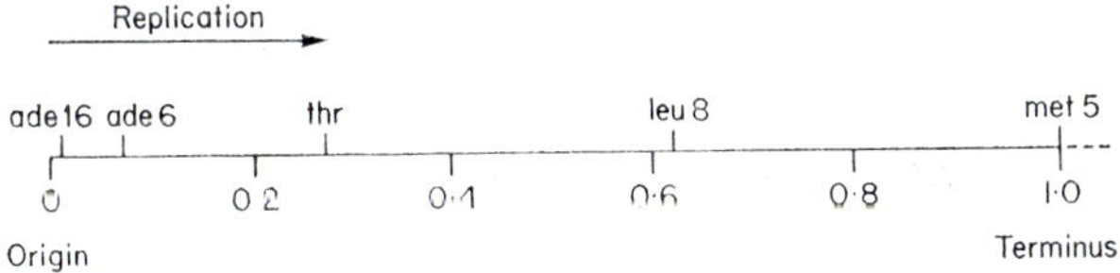

FIG. 7. An abbreviated linear map of the *Bacillus subtilis* chromosome depicting the *ade*16 to *met* replicon. From Matsushita *et al.* (1971).

clear hybrid peak after dBrUTP labelling and CsCl centrifugation, they showed that the hybrid peak separated into two peaks on an alkaline CsCl gradient, a newly synthesized heavier ^{3}H-labelled strand and a lighter ^{14}C-parental strand. They also demonstrated that the DNA synthesized was biologically active since the transformation activity of the hybrid DNA compared favourably with that in the parental peak. Furthermore synthesis proceeded only at replication forks existing in the cells prior to toluenization. To show this, a *dna* initiation mutant (White and Sueoka, 1973) was incubated at a non-permissive temperature to allow completion of the chromosomes without any new initiations. After restarting replication, the synchronized cells were subjected to parallel *in vivo* and *in vitro* density-transfer studies. For an early time sample, only the early gene *ade*16 (Fig. 7) was synthesized semiconservatively and the *leu* and *met* markers, representing the middle and terminus of the chromosome, respectively, were not synthesized at all. If replication was allowed to proceed further before toluenization, semi-conservative synthesis of the next later gene (*ade*6) could be detected. Thus sequential semiconservative replication of genes in their proper mapped order is the dominant mode of DNA synthesis in toluenized cells. This removed any doubt that DNA synthesis in toluenized cells might represent extensive random repair. Burger (1971) also presented evidence, using toluenized *E. coli*, that *in vitro* synthesis proceeds only at pre-existing replication forks. Both of these studies were performed in *polA*$^+$ cells, showing the usefulness of this system even in the presence of DNA polymerase I.

Further studies with toluene-treated *B. subtilis* cells showed that the ATP-dependent synthesis represents, specifically, elongation and that there is no chromosome initiation (Sueoka *et al.*, 1973; K. White, T. Matsushita and N. Sueoka, unpublished results). In a non-synchronized population of cells, the very early (near the origin) marker *ade*16 was synthesized in toluenized cells in low amounts compared with the parallel *in vivo* sample (Fig. 8). If there is no initiation in toluenized cells, relatively few preformed forks would exist in front of *ade*16 as compared with later markers, resulting in low amounts of *ade*16 synthesis as compared with other markers. This loss of initiation was a permanent lesion since cells washed free of toluene still showed hybrid *ade*16 in low amounts (S. E. Winston and T. Matsushita, unpublished results) and suggested permanent denaturation of the membrane/initiation complex or an irreversible lesion that blocked synthesis of some "initiation precursor".

FIG. 8. Evidence for the absence of initiation in toluenized cells. *Bacillus subtilis* strain 168TT (*thy try*) was grown at 30°C to a concentration of 7×10^7 cells/ml in 50 ml of medium C$^+$, containing 12·5 μCi of ^{14}C thymine, 3 μg cold thymine/ml, and 50 μg tryptophan/ml. Cells, 25 ml were collected, washed, resuspended in phosphate buffer, and agitated for 10 minutes at 25°, and 5 minutes at 4° in 1% toluene. The complete reaction mixture (1 ml) contained 70 mM KH_2PO_4 (pH 7·4), 13 mM Mg SO_4, 1·3 mM ATP, 33 μM dGTP, dCTP, dATP, dBrUTP, 1 μM ^{3}H-dATP, 4 mM phosphoenolpyruvate, 13 I.U. pyruvate kinase/ml and 8×10^8 toluene-treated cells. The remaining 25 ml of cells for the *in vivo* control (a) were collected by filtration and resuspended in 25 ml of C$^+$ media containing 50 μg tryptophan/ml and 10 μg bromouracil/ml. This sample was grown for 20 minutes at 37°C. Cells were collected by filtration and resuspended in 5 ml of 0·1 M KH_2PO_4 (pH 7·4). Lysates of 1 ml of this *in vivo* suspension and 1 ml of the *in vitro* reaction mixture were prepared by separately adding 0·6 ml of 0·3 M NaCl, 0·2 M EDTA (pH 8·2) and 0·15 ml of 5 mg lysozyme/ml, and incubated for 30 minutes at 37°C. Sodium dodecyl sulphate (0·042 ml of 20%) was added and incubated for an additional 15 minutes, added to 1·74 ml of the lysate and shaken with 4·6 g of CsCl were 1·95 ml of 0·01 M Tris containing 0·001 M EDTA (pH 8·4). After 72 hours centrifugation at 35,000 rpm (25°C) in a Spinco SW 50·1 rotor, three-drop fractions were collected, precipitated, and counted. Transformations were performed by adding 0·1-ml aliquots to 1 ml of competent cells, shaking at 37°C for 40 minutes, and plating on selective plates. Recipient 168 *leu*8-*met*5-*ade*16 was made competent by the method of Bott and Wilson (1967). (a) Fractions 12 to 17 represent hybrid DNA, and fractions 18 to 23 represent parental DNA hybrid *ade*16, 22%; *leu*8, 15%, *met*5, 15%. (b) Fractions 15 to 19 represent hybrid DNA, and fractions 22 to 26 represent parental DNA hybrid *ade*16, 6%; *leu*8, 29%; *met*5, 29%. Adapted from Sueoka *et al.* (1973).

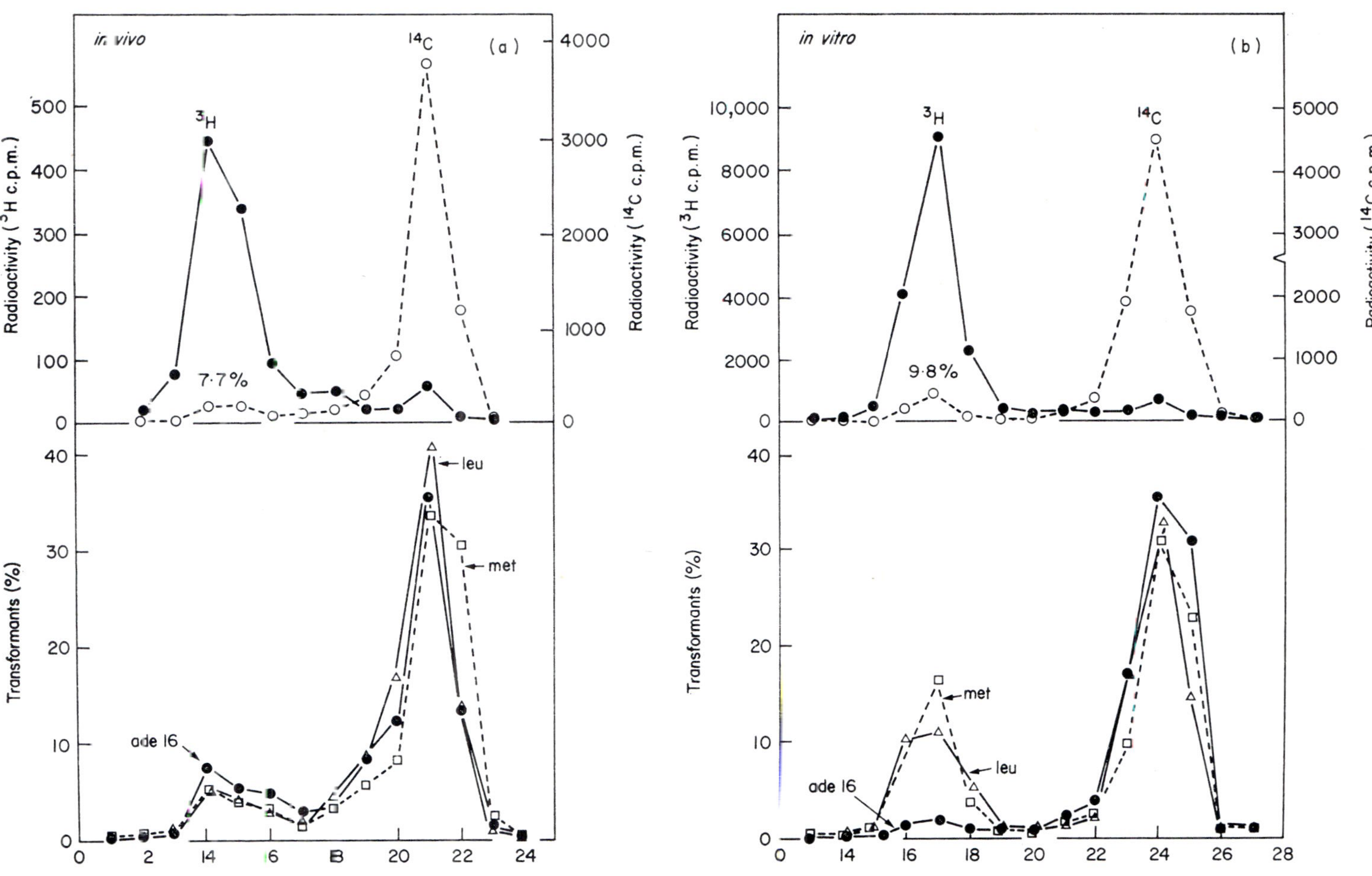

in vivo
(a)
in vitro
(b)
Radioactivity (^{3}H c.p.m.)
Radioactivity (^{14}C c.p.m.)
Transformants (%)
^{3}H
^{14}C
7·7%
9·8%
leu
met
ade 16

Matsushita and Sueoka (1974) showed that toluenized *B. subtilis* cells were also permeable to DNAse I, by inhibiting replication in a *polA*$^+$ strain with *p*-chloromercuribenzoate (pCMB) and stimulating DNA repair with exogenous DNAse I. This *p*CMB-resistant DNAse I-stimulated DNA synthesis did not occur in *polA*$^-$ strains, and indicated a DNA polymerase I dependency for this synthesis. Furthermore this "repair" synthesis remained associated with the toluenized cell when samples were subjected to vortex agitation and centrifugation before precipitation with trichloroacetic acid. This was a further indication of the penetration of DNAse I into the non-lysed toluenized cell, rather than incorporation by the action of polymerase I on DNA released from lysed cells.

(c) *Some contributions of the toluenized cell system:* The toluenized cell system gave a clear indication that ATP was directly involved in DNA replication. Moreover, in *B. subtilis,* toluenized cells have been extremely useful in studying the mechanism of action of 6-(*p*-hydroxyphenylazo)-uracil (HPUra), a drug which specifically inhibits DNA replication (see Section III, C.4(c), p. 294). Toluenized-cell studies implied that the reduced form of HPUra was the active inhibitor, and also indicated that deoxyguanosine triphosphate (dGTP) competitively reversed the action of the drug. Thirdly, the use of toluenized HPUra-resistant mutants helped establish that DNA polymerase III is the essential replicating enzyme in *B. subtilis* (Section III, C.4(c), p. 294). A fourth contribution was that sulphydryl reagents appeared to act on replication specifically, and indicated that sulphydryl reagent sensitivity might be a characteristic of an essential DNA polymerase (polymerase III has this characteristic; see p. 292) In addition, Fleischman and Richardson (1971) have shown that, if hydroxymethyldeoxycytidine 5-triphosphate (dHMCTP) replaces dCTP, then dTTP is not incorporated into DNA in toluenized cells restrictive to non-glucosylated T-even phage, but is incorporated using permissive *E. coli*. This indicates that phage enzymes are not necessary for dHMCTP incorporation, and the restrictive mechanism, present in the host before infection, recognizes HMC residues in its own DNA as well as in phage DNA. Another important contribution has been the demonstration of the RNA-linked nascent DNA fragments, and the partial sequencing of the RNA-DNA junction in toluenized *E. coli* cells (see Section III, B.1, p. 284). Recently, toluenized cells have been useful for studying repair. The usual method for studying repair in toluenized cells is to inhibit replication by either UV (Bowersock and Moses, 1973; Masker and Hanawalt, 1973), use of conditionally lethal *dna* mutants (Masker and Hanawalt, 1973; Masker *et al.*, 1973), treatment with sulphydryl group reagents (Matsushita and Sueoka, 1974), or treatment with HPUra (T. Matsushita, unpublished results). Repair is stimulated by prior UV treatment or by addition of exogenous DNAse I. Although repair is thought to be mainly mediated by DNA polymerase I (Matsushita and Sueoka, 1974), repair studies in toluenized cells have revealed an ATP-dependent DNA polymerase III-mediated repair (Bowersock and Moses, 1973; Youngs and Smith, 1973). Also ATP-dependent DNA polymerase II-mediated repair has been found in toluenized *E. coli* (Masker and Hanawalt, 1973; Masker *et al.*, 1973) and in toluenized *B. subtilis* (T. Matsushita, unpublished results; see Section III, C.6, p. 302). The ability of DNA polymerase III, an essential replication enzyme, to participate in the repair of UV-damage in the absence of polymerase I, emphasizes the importance of studying the possible relationships between replication and repair. The toluenized cell systems are useful for studying these relationships and the common enzymes and proteins involved.

2. *The Cellophane* (*Lysed Cell*) *System*

Schaller *et al.* (1972) immobilized concentrated *E. coli* cell suspensions on a cellophane surface, lysed the cells gently in the cold, and dried the lysate on the cellophane. The reaction mixture of dNTPs, ATP and Mg^{2+} was then added to the lysate containing the intact, functioning replication complexes. This cellophane system permits diffusion of small molecules through the disc (similar to dialysis tubing) to the replication complex while retaining the large molecules involved in replication. The characteristics of this *in vitro* system are quite similar to those of toluenized-cell replication. Synthesis is semiconservative over a period of 60 minutes, and proceeds at about 20% of the *in vivo* rate. The synthesis is ATP dependent, and sensitive to N-ethyl maleimide (NEM) and *p*HMB. With lysates of conditionally lethal *dna* mutants, synthesis is decreased at a restrictive temperature. Okazaki pieces (see p. 281) are formed and joined subsequently to form larger DNA intermediates. The principal advantage of this system over the use of toluenized-cell preparations is that it is accessible to molecules of any size; therefore complementation is possible. However the system is, in essence, unpurified from the *in vivo* state and there is still the need for a simpler, well-defined, bacterial replication system.

(a) *Some contributions of the cellophane system.* The cellophane system was used to purify *dna*E crude extracts and confirmed the important finding that the *dna*E gene product was DNA polymerase III (see Section III, C.4a, p. 292). DNA polymerase III was the first enzymic activity to be identified with a *dna* mutant, and therefore the first essential replicating protein to be identified (Gefter *et al.*, 1971). The cellophane system has also been used to purify *dna*G protein (Klein *et al.*, 1973; Section III, C.8f, p. 307). This protein also has been purified using the phage ϕX174 SS-RF system (Wickner, S. *et al.*, 1973b; Section III, A.3, p. 279). In addition, the cellophane system has been used to describe the function of the *dna*G locus during discontinuous DNA synthesis (Lark, 1972b; Section III, C.8f, p. 307). Detailed studies of Okazaki-piece formation in this system helped to characterize discontinuous synthesis in *E. coli*. Replication is usually discontinuous for both strands in *E. coli*, but in this system one strand is replicated discontinuously and the other continuously. Olivera and Bonhoeffer (1972b) suggest that this is a consequence of strand polarity (see Section III, B.3, p. 285). Lundquist *et al.* (1974) also showed that dCTP increases the frequency of initiation of Okazaki pieces in the cellophane system, a finding which fits with the known sequence of the RNA primer-DNA junction (Sugino and Okazaki, 1973; Section III, B.1, p. 284). Olivera and Bonhoeffer (1972a) showed the conversion of phage ϕX174 SS–RF DNA synthesis using *E. coli* lysates in the cellophane system. This correlates with the SS–RF phage *in vitro* systems which show dependency on host proteins for DNA replication (Wickner, W. *et al.*, 1972; see following section).

3. *The Single-Stranded DNA Phage System*

The life cycle of bacteriophages M13 and ϕX174 in *E. coli* consists of three phases: (1) synthesis of complementary strand to the virus plus-strand template, resulting in a double-stranded replicative-form parent molecule (SS–RF phage DNA synthesis); (2) synthesis of replicative-form (RF) progeny from the RF parent; (3) the asymmetric replication of RF to produce single-stranded progeny. Of the three phases, the second is the most similar to bacterial semiconservative replication. However, the first phase is readily reproduced *in vitro* and, since the ϕX174 *in vitro* process is dependent on many of the host *dna* gene functions, it has been very useful for studying the function of bacterial replicating proteins.

There are two kinds of experiments performed with the SS–RF system, those done with crude extracts, and those using semipurified components. In experiments with crude extracts, a high-speed supernatant enzyme fraction, from gently lysed cells, converts M13 and ϕX174 to the RF form (Wickner, W. *et al.*, 1972). The conversion of M13-DNA but not ϕX174-DNA was inhibited by rifampicin, a specific inhibitor of RNA polymerase. This paralleled the *in vivo* drug effects (Brutlag *et al.*, 1971; Silverstein and Billen, 1971). The Kornberg group later showed that RNA served as a primer for initiating a DNA chain. Priming with RNA has also been implicated in chromosome initiation

(Lark, 1972a), in events prior to lambda DNA replication (Dove *et al.*, 1969, 1971), the replication of colicinogenic factor Col-El (Clewell *et al.*, 1972; Blair *et al.*, 1972) and in the "initiation" of Okazaki-piece formation (Sugino *et al.*, 1972; Section III, B.1, p. 282). While RNA polymerase is not needed for the replication of ϕX174, there is some evidence that RNA made by a different synthetic system may prime its DNA synthesis.

Schekman *et al.* (1972) discovered two other requirements for the ϕX174 and M13 replicating systems, namely ATP and the polyamine spermidine. Further analysis revealed that the conversion of SS to RF was dependent on all known *E. coli dna* gene products (*dna*A, B, C[D], G and E; Schekman *et al.*, 1972; Wickner, R. B. *et al.*, 1972b) except for *dna*F. Interestingly, phage M13 and fd (a very similar phage) *in vitro* systems only required the *dna*E (DNA polymerase III) gene product among the *dna* mutants. Thus these two phage classes of SS to RF conversion are quite distinct in at least five components, namely *dna*A, B, C(D) and G, and RNA polymerase.

Since *dna*E was responsible for replication of both M13 and ϕX174, Kornberg's group purified DNA polymerase III for use in these phage *in vitro* systems (Wickner, W. *et al.* 1973), leading to a second series of experiments using the phage system with semipurified components. Initially they found that polymerase III was inactive in their systems and instead isolated a new form of polymerase III which they called Pol III*. They also found a factor which they called Copol III* which was necessary for chain growth. This semipurified system will be described in detail in the section on DNA polymerase III (p. 293). In summary, the SS to RF *in vitro* system can be utilized in two kinds of experiments; firstly, those involving *E. coli* crude extracts, and secondly a semi-purified system using a different form of DNA polymerase III (Pol III*). These *in vitro* systems are certainly more well defined and purified than any obtained from bacteria.

(a) *Some contributions of the SS to RF system.* This system yielded one of the first clear indications that RNA could be involved as a primer molecule for DNA polymerization, and RNA priming is a solution to the inability of DNA polymerases to self-prime. Further, a new mechanism for DNA polymerase elongation (pol III*) was discovered which could be important for bacterial as well as phage DNA replication (see Section III, C.4(b), p. 293). Lastly the dependency of the phage ϕX174 system on host *dna* genes allowed the purification of the *dna*B (Wright *et al.*, 1973), *dna*C(D) (Wickner, S. *et al.*, 1973a) and *dna*G (Wickner, S. *et al.*, 1973b) gene products. The SS to RF system should prove valuable for studying the role of these gene products, much in the same way as the *dna*E gene product has been studied.

B. THE DISCONTINUOUS MODE OF ELONGATION

There are many models for DNA replication (see Smith, 1973; Gross, 1972). However, we will review only the discontinuous model for two reasons. Firstly, this model describes events at the replication fork at our third, biochemical, level of discussion and does not involve mechanisms at the whole chromosome or intermediate levels. Therefore it is convenient for categorizing the replication proteins and for giving an overall view of the biochemical events. Secondly, there is considerable evidence that discontinuous synthesis is the actual mode of elongation and that it is an accurate description of the replication fork events.

The basic concept of discontinuous synthesis is simple, and can be viewed as resolving the question of whether DNA elongation is one continuous polymerization reaction or a series of simultaneous polymerizations. Simultaneous polymerizations are advantageous for increasing rate of DNA synthesis without increasing replication-fork travel speed. Sakabe and Okazaki (1966) provided evidence for the latter alternative (Fig. 9) when they demonstrated synthesis of small DNA intermediates (known popularly as "Okazaki pieces"). There is evidence for Okazaki pieces occurring in *E. coli* (Sakabe and Okazaki, 1966), *B. subtilis* (Oishi, 1968), bacteriophage T4 (Sugino and Okazaki, 1972); and in many eukaryotic systems inclding Hela cells (Painter and

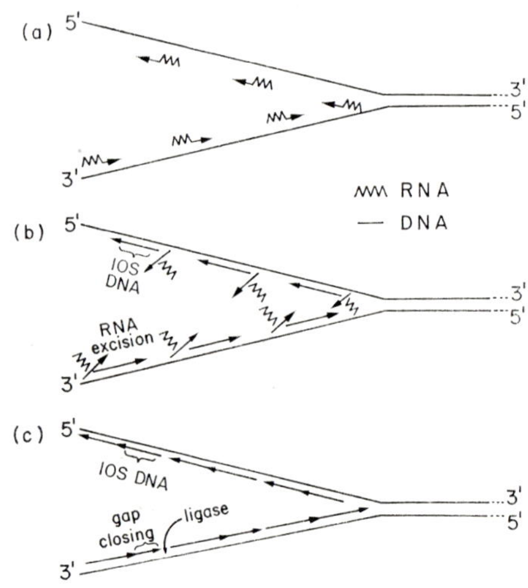

FIG. 9. The discontinuous mode of elongation. (a) RNA primer formation, (b) formation of the 10S DNA intermediate and excision of the RNA primer, and (c) gap closing between Okazaki pieces and final ligase sealing. Adapted from Okazaki *et al.* (1968b) and Hirose *et al.* (1973).

Schaefer, 1969; Habener *et al.*, 1970); Chinese hamster fibroblasts (Schandl and Taylor, 1969); Ehrlich ascites cells (Sato *et al.*, 1970); regenerating rat liver (Berger and Irvin, 1970) and human lymphocytes (Fox *et al.*, 1973). The basic experiment is to pulse-label DNA very briefly with ^{3}H-thymidine and observe these small single-stranded 4-10S DNA pieces by separation on sucrose gradients, or by hydroxyapatite column chromatography or nitrocellulose binding. The single-stranded fragments are then chased into higher molecular-weight double-stranded DNA using longer pulse times (Fig. 10). Thus the discontinuous model of replication invokes a series of short single-stranded DNA intermediates which are synthesized in the 5′ to 3′ direction along one or both template strands at the replication fork (Fig. 9).

1. RNA Primer Formation

There is evidence for an RNA synthesis step during bacterial chromosome initiation (Lark, 1972a; Laurent, 1973). However RNA synthesis did not appear to be required for bacterial chain elongation since this latter process is resistant to rifampicin, an inhibitor of classical RNA polymerase. For instance, DNA replication in toluenized cells is not inhibited by high concentrations of rifampicin in *B. subtilis* (Sueoka *et al.*, 1973) or in *E. coli* (T. Matsushita, unpublished results). Therefore it was surprising when Okazaki and his coworkers reported that very early (nascent) Okazaki pieces contained RNA that was covalently linked to DNA (Sugino *et al.*, 1972; Okazaki *et al.*, 1973). This suggested that Okazaki pieces are formed by extension of even shorter RNA chains during the elongation process.

The RNA-linked nascent DNA fragments were demonstrated in *E. coli* as follows. The pulse-labelled DNA was denatured by heat, formamide or hot formaldehyde, and the single-stranded DNA was subjected to $CsSO_4$ equilibrium sedimentation. Nascent fragments were observed at an intermediate density between the RNA and DNA marker peaks (Fig. 11a) and were chased into the DNA marker density after longer pulse times. When treated with alkali and RNAse (conditions which would remove RNA covalently linked to DNA), parallel samples of the nascent single stranded fragments showed a density equivalent to the DNA marker (Figs. 11b and c); however sonication of the nascent samples did not shift the density back to DNA. These results are all consistent with a covalently linked RNA and DNA molecule. These RNA–DNA fragments were found under growth conditions (14°C) giving a doubling of 400–600 minutes and for a 15–30 seconds pulse of tritiated thymidine. Since the RNA could not be detected with slightly longer pulse times, the formation and removal of the RNA primer appears to be a fast process.

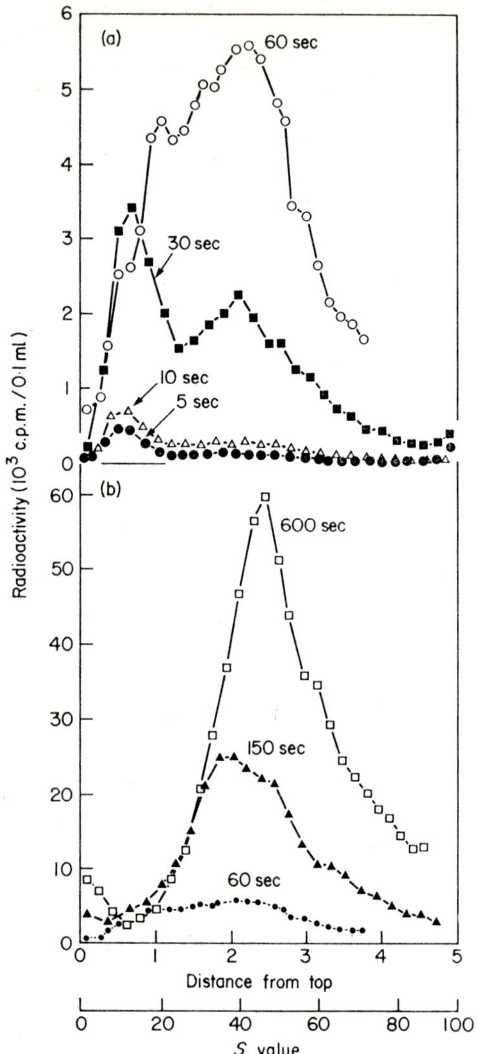

FIG. 10. Alkaline sucrose-gradient sedimentation of pulse-labelled DNA from *Escherichia coli* B. Cells were grown in a glucose-salt medium at 37°C to a titre of 3×10^8 cells/ml and then at 20°C to 5×10^8 cells/ml. The culture was pulse-labelled with 10^{-7} *M* ^{3}H-thymidine at 20°C for the indicated time. DNA was extracted by NaOH-EDTA treatment and sedimented in a SW 25·3 rotor for 10 hours at 22,500 rpm and 4°C. From Okazaki *et al.* (1968a).

The formation of the RNA primer was resistant to 200 μg of rifampicin/ml which is consistent with the rifampicin-resistance found in elongation. Therefore the RNA primer formation could be synthesized by the classical

RNA polymerase which is interacting with some replication factor that confers rifampicin-resistance on the enzyme; or this RNA primer could be synthesized by an unknown non-classical rifampicin-resistant RNA polymerase.

To demonstrate further the probable role of nascent RNA–DNA pieces in elongation, Okazaki's group took advantage of the toluenized

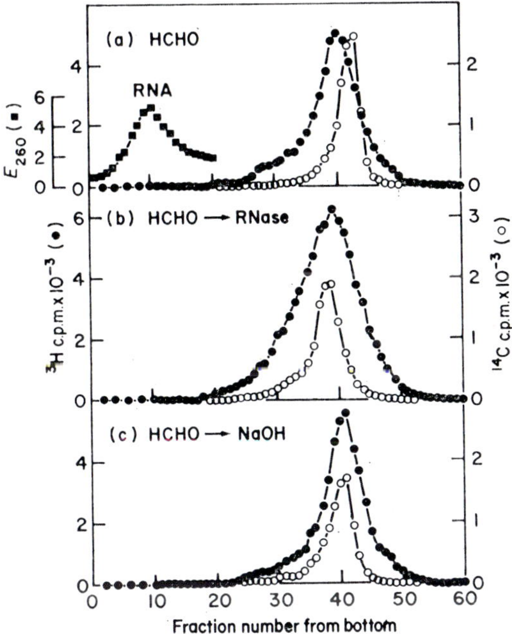

FIG. 11. Effect of ribonuclease and alkali treatment on the density of nascent DNA. Nucleic acid, extracted from *Escherichia coli* Q13 pulse-labelled with ^{3}H-thymidine for 15 seconds at 14°C and heated in 12% formaldehyde at 100°C for 10 minutes, was subjected to Cs_2SO_4 equilibrium centrifugation in a SW50L rotor without further treatment (a); or after incubation for 1 hour at 37°C in SSC containing 50 μg ribonuclease IA/ml and 10 μg of ribonuclease T_1/ml (b) or after incubation in 0·3 *M* NaOH at 37°C for 15 hours; (c) denatured *E. coli* ^{14}C-DNA was included as marker. From Sugino *et al.* (1972).

cell system (see p. 273). This *in vitro* system performs normal semi-conservative replication in the presence of ATP and dNTPs. Since DNA synthesis in toluenized cells is only 7% of the *in vivo* rate at 37°C (Matsushita *et al.*, 1971), these conditions result in a decreased replication rate. Thus Sugino and Okazaki were able to isolate RNA-linked DNA fragments after pulse-labelling toluenized cells for 15–30 seconds. They also demonstrated that a transfer of ^{32}P label occurred from the deoxyribonucleotides to ribonucleotides after digestion with alkali or pancreatic RNAse. These conditions split the bond between the 3′ ribonucleotide and the 5′ position of the adjacent deoxyribonucleotide to yield a terminally ^{32}P-labelled (2′ or 3′ phosphate) ribonucleotide or oligoribonucleotide. These ^{32}P-transfer experiments also revealed a unique structure at the RNA–DNA junction, namely . . . p(rPy)p(rA)p-(rU or rC)p(dC)p . . ; Sugino and Okazaki (1973) suggest this sequence may be a universal sequence of the RNA–DNA junction in *E. coli* Okazaki pieces and serve as a specific signal for termination of synthesis of the RNA segment and/or the "initiation" of the DNA segment. This "initiation" of the formation of the 10S Okazaki DNA piece should not be confused with initiation of synthesis of the whole chromosome. Initiation of the whole chromosome occurs only once per cell cycle as opposed to the frequent "initiations" of the many 10S Okazaki pieces required for the synthesis of one chromosome. Further evidence for specific "initiation" points along the chromosome was the finding that *in vivo* RNA–DNA fragments also have cytosine as the specific 5′-terminal base of the DNA segment, and the RNA–DNA juncture is also . . .p(rPy)p-(dC)p . . . (Hirose *et al.*, 1973).

2. *The Present Model for Discontinuous Synthesis*

Our present knowledge of the discontinuous mode of elongation can be summarized as follows. After the replication fork has formed (chromosome initiation), it starts the process of elongation (Fig. 9, p. 281). A rifampicin-resistant RNA polymerase forms the RNA primer which "initiates" the formation of Okazaki pieces at initiation sites along the chromosome. With the formation of this primer molecule, DNA polymerase III extends polymerization in a 5′–3′ direction until the 10S intermediate is formed. At some early time before joining of the fragments, the RNA primer is excized. The gaps between the Okazaki pieces are then closed by DNA polymerase I and sealed by DNA ligase to form a continuous DNA strand.

3. *Discontinuous and Continuous Strand Synthesis*

An oddity in piece formation in *B. subtilis* is that only one replicated strand is synthesized in small pieces (Kainuma and Okazaki, 1970) whereas the synthesis of the other strand seems to be continuous. In *E. coli* both strands appear to be synthesized discontinuously since all DNA is synthesized in short fragments *in vivo* (Okazaki *et al.*, 1968a, b). However, Olivera and Bonhoeffer (1972b) were able to detect two distinct size classes of DNA intermediates in the *E. coli* cellophane system, namely 9S and 38S, in equal quantity. Furthermore, Herrman *et al.* (1972) found that hybridization gave complementarity between the two classes but no complementarity within either class alone. This suggested that the synthesis of Okazaki pieces in the *in vitro E. coli* system was similar to that in *B. subtilis* with continuously and discontinuously replicating daughter strands. Therefore Olivera and Bonhoeffer (1972b) proposed that the size of Okazaki pieces is determined by competition between the "initiation" of Okazaki-piece formation and subsequent elongation of the pieces. When initiation rate is low, an initiation site is by-passed and overgrown by the 5′–3′ elongation of the Okazaki piece initiated from a neighbouring site. Because of the opposite polarity of the sequentially replicating strands, one strand would have its initiation site available for a shorter time. Thus DNA intermediates would be much larger on the 3′ left-ended template strand than on the 5′ left-ended template strand for a replication fork travelling from left to right (Fig. 9, p. 281). In *E. coli*, where discontinuous synthesis occurs on both strands, Okazaki-piece elongation relative to initiation could increase under certain conditions and result in larger intermediates synthesized on the 3′ ended template strand. If Okazaki-piece elongation is decreased by lowering the concentration of deoxyribonucleotides *in vitro*, the size of the large 38S intermediates is decreased (Olivera and Bonhoeffer, 1972b). This argues for their mechanism of sizing of DNA intermediates by competition between Okazaki-piece initiation and elongation.

4. *A Very Early DNA Intermediate in* Bacillus subtilis

Okazaki pieces also participate in *B. subtilis* DNA replication. Wang and Sternglanz (1974) have confirmed that Okazaki pieces are intermediates in *B. subtilis* DNA replication and can be labelled with thymine or thymidine. However thymine pulse-labelling resulted in the detection of a smaller 2S DNA intermediate which was labelled by very short pulses. This 2S intermediate looked like an early replication intermediate since it could be chased into larger DNA segments, and its synthesis inhibited by HPUra (Section III, C.4(c), p. 294) or by nalidixic acid, two replication inhibitors. But no RNA could be detected on the 2S piece which, in view of the *E. coli* work, argues against its being an early intermediate in replication. Wang and Sternglanz suggest two possibilities for the 2S piece: (1) it is a DNA replication intermediate; or (2) it is involved in DNA repair processes away from the replication fork. However an RNA primer role in elongation must be defined in *B. subtilis* (there is presently no evidence for this) before the role of the 2S piece in repair and/or replication can be resolved.

5. *Are Okazaki Pieces Artifacts?*

Some earlier observations (Werner, 1971) raised the possibility that the short pieces observed by Okazaki and others were artifacts of thymidine pulse-labelling. However, Okazaki *et al.* (1971) observed no difference when thymine or thymidine was used during pulse-labelling. With the finding of covalently linked RNA in nascent Okazaki pieces (Sugino *et al.*, 1972), the likelihood of these small intermediates being artifacts of DNA degradation and/or repair intermediates is considerably diminished. There is no evidence for covalently linked RNA and DNA existing in completed bacterial chromosomes, a situation which would exist were these degradation products. In addition, there is no direct or implied evidence at present for any RNA requirement in repair synthesis. However, *in vitro* studies show that RNA polymerase covalently attaches ribonucleotides to the 3′ OH end of DNA (Wickner, S. *et al.*, 1972; Nath and Hurwitz, 1974). These DNA–RNA covalent hybrids could also act as primers for DNA synthesis. Therefore, Nath and Hurwitz (1974) suggest that isolation of covalent RNA–DNA hybrids cannot be taken as evidence that RNA has participated solely as an initiator of DNA synthesis.

C. Replication Enzymes and Proteins

1. *A Biochemical Model*

The study of bacterial DNA replication at the biochemical level falls into two major categories: (1) chromosome initiation (formation of the replication fork); and (2) elongation (advancement of the replication fork). Since little is known about the initiation proteins, we will concentrate mainly on elongation proteins. To simplify the discussion, a biochemical model, or framework, is presented in Table 3 which includes suspected replication proteins and the *dna* gene products. Although many compounds in this table are necessarily speculative, the approach is useful because it emphasizes where the gaps in knowledge are. These gaps will be considered in detail in the discussion of the individual proteins.

2. *Does Chromosome Initiation Involve Any Elongation?*

Chromosome initiation is biochemically and genetically distinct from elongation, but little is known about what fork formation actually represents. Some thymine starvation experiments raise the possibility that a small amount of elongation could be involved as part of the initiation process. Thymine starvation induces premature initiations, presumably by increasing the pool of initiator proteins while stopping DNA synthesis. The chromosomes have been thought to initiate *after* release from a thymineless condition. However, Sueoka *et al.* (1973) showed that toluenized *B. subtilis* cells (which do not initiate) synthesize

TABLE 3. A Biochemical Model of Bacterial DNA Replication

Replication Step	Replication Protein
A. *Chromosome Initiation*	
1. Membrane attachment	*dna*C?
2. Derepression of protein synthesis?	*dna*A?
3. Synthesis of initiation proteins	
4. RNA synthesis	classical RNA polymerase
5. Replication fork formation, (a small amount of elongation)?	?
6. RNA excision	?
B. *Elongation*	
1. Unwinding	*E. coli* "unwinding" protein *rec*BC enzyme?
a. Swivel	omega protein
2. "Initiation" of Okazaki-piece formation	
a. RNA primer formation	non-classical RNA polymerase (*dna*G) or classical RNA polymerase + *dna*G
3. 10S Okazaki-piece formation	
a. Pol III* complex formation?	
(1) ATPase	*dna*B?
(2) Copol III*	
(3) DNA polymerase III	*dna*E
b. DNA polymerization	
(1) Pol III* ?	*dna*E
c. RNA primer excision	pol I 5′-3′ exonuclease?
4. Gap closing between Okazaki pieces	DNA polymerase I
5. Sealing of DNA fragments	DNA ligase

large amounts of origin marker *ade*16 (see Fig. 12; *in vitro* hybrid peak) when they were thymine-starved before toluenization. This synthesis results from continuation of the replication forks already formed *during* thymine starvation. A small, completely heavy peak was also observed and indicated that two initiation cycles had occurred during thymine starvation. These results are compatible with a small amount of elongation during the period of thymine starvation. If the *dna*-1 initiation mutant is of thymine and given a pulse of bromouracil, at 45°C, semi-conservative synthesis of origin *ade*16 occurs if thymine starvation had been imposed at permissive temperatures, but not if it had been imposed at 45°C (a non-permissive temperature; K. White, T. Matsushita and N. Sueoka, unpublished results). These *in vivo* studies again show that initiation is completed *during* the thymine starvation period, since initiation would not occur during the subsequent BU pulse at 45°C.

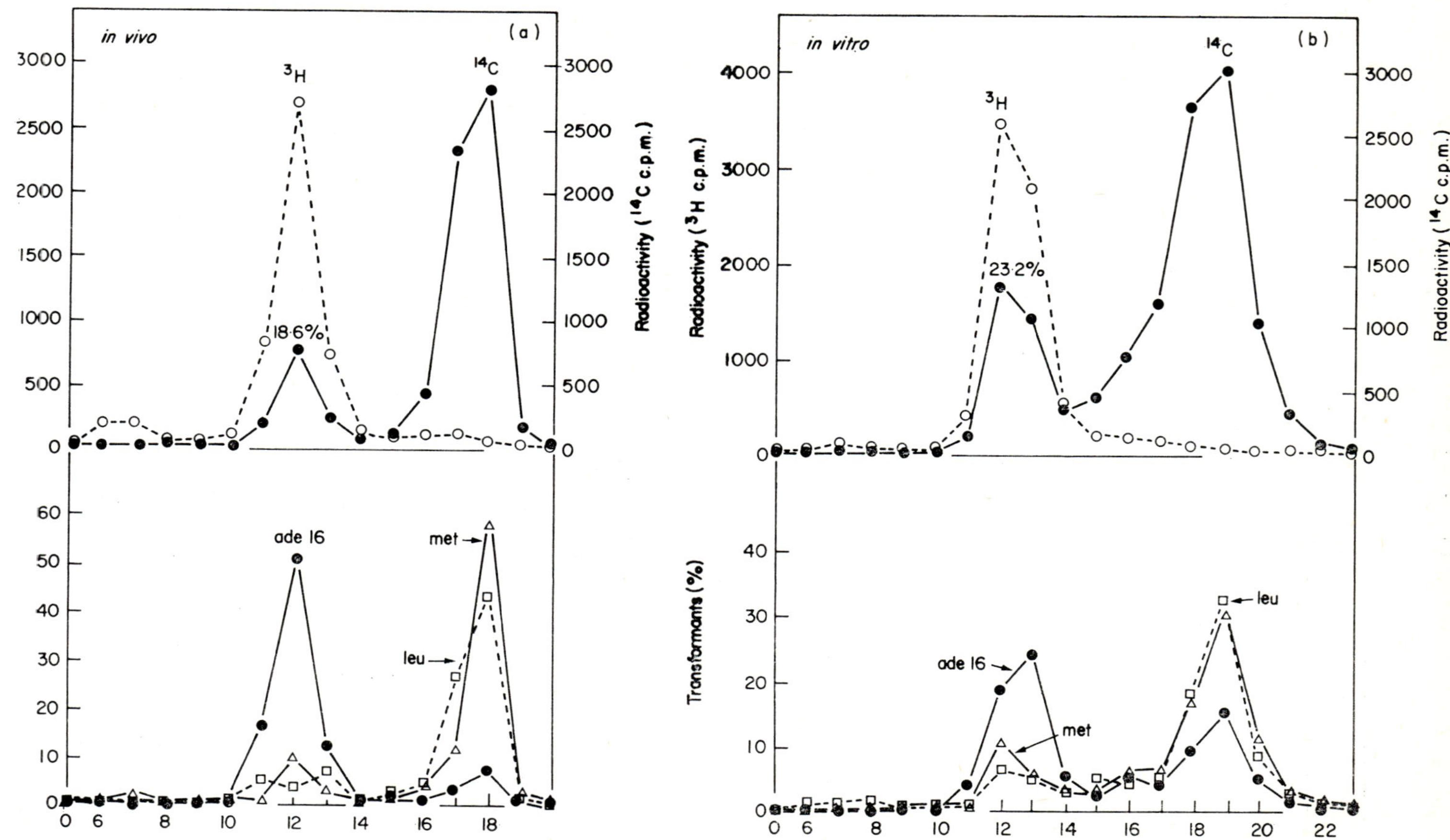

in vivo
(a)
3H
14C
18·6%
Radioactivity (14C c.p.m.)
ade 16
met
leu
in vitro
(b)
23·2%
Radioactivity (3H c.p.m.)
Transformants (%)

To summarize, the thymine starvation data indicate that initiation might include some small amount of elongation to complete the process. Alternatively, the initiation could be completed during thymine starvation, and any slight fork movement (up to a point where a thymine residue needs to be incorporated) might represent the start of elongation. Further work must be done to determine what constitutes the process of replication fork formation and what differentiates the later steps of initiation from elongation.

3. *Unwinding Proteins*

(a) *"Unwinding" protein of* Escherichia coli. Although the unwinding mechanism during DNA replication is still unknown, proteins which lower the stability of the helix by tight co-operative binding to exposed single-strands of DNA have been isolated from prokaryotes (Alberts and Frey, 1970; Sigal *et al.*, 1972), mammalian cells (Tsai and Green, 1973; Herrick, 1973) and virus-infected mammalian cells (Van der Vleit and Levine, 1973). The first protein of this type was found in *E. coli* infected with T4 phage, and was identified as the product of T4 gene-32 (Alberts *et al.*, 1968; Alberts and Frey, 1970; Alberts, 1971; Delius *et al.*, 1972). This gene product is required in both DNA replication (Epstein *et al.*, 1963; Kozinski and Felgenhauer, 1967) and recombination (Tomizawa *et al.*, 1966; Broker and Lehman, 1971) in T4 phage. The "unwinding" protein stimulates *in vitro* DNA synthesis, catalysed by T4 DNA polymerase on single-stranded DNA templates, some five- to ten-fold. In addition, gene-32 protein promotes denaturation of double-stranded DNA, as indicated by the lowering of the melting temperature (Alberts and Frey, 1970). However gene-32 protein also promotes renaturation of DNA, possibly by preventing intrastrand hydrogen bonding of single-stranded DNA. Its specificity for single-stranded DNA suggests that this

FIG. 12. Excess initiation during thymine starvation in *Bacillus subtilis*. A culture of 50 ml of 168TT (*thy try*) was uniformly labelled at 30°C with ^{14}C-thymine in a C^+ medium plus *L*-tryptophan (50 μg/ml) and thymine (5 μg/ml) and quick-filtered. Cells were washed and shifted to 37°C in C^+ medium without thymine. At the end of 30 minutes the two 25 ml samples were withdrawn. One sample (for *in vivo*) was suspended in BU-C^+ medium (C^+ medium plus 30 μg bromouracil/ml, 3 μg thymine/ml, 2 μCi ^{3}H-thymine/ml, and 50 μg *L*-tryptophan/ml) and incubated for 20 minutes at 37°C, before the lysate was prepared. The other sample (for *in vitro*) was toluenized and incubated for 45 minutes at 37°C in the reaction mixture described in Fig. 8. Both lysates were centrifuged in 55% CsCl for 68 hours at 35,000 rpm at 25°C. Fractions were collected from the bottom of the tube and analysed for radioactivity and transforming activities for *ade*16, *leu*8, and *met*5 markers. Fractions 5 to 9 represent double-density labelled DNA, fractions 10 to 15 hybrid DNA, and fractions 16 to 22 parental DNA. From Sueoka *et al.* (1973).

protein is not the only one involved in the unwinding process. The precise role of gene-32 protein in T4 DNA replication is still unknown.

A similar protein has been isolated from uninfected *E. coli* (Sigal *et al.*, 1972). This *E. coli* "unwinding" protein (molecular weight of 22,000 daltons) shows the same tight co-operative binding to single-stranded DNA, but much less binding to RNA or double-stranded DNA. Also, like the gene-32 protein, it depresses the melting temperature of double-stranded DNA and preferentially melts adenine-thymine base pairs. A puzzling result is that this protein stimulates *in vitro* DNA synthesis using DNA polymerase II but not with DNA polymerases I or III. Since none of the *dna* mutants shows any altered denaturation protein activity, Sigal *et al.* (1972) suggest that the "DNA-unwinding" protein they have isolated may be involved in recombination rather than DNA replication. Thus there may be several of these proteins in *E. coli* with separate roles in replication and recombination.

(b) *Omega protein.* Another *E. coli* protein which could be involved in the unwinding process is the omega protein (Wang, 1971). This high molecular-weight protein (100,000 daltons) can remove superhelical twists from negatively twisted closed circular DNA. It also binds preferentially to single-stranded DNA. Although the mechanism for this reaction is not known, Wang (1971) suggests that omega first nicks the DNA and then covalently bonds to the 5′-phosphoryl group at the nick. After unwinding the supertwists from the DNA, the nick is sealed and omega released. Thus omega could be acting as a swivel protein for unwinding during DNA replication.

(c) *The* rec*BC enzyme* (*exonuclease V*). An ATP-dependent exonuclease activity is absent from *rec*B and *rec*C mutants, correlating with the recombination deficiency and radiation sensitivity of these mutants (Oishi, 1969; Wright and Buttin, 1969; Barbour and Clark, 1970; Wright *et al.*, 1971). The wild-type enzyme has both exonuclease and endonuclease activities, the exonuclease digesting single-strand and duplex DNA, and the endonuclease degrading single-strand DNA (Goldmark and Linn, 1970, 1972; Karu *et al.*, 1973). The endonuclease activity is stimulated by ATP, but is not absolutely ATP-dependent. The *rec*BC enzyme also has an ATPase activity which is only active in the presence of polydeoxyribonucleotides. However, the ATPase activity is not coupled to DNA hydrolysis since its activity can be demonstrated in systems containing non-degradable polymers or duplex DNA with interstrand crosslinks (Karu *et al.*, 1973; Karu and Linn, 1972). This suggests that the ATP might be needed for unwinding native DNA in order for the *rec*BC nucleolytic activities to function.

Friedman and Smith (1973), and later Linn and MacKay (1974), have found evidence for this unwinding by isolating *rec*BC degradation products which are of two classes: duplex molecules with single-stranded tails of several thousand nucleotides (the unwound, partially degraded remnant DNA) and single-stranded fragments of several hundred nucleotides (degradation products).

The ATP-dependent DNAse also appears to be involved in recombination in *Diplococcus pneumoniae* (Vovis and Buttin, 1970) and *Haemophilus influenzae* (Greth and Chevallier, 1973; Friedman and Smith, 1973). Recently an ATP-dependent DNAse has been isolated from *B. subtilis* (Sueoka *et al.*, 1973; Ohi and Sueoka, 1973). In contrast to *E. coli*, the ATP-dependent DNAse required ATP for hydrolysis of double-stranded DNA but not for single-stranded DNA. Since both activities reside in the same enzyme, they also suggested that ATP consumption might be needed for unwinding of native DNA before hydrolysis. In an attempt to visualize this unwinding, Ohi *et al.* (1974) examined DNA-enzyme complexes by electron microscopy and found, unexpectedly, that the enzyme brought DNA molecules into close proximity and caused non-specific "pairing" of native DNA molecules with, or without, ATP. Although this "pairing" phenomenon is apparently unrelated to unwinding, this property of the *B. subtilis* enzyme could have special significance in the mechanism of recombination.

In *E. coli* (Linn and MacKay, 1974), *B. subtilis* (Ohi and Sueoka, 1973) and *Haemophilus influenzae* (Friedman and Smith, 1973), an ATP role in unwinding double-stranded DNA has been proposed for the *rec*BC enzyme. Therefore it seems reasonable to suggest that this enzyme is involved in DNA replication, as well as recombination, as an unwindase. Although the non-specific pairing observed by Ohi *et al.* (1974) conflicts with an unwinding function, they suggest two modes of DNA binding by this enzyme: one which enables the DNAse to work in unwinding and hydrolysis, and one which causes the non-specific pairing of DNA. In summary, the *rec*BC enzyme is multifunctional, with nucleolytic, ATPase, non-specific DNA pairing and perhaps unwinding activities. Whether it serves a role in DNA replication as well as recombination still remains to be proved.

There are many other nucleases in the bacterial cell, but the role of most of these enzymes in DNA replication is unknown at present. For extensive reviews of these nucleases the reader is referred to Lehman (1963, 1967); Richardson (1969); Radding (1969); Arber and Linn (1969); Koerner (1970); Echols (1971); Goulian (1971); Signer (1971); Smith (1973).

4. *DNA Polymerase III*

(a) *DNA polymerase III of* Escherichia coli. DNA polymerase III (*pol*C) was first discovered by T. Kornberg and M. L. Gefter during an investigation of residual polymerase activity in the DeLucia and Cairns (1969) *pol*A mutant. Because of the temperature lability of *pol*C, early isolations from the *pol*A mutants had revealed only DNA polymerase II (Moses and Richardson, 1970a, b, c; Kornberg and Gefter, 1971; Wickner, R. B. *et al.*, 1972a). During their purification of DNA polymerase II, Kornberg and Gefter (1971) noticed another activity ("peak A" on a phosphocellulose column) which appeared biochemically distinct from DNA polymerase II ("peak B"). This peak A activity was inhibited by low concentrations of ammonium sulphate whereas there was a two-fold stimulation of DNA polymerase II by 10 m*M* ammonium sulphate. Peak A (*pol*C) was three-times more sensitive to N-ethylmaleimide (NEM) and heat inactivation, than was DNA polymerase II. Both activities were similar however in their reaction requirements and insensitivity to anti-DNA polymerase I antibody.

The identification of *pol*C as an enzyme distinct from DNA polymerase II was established by showing that the peak A activity was the gene product of several of the *dna*E temperature-sensitive mutants (Gefter *et al.*, 1971). Each of the known *dna* mutants (in a *pol*I$^-$ background) were checked for temperature-sensitive polymerases. In all strains, DNA polymerase II was at wild-type levels, but *pol*C appeared to be temperature-sensitive in two of the four *dna*E mutants tested. Thus *pol*C was the first enzymic activity identified with a *dna* mutant, and appears to be an essential enzyme for DNA replication in *E. coli*. This result was also confirmed by Nusslein *et al.* (1971), using their cellophane *in vitro* DNA synthesis system (Schaller *et al.*, 1972) to test *dna* mutants. They were able to identify four *dna*E mutants which contained a soluble temperature-sensitive factor for replication in their *in vitro* system. Purification of this factor, by using complementation of the mutant *in vitro* system, revealed a protein similar in properties to the *pol*C activity found by Gefter *et al.* (1971).

Although DNA polymerase III is an essential replicating enzyme, its polymerization reaction is remarkably similar to polymerases I and II. Polymerase III does not self-prime, synthesizes DNA in the 5′ to 3′ direction, and requires Mg^{2+} and dNTPs but not ATP *in vitro*. However, it is both biochemically (Table 3) and genetically distinct from the other DNA polymerases. Polymerase III presumably is responsible for 10S Okazaki piece formation, and could also be involved in the gap-closing step of the discontinuous mode of replication. At present, neither replication function has been identified exclusively with *pol*C, or with any other gene product.

(b) *Pol III* and Copolymerase III* of* Escherichia coli. Recently Kornberg and his coworkers have discovered a new form of DNA polymerase III, termed Pol III* (Wickner, W. *et al.*, 1973), which is necessary for *in vitro* synthesis of bacteriophage M13 and ϕX174 replicative forms (RF). During their purification of *E. coli* extracts, Wickner, W. *et al.* (1973) found purified polymerase III to be inactive in the phage systems, even though the system required the *dna*E gene product. Instead a novel, more complex form of polymerase III, was found to be necessary for chain growth, namely Pol III*. Although the enzymes are readily separable by gel filtration, Pol III* was easily identifiable as a form of polymerase III because its activity was temperature-sensitive when isolated from a *dna*E mutant. In addition Wickner, W. *et al.* (1973) found that another protein, called copolymerase (Copol) III*, was needed for pol III* to replicate DNA in the bacteriophage M13 and ϕX174 systems *in vitro.*

Further studies of the mechanism revealed that, after the template is primed by RNA (see Section A.3, p. 279), Copol III* and ATP participate in an early stage of the synthesis (Wickner and Kornberg, 1973). Pre-incubation of Copol III* with anti-Copol III in the absence of dNTPs showed no subsequent DNA synthesis when dNTPs were added because the antibody had prevented formation of an "initiation" complex. However, when anti-Copol III* was absent during pre-incubation, but added with the dNTPs, subsequent rapid DNA synthesis took place. Thus there appear to be two stages of DNA synthesis after RNA priming—a Copol III*-dependent step and a Copol III*-independent step. Copol III* appears to be involved in the formation of an "initiation" complex of RNA-primed template, pol III* and spermidine. The formation of the complex requires Copol III*, ATP, and Mg^{2+}. After the formation of the complex, the second chain-elongation stage takes place which is now independent of Copol III* and ATP. Hurwitz *et al.* (1973) have also isolated a stimulatory protein for the fd (a filamentous phage very similar to M13) and ϕX174 *in vitro* replicating systems which they call Factor I; this protein may be identical with Kornberg's Copol III*. Hurwitz and Wickner (1974) have also isolated a second factor, Factor II, which copurifies with polymerase III but appears to be distinct from *pol*C polymerizing activity since its stimulatory activity is not temperature-sensitive when isolated from a *dna*E mutant. Factor II plus polymerase III may be equivalent to Pol III*. Hurwitz and Wickner (1974) confirm the ATP dependency of the fd phage replicating system and also find that dATP can satisfy this nucleoside triphosphate requirement. These *in vitro* phage replication systems have elucidated at least two functions for ATP in DNA synthesis: (1) ATP (along with the other ribonucleoside triphosphates) is a precursor in the synthesis of the RNA-priming fragment; and (2) ATP is needed for the operation of the pol III system. However there is no evidence that these functions account for the ATP requirement seen in the bacterial *in vitro* systems. These uncertainties emphasize that synthesis of double-stranded RF from a viral single strand is certainly different from bacterial DNA replication. Although there is no evidence at present for the pol III* system functioning in the host bacterial DNA replication process, this system does exist in *E. coli* and could be the functioning replicating system *in vivo*. This possibility must be considered in future studies on the role of DNA polymerase III in bacterial replication. These cofactors of polymerase III are prime examples of non-enzymic control proteins, and emphasize the interesting complexities we are finding in the study of DNA replication.

(c) *DNA polymerase III of* Bacillus subtilis. Several laboratories have investigated residual DNA polymerase activities in *B. subtilis pol*A$^-$ mutants (Gass *et al.*, 1971; Gass and Cozzarelli, 1973a; Neville and Brown, 1972; Bazill and Gross, 1972; Ganesan *et al.*, 1973). There appear to be many similarities and some differences to the correspondingly numbered *E. coli* polymerases (see Table 4, p. 296). The function of DNA polymerase I (*pol*A) seems to be primarily DNA repair, but may include replication and recombination in both organisms. DNA polymerase II is capable of ATP-stimulated DNA repair in the absence of polymerases I and III (see Section III, C.6, p. 301) but thus far its specific biological role is unknown in both *E. coli* and *B. subtilis*. DNA polymerase III is essential for DNA replication in both organisms. As mentioned above, the requirement for polymerase III in *E. coli* was established by finding temperature-sensitive polymerase III in *dna*E mutants. However the evidence for the requirement of polymerase III in *B. subtilis* relies upon conditionally lethal *pol*C mutants which do not make DNA at the non-permissive temperatures, and on the unique

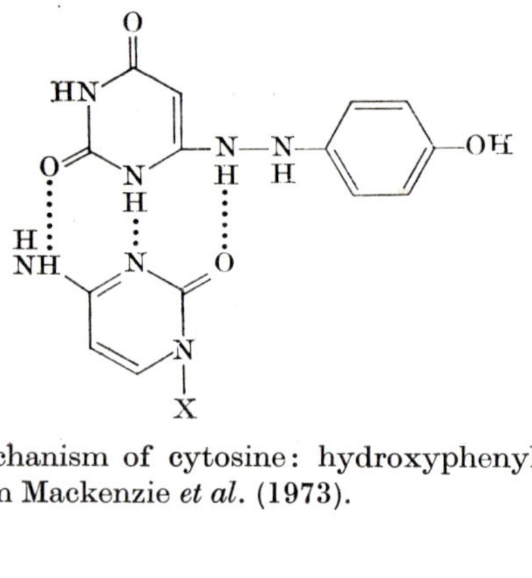

FIG. 13. Proposed mechanism of cytosine: hydroxyphenylhydrazinopyrimidine (HPUra) pairing. From Mackenzie *et al.* (1973).

sensitivity of this enzyme to 6-(*p*-hydroxyphenylazo)-uracil (HPUra; see Fig. 13), a drug selectively toxic for Gram-positive bacteria.

Brown (1970) first established that HPUra inhibited DNA synthesis but did not affect RNA or protein synthesis in log-phase cultures of *B. subtilis*. The effect on DNA synthesis was immediate, but readily reversible when the culture was washed free of the drug. At the same time, Brown (1970) also demonstrated that HPUra had no effect on the replication of the DNA phages SP3 or SP02C_1, indicating that the HPUra-sensitive site was unique to the bacteria. Through further *in vivo* studies, Brown (1971) was able to determine that HPUra inhibition of DNA synthesis was confined specifically to semi-conservative replication and that HPUra did not inhibit DNA repair. By performing density-transfer studies on DNA from UV-irradiated cells, Brown showed that HPUra inhibited only hybrid peak DNA formation (semiconservative synthesis) and did not affect the parental peak incorporation (repair synthesis). Thus these *in vivo* studies clearly indicated that HPUra was a specific DNA replication inhibitor. This was verified when ATP-stimulated DNA synthesis in toluene-treated *B. subtilis* cells was shown to be inhibited by HPUra (Brown *et al.*, 1972).

The first indications that HPUra might be acting directly on a DNA polymerase came from studying residual polymerase activity in *pol*A mutants of *B. subtilis*. At this time, the characteristics of *B. subtilis* DNA polymerases were not well-defined, but Neville and Brown (1972) found some indication in the partially purified extracts of a *pol*A mutant that there existed at least one drug-sensitive polymerase. Bazill and Gross (1972) independently detected three chromatographically distinct polymerases in the residual activity of a *pol*A$^-$ mutant of which two peaks were HPUra-sensitive. In a further study of the residual polymerases in *polA* mutants, Cozzarelli and his coworkers established a remarkably close parallel between *B. subtilis* DNA polymerases and the already known *E. coli* DNA polymerases I, II and III (Gass *et al.*, 1971; Gass *et al.*, 1973; Gass and Cozzarelli, 1973a, b). One of the enzymes was designated DNA polymerase III because of its heat lability and salt and sulphydryl-reagent sensitivity—all characteristics similar to *E. coli* DNA polymerase III (see Table 4). Subsequently Gass *et al.* (1973) showed that HPUra specifically inhibited the enzyme they had designated polymerase III. This was independently confirmed by Mackenzie *et al.* (1973) and was strong evidence that DNA polymerase III was an essential replicating enzyme in *B. subtilis*. Further genetic evidence of a requirement for polymerase III for replication *in vivo* was that HPUra-resistant mutants contained a drug-resistant polymerase III (Cozzarelli and Low, 1973) and, when toluenized, showed drug-resistant cell replication. The increased mutation rate in *polC* mutants (Gass and

TABLE 4. Comparison of *Bacillus subtilis* and *Escherichia coli* DNA Polymerases. Adapted from Gass and Cozzarelli (1973a)

Property	*Bacillus subtilis*	*Escherichia coli*
Biological function:		
I	DNA repair, replication	⟶
II	Unknown	⟶
III	DNA replication	⟶
Heat lability	III most sensitive (III, II, I)	⟶
Salt sensitivity	III most sensitive (III, II, I)	⟶
Sulphydryl reagents sensitivity	III most sensitive (III, II, I)	⟶
AraCTP sensitivity	II most sensitive	⟶
Order DEAE cellulose elution	II, I, III	⟶
Order phosphocellulose elution	III, I, II	⟶
Poly dAT as template	I good template	⟶
	II 25% of I	Poor template
	III 1% of I	
Long single-stranded template regions	Inhibits II and III	⟶
Antiserum *Escherichia coli* pol I	Cross reacts weakly	Strong reaction
Mol. at. pol I	75,000 daltons	109,000 daltons
Associated nuclease	I none	Yes
	II none	Yes
	III yes	Yes (Otto *et al.*, 1973)
HPUra sensitivity	I none	None
	II slight inhibition (500 fold less sensitive than III)	None
	III sensitive	None

Cozzarelli, 1973a; Bazill and Gross, 1973) also suggested a direct role for polymerase III in replication. Conditionally lethal *polC* mutants have been isolated (Gass and Cozzarelli, 1973a) which are HPUra-resistant and temperature-sensitive at 51°C, and have a greatly decreased polymerase III activity. The correlation of HPUra-resistance with genetic effects, such as increased mutation rate and conditional lethality, confirm that the site of action of HPUra *in vivo* is the specific inhibition of polymerase III.

The study of the action of HPUra on DNA polymerase III has proved to be quite interesting. The drug appears to be active only when the azo linkage is reduced. The first evidence for this came from studies of HPUra in toluene-treated cells where it was shown that $NADPH_2$, $NADH_2$ or dithiothreitol was necessary for the drug to act (Brown *et al.*, 1972). The reduced drug was confirmed as the active form by the demonstration that it, but not the parent azo form, inhibited polymerase III *in vitro* (Gass *et al.*, 1973). The structure of the reduced drug was shown to be the hydrazino derivative by nuclear magnetic resonance analysis of dithionite-treated HPUra (Mackenzie *et al.*, 1973), and by X-ray crystallography (C. L. Coulter and N. R. Cozzarelli, personal communication).

Another interesting finding from studies on toluenized cells was that HPUra inhibition was abolished by high concentrations of dGTP (Brown, 1972a, b). This reversible effect of dGTP on drug action was also demonstrated with polymerase III (Gass *et al.*, 1973). The dGTP effect immediately suggested specific hydrogen bonding to template cytosine residues. Further evidence for this specific hydrogen bonding between HPUra and cytosine came from nuclear magnetic resonance analysis and X-ray data. All four bases and their derivatives were tested for their effects on the nuclear magnetic resonance spectra of reduced drug solutions (Mackenzie *et al.*, 1973) and the only significant peak shifts were observed with cytosine. Also the magnitudes of binding constants for cytosine binding to both reduced HPUra and deoxyguanosine were comparable. On this basis Mackenzie *et al.* (1973) proposed a model for drug:base pairing that involved three specific hydrogen bonds (Fig. 13, p. 294). This model has since been supported by X-ray crystallography on oxidized (Coulter and Cozzarelli, 1974) and reduced forms of HPUra (C. L. Coulter and N. R. Cozzarelli, personal communication). These studies have shown that the uracil moiety in the reduced drug exists as the diketo form and is *not* enolized. Secondly, the atoms of reduced HPUra, in the proposed hydrogen bonding scheme, are all coplanar due to resonance energy stabilization, directly analogous to the electron delocalization that maintains peptide bond planarity. Only drug crystals in the above-described configuration are active against DNA polymerase III; the oxidized form is completely inactive.

Although HPUra appears to hydrogen-bond to cytosine residues on the template DNA, it is not a simple competitive inhibitor of dGTP. This comes from the observation that the incorporation of all four dNTPs is diminished by HPUra (Gass *et al.*, 1973). Since only the triphosphate form of deoxyguanosine reverses drug inhibition, they proposed the HPUra was binding to the enzyme at the dNTP binding site. A third observation showed that polydA: oligo-dT primed incorporation of dTTP, by polymerase III, was not inhibited by HPUra until a minute amount of gapped DNA was added. Thus the formation of an HPUra–DNA template and polymerase III complex leads to scavenging of free polymerase III, trapping the enzyme in an inactive state.

Confirmation of this model of ternary-complex scavenging comes from the isolation of the ternary complex, by use of agarose column chromatography, which effectively separates the complex from unbound enzyme (Low *et al.*, 1974). Only DNA with 3′-hydroxyl terminated breaks leads to complex formation; single-stranded or native DNA does not form the ternary complex. This formation requires divalent or polyvalent cations, is inhibited by sulphydryl reagents and is prevented, or reversed, by dGTP addition. The complex dissociates slowly with a half-life of several minutes at 4°C. An interesting observation which further supports this ternary complex-scavenging model is the finding of an intrinsic exonuclease in the polymerase III molecule. This exonuclease is single-strand specific and generates 5′-mononucleotides. It is inhibited by HPUra only when 3′-hydroxyl-terminated gapped DNA is used, and has the same requirement of divalent cation, is inhibited by sulphydryl reagents and is reversed by dGTP. This provides further evidence for ternary complex involvement in the inhibition of both polymerase III activities.

In summary, Cozzarelli and his coworkers suggest that HPUra is bound to the DNA template by a hydrogen bond-stabilized base pair, and to polymerase III in a position overlapping, but not identical with, the triphosphate binding site. This ternary complex of HPUra, DNA and enzyme dissociates slowly, trapping polymerase III in an inactive state. Thus HPUra appears to inhibit replication of DNA by acting not only as a competitive inhibitor of dGTP incorporation by polymerase III, but also as a scavenger for DNA polymerase III.

5. *DNA Polymerase I*

The properties of DNA polymerase I (pol I) have been reviewed extensively, and the reader is referred to recent reviews (Englund *et al.*, 1968; Richardson, 1969; Kornberg, 1969; Goulian, 1971; Smith, 1973) for more detailed information. In *E. coli* this enzyme has a molecular weight of 109,000 daltons (Jovin *et al.*, 1969). There are five enzymic activities associated with DNA polymerase I: (1) DNA polymerization; (2) 3′ to 5′ exonuclease; (3) 5′ to 3′ exonuclease; (4) 3′ to 5′ pyrophosphorolysis; and (5) pyrophosphate-triphosphate exchange. The first three activities will now be discussed, and their possible roles in DNA replication examined.

(a) *DNA polymerase I polymerization in* Escherichia coli. Although there is general agreement about *polA* participation in DNA repair, there is considerable debate about the essential role of DNA polymerase I (Pol I) polymerization activity in DNA replication. If Pol I polymerization were totally absent from *polA* mutants, one could argue that the enzyme is not essential since DNA replication can occur with only polymerases II and III available. However, Lehman and Chien (1973) re-examined the *E. coli polA* mutants and found residual Pol I polymerization varying from 0·6 to 35% of the wild-type levels. Since this residual activity could account for enough polymerization to supply the essential replication function, no positive conclusions can be drawn regarding the essential nature of the Pol I polymerization function.

Whether or not it is essential, Pol I polymerization contributes to some step(s) of replication since elongation is affected in *polA* mutants. Okazaki pieces accumulate in *polA* mutants (Kuempel and Veomett, 1970; Okazaki *et al.*, 1971) due to slower closing of gaps between the DNA intermediates. This gap-closing step is essential for replication to proceed, but the slow joining in *polA* mutants could be due to the residual Pol I activity or alternatively to DNA polymerase II and/or III. Another indication that Pol I polymerization is performing some role in replication is a comparison of the ATP-stimulated DNA synthesis in toluene-treated wild-type and *polA* cells. Although *E. coli* and *B. subtilis* grow as well as wild type, the ATP-stimulated synthesis is some 20–30% lower in *polA* strains (T. Matsushita, unpublished results). Thus the toluenized cell-elongation process seems to be less efficient in the partial absence of Pol I polymerization.

In summary, Pol I polymerization contributes to replication by acting at a late step in elongation during the closing of gaps between Okazaki pieces. The less efficient replication observed in *polA* mutants may be due to less efficient gap closing by polymerases II and III, or alternatively to gap closing by the essential, but residual, Pol I activity. The isolation of a more complete (or less leaky) *polA* mutant is probably necessary to resolve this question.

(b) *DNA polymerase I (5′–3′)-exonuclease in* Escherichia coli. In their re-investigation of enzymic activities in *polA* mutants, Lehman and Chien (1973) found nearly normal DNA polymerase I (Pol I)-associated

(5′–3′)-exonuclease activity in all mutants. Recently Konrad and Lehman (1974) have isolated a new mutant, *E. coli* K-12 *polAexl*, that has a wild-type level polymerizing activity but only 2% of normal (5′–3′)-exonuclease activity in partially purified extracts. This mutant is similar to other *polA* mutants in its MMS- and UV-sensitivity and in its accumulation of Okazaki pieces. However, unlike other *pol A* mutants, this strain is conditionally lethal and does not grow at 43°C. Revertants to the parental phenotype show normal (5′–3′)-exonuclease activity. These results indicate that the Pol I (5′–3′)-exonuclease may be essential for DNA replication in *E. coli*. Perhaps the nuclease removes the RNA primer of the adjoining Okazaki piece by a "nick translation" mechanism, with simultaneous 5′–3′ degradation and polymerization performed by the single polymerase exonuclease (Kelly *et al.*, 1969, 1970). The *polAexl* mutant should prove useful for determining the function of the Pol I (5′–3′)-exonuclease in DNA replication.

(c) *Pol I (3′–5′)-exonuclease in* Escherichia coli. The Pol I (3′–5′)-exonuclease appears to serve a "proof-reading" function during DNA polymerization (Brutlag and Kornberg, 1972). If the 3′ terminal nucleotide on the primer strand does not pair with the template strand, DNA polymerases will not extend the chains until the mispaired nucleotides have been removed. The Pol I (3′–5′)-exonuclease acts specifically against mispaired nucleotides thereby increasing the fidelity of template copying and minimizing mutations. This activity also ensures that errors in polymerization will not stop DNA replication since the nuclease removes this blockage and permits further chain extension by the DNA polymerases.

(d) *DNA polymerase I in* Bacillus subtilis. The characteristics of Pol I polymerization in *B. subtilis* are similar to those in *E. coli*. Early studies by Okazaki and Kornberg (1964) showed that *B. subtilis* Pol I requires a DNA template, Mg^{2+} and dNTPs. The DNA primer requirements are similar to those in *E. coli* with dAT copolymer resulting in 20-fold faster rates of synthesis than DNA. The efficiencies of incorporation of analogues such as 5-bromo- and 5-fluoro-uracil are indistinguishable from those with *E. coli*. *PolA* mutants have also been isolated in *B. subtilis* (Gass *et al.*, 1971; Neville and Brown, 1972; Ganesan *et al.*, 1973). The *polA* gene was mapped between *leu* and *argA* both by transduction and transformation data (Gass and Cozzarelli, 1973a).

There are two significant differences between the *E. coli* and *B. subtilis polA* gene products. There is no nuclease activity associated with the *B. subtilis* enzyme (Okazaki and Kornberg, 1964) whereas the *E. coli* enzyme has two exonuclease activities but no endonuclease (see

p. 299). In addition, the *B. subtilis* Pol I does not utilize NTPs whereas the *E. coli* enzyme incorporates these compounds (Berg *et al.*, 1963).

The biological role of Pol I in *B. subtilis* is similar to that in *E. coli* since it appears to function in repair and replication. *PolA* mutants show increased sensitivity to UV, MMS and X-rays (Ganesan *et al.*, 1973), and digest their DNA at an increased rate after treatment with either UV or MMS (Gass *et al.*, 1971). Also "repair synthesis" (that is, that synthesis stimulated by exogenous DNAse I when replication is inhibited in toluenized cells) is absent from toluene-treated *B. subtilis polA* (Matsushita and Sueoka, 1974). The role of Pol I in *B. subtilis* DNA replication is similar to that in *E. coli*—i.e., the closing of gaps between Okazaki pieces. Okazaki fragments are joined very slowly in *polA* mutants (R. Okazaki, personal communication) and replication in toluenized *B. subtilis* cells is 20–30% less efficient than in wild-type (T. Matsushita, unpublished results). As for *E. coli*, determination of the essential requirement of Pol I polymerization for DNA replication in *B. subtilis* will necessitate the isolation of completely clean *polA* mutants.

Since there is no nuclease activity in purified Pol I, the *pol*A gene in *B. subtilis* cannot perform nick translation; such translation requires a single polymerase-exonuclease (Kelly *et al.*, 1969, 1970) or a proof-reading function as served by the *E. coli* Pol I (3′–5′)-exonuclease (Brutlag and Kornberg, 1972). Although there is no evidence at present for RNA primer molecules in *B. subtilis* (Wang and Sternglanz, 1974), the lack of nuclease activity also eliminates the *polA* gene from serving any RNA primer excision function, a possibility in *E. coli*. Whether this function, presumably essential for DNA replication in *B. subtilis*, can be performed by a polymerase-independent nuclease still remains to be proven.

6. DNA Polymerase II

When the Delucia and Cairns (1969) *polA* mutant was examined for residual DNA polymerase activity, DNA polymerase II was discovered before DNA polymerase III. Three separate groups have purified and characterized DNA polymerase II: Moses and Richardson (1970a, b, c), Kornberg and Gefter (1971), and Wickner, R. B. *et al.* (1972a). Polymerase II activity in wild-type cells was estimated to be around 5–10% (Moses and Richardson, 1970c). Polymerase II requires a nicked double-stranded DNA substrate, four dNTPs, Mg^{2+}, and is not stimulated by ATP *in vitro*. Polymerase II is sensitive to salt and sulphydryl reagents and unable to use polyd(A-T) as primer-template. These properties, along with insensitivity to anti-DNA polymerase I, distinguish this enzyme from polymerase I. The enzyme also has a 3′–5′-exonuclease activity against single-strand DNA but no 5′–3′-nuclease activity has been found. The molecular weight has been estimated to be around 120,000 daltons with about 17 molecules occurring per bacterium (Wickner, R. B. *et al.*, 1972a).

So far, no essential biological function has been determined for DNA polymerase II. This enzyme is sensitive to ara-CTP *in vitro*, but this inhibitor has no effect on polymerase I and high concentrations are needed to inhibit polymerase III. In addition, ara-CTP inhibits ATP-stimulated DNA synthesis in toluenized cells (Rama Reddy *et al.*, 1971; Masker and Hanawalt, 1974). These data argue for polymerase II being a replicative polymerase. However Masker and Hanawalt (1974) have found that, in a mutant deficient in polymerases I and III (ts), semi-conservative synthesis is more severely inhibited by ara-CTP than is the repair synthesis induced by UV irradiation. This suggests that ara-CTP inhibits DNA replication in toluenized cells by a mechanism other than specific polymerase II inhibition.

Masker and Hanawalt (1973) have established that polymerase II is responsible for UV-induced repair in toluenized cells when polymerases I and III are absent. The polymerase II role was confirmed when Masker *et al.* (1973) used a triple mutant, deficient in polymerases I, II and III(ts), and found UV-induced repair was absent from toluenized cells. Additionally they found this repair to be ATP-dependent. However, the ATP-dependency differs from the specific ATP requirement found in replication since GTP can substitute for ATP during UV-induced repair (W. E. Masker and P. C. Hanawalt, personal communication). This NTP requirement prompted Masker and Hanawalt to examine strains deficient in the ATP-dependent *rec*BC DNAse (Oishi, 1969; Wright *et al.*, 1971; Goldmark and Linn, 1972). Since NTP-dependent UV-stimulated repair also occurred in these *rec*BC mutants, the ATP effect in UV-induced repair in toluenized cells is probably not due to nucleolytic action of the *rec*BC nuclease.

DNA polymerase II-mediated repair has also been demonstrated in toluenized *B. subtilis* cells by a different approach (T. Matsushita, unpublished results). In this organism, DNA replication and DNA polymerase III can be inhibited by HPUra (see p. 294). By using a *polA* mutant in conjunction with HPUra, polymerase I and III activities are completely inhibited. When exogenous DNAse I is added under these conditions, toluenized cells show an ATP-dependent polymerase II-mediated repair. This repair can be completely inhibited by *p*HMB, a result which precludes any involvement of residual polymerase I.

Masker *et al.* (1973) suggest that the polymerase II-mediated repair which they observed in toluenized *E. coli* cells is similar to "long patch" excision repair *in vivo* (Cooper and Hanawalt, 1972a, b). Long-patch repair occurs in increased amounts in *polA* cells, but is absent from a *rec*A *rec*B double mutant. Therefore Cooper and Hanawalt proposed that DNA polymerase I performs short-patch repair of damaged DNA, and that the *rec* system also participates in excision repair but produces predominantly long patches of DNA. The relationship of polymerase II to these repair and recombination processes should prove to be an interesting area of research. At present there is no indication that polymerase II is involved in any aspect of DNA replication except perhaps in gap closing between Okazaki pieces in *polA polC* mutants (Tait and Smith, 1974).

7. *DNA Ligase*

The DNA ligases, and their reactions, also have been reviewed earlier (Richardson, 1969; Goulian, 1971; Smith, 1973). In a discontinuous mode of replication, DNA ligase is thought to seal the juxtaposed ends of Okazaki pieces that have finished elongating except for the last step. That is, this enzyme forms the final phosphodiester bond between the 3′-OH group and the 5′-phosphoryl group at the ends of adjacent completed Okazaki pieces. The reaction mechanism appears to involve three steps. In *E. coli*, the NAD cofactor is joined covalently to the enzyme in the first step, to form an enzyme-AMP complex plus NMN (Olivera *et al.*, 1968b; Little *et al.*, 1967). In the second step, the AMP moiety of the complex is transferred to the 5′-phosphoryl group of the DNA substrate via a pyrophosphate linkage (Olivera *et al.*, 1968a). The third step is the ligase-catalysed formation of the phosphodiester bond with release of AMP (Hall and Lehman, 1969).

The need for a DNA ligase to repair nicks in chromosomes during DNA replication and repair processes is obvious. The essential nature of this enzyme for survival of *E. coli* has been demonstrated to some degree. Pauling and Hamm (1968) isolated a mutant of *E. coli* TAU called ts-7, which exhibits a two-fold increased sensitivity to UV irradiation at 40°C compared with that at 27°C. At 40°C, short DNA fragments were not joined suggesting inactivation of a DNA ligase in this ts-7 mutant. Subsequent studies revealed that ts-7 indeed contained a temperature-sensitive DNA ligase (Modrich and Lehman, 1971). Since ts-7 cells also die at the non-permissive temperature, it appears that DNA ligase is essential for cell survival because of its action in DNA replication and repair, although DNA synthesis does continue at a diminished rate. In the presence of BU, the DNA synthesized had a normal buoyant density, indicating an abnormal DNA synthesis, perhaps due to nick translation by polymerase I at the unsealed nicks. Gellert and Bullock (1970) isolated several mutants containing smaller amounts of DNA ligase and these were not able to support the growth of T4 (ligase-deficient) phage. Okazaki pieces did not appear to accumulate in these mutants (in contrast to ts-7 mutants) but these strains exhibited 2–10% residual activity which could perform the ligase-sealing step in the DNA replication of these mutants. Mutants isolated later with less residual activity (that is, less than 1%) at 42°C did show a five- to ten-fold lower activity in the joining of Okazaki pieces. However these strains grew normally at 42°C. Although the ligase marker for phage growth was mapped near the *ctr* locus (Gellert and Bullock, 1970), the ts-7 mutant has not been mapped because it is a derivative of *E. coli* $\overline{\text{TAU}}$. However, *lig* ts-7 has been transferred by P1 transduction into *E. coli* K12 (Konrad *et al.*, 1973). The resulting *E. coli* K12 *lig* ts-7 is temperature sensitive and displays the same characteristics as $\overline{\text{TAU}}$ ts-7. The problem of survival of the T4 ligase mutant may be similar to that discussed previously for *polA*, where a completely "clean" mutant is needed to show an essential function. Part of the problem may also lie in the difficulty of the ligase assay. Recently a method has been reported (Karkas, 1974) which consists of performing two parallel DNA polymerase assays with poly(dA):oligo(dT) as template-primer but supplementing one assay with NMN and the other with NAD. The difference in dTMP incorporation between the two assays provides a measure of the activity of the ligase. Perhaps when a simpler assay of this sort is available, the residual enzyme levels of all ligase mutants can be compared accurately and the essential role of DNA ligase in DNA replication further substantiated.

8. *The Gene Products of* Escherichia coli dna *Mutants*

At present the *dna* mutants of *E. coli* appear to reside in six cistrons, *dna*A, B, C(D), E, F and G. The genetics and mapping of these mutants have been extensively reviewed by Gross (1972) and by Smith (1973). We will not discuss the mapping data except to mention that on the basis of *in vivo* complementation tests the *dna*C and D loci appear to be one gene (Wechsler, 1973). The *dna*A and C(D) groups appear to be gradual shutoff (initiator) mutants, while *dna*B, E and G are immediate stop (elongation) mutants. A brief summary of the currently known facts about the *dna* mutants follows.

(a) dna*A*: Both the gradual stopping of DNA synthesis at nonpermissive temperatures and "integrative suppression" provide evidence for the involvement of *dna*A in chromosome initiation. Integration of an F-factor into the bacterial chromosome results in the suppression of *dna*A temperature sensitivity (Nishimura *et al.*, 1971). This arises because the F-factor, as an independent replicon (Jacob *et al.*, 1963), has its own initiation system. Prophage can also show integrative suppression in *dna*A (Lindahl *et al.*, 1971; Hirota *et al.*, 1972).

An indication of the role of the *dna*A gene comes from the dependency of colicin E1 plasmid DNA synthesis on the *dna*A and C genes (Goebel, 1970, 1973, 1974; Spratt, 1972). This dependency is at first puzzling when one considers that ColEl is an independent replicon with its own initiation system. However, the addition of low concentrations of chloramphenicol (CM) eliminates the inhibition of ColEl synthesis in the *dna*A mutant CRT46 at 43°C (Goebel, 1974). Goebel suggests that a low concentration of CM blocks the synthesis of a repressor specific for plasmid initiation. The *dna*A gene would then serve as an antagonist of this plasmid repressor, that is, an "antirepressor". At non-permissive temperatures, ColEl does not initiate DNA synthesis in a *dna*A mutant, because the plasmid repressor blocks initiation by inactivation of the *dna*A antirepressor. Whether the *dna*A gene serves a similar antirepressor function in the initiation of the *bacterial* chromosome remains to be seen.

(b) dna*B*: The *dna*B gene product has been partially purified using the phage ϕX174 system (Wickner, S. *et al.*, 1974). This protein contains both DNA-dependent and DNA-independent ATPase activities. Furthermore, there are indications that the *dna*B gene product is heterogeneous, and the ATPase activity may only be one of several functions associated with this gene. Because of the lability of the protein, they were unable to demonstrate the temperature sensitivity of the *dna*B ATPase. The ϕX174 SS–RF "semipurified" replicating system can be divided into three stages, namely initiation by RNA primer formation, formation of the pol III* complex, and elongation by pol III*. The *dna*B ATPase activity could be involved in complex formation which requires both ATP and Copol III*. Wickner, S. *et al.* (1974) suggest the *dna*B gene product also could function in RNA primer formation since the *dna*B gene product was not needed when phage ϕX174 DNA:RNA hybrids were used in place of single stranded ϕX174 DNA. Since *dna*C and *dna*G were also bypassed by use of ϕX174 DNA:RNA hybrids, *dna*B, C(D) and G might also facilitate ϕX174 RNA primer formation (Schekman *et al.*, 1972).

A unique 4S RNA (*oop* RNA) has been isolated and implicated in the formation of the left replication fork during lambda phage DNA replication (Hayes and Szybalski, 1973a, b). The synthesis of *oop* RNA primer depends on three lambda genes and the host genes (*dna*B and *dna*C). Although the *dnaB* gene facilitates RNA primer formation in both lambda and ϕX174 bacteriophages, the specific biochemical function(s) of the *dna*B gene product is still very much in doubt.

(c) dna*C*(*D*): The *dna*C gene product has also been purified (Wickner, S. *et al.*, 1973a) using the same phage ϕX174 SS–RF replication system. They demonstrated temperature sensitivity of the mutant gene product but no enzymic activity could be found. There was no incorporation of dNTPs or NTPs, and no ATPase, DNAse or RNAse activities. The *dna*C mutation does appear to be involved in initiation, however, since the amount of residual synthesis at 43°C is comparable to that observed at the permissive temperature in the absence of protein synthesis (Beyersmann *et al.*, 1971; Hirota *et al.*, 1970; Carl, 1970). As with *dnaA* mutants, integrative suppression by F-factors can be demonstrated in some *dna*C mutants.

An obligatory transcriptional step has been suggested for chromosome initiation in *E. coli* (Lark, 1972a; Messer, 1972) and in *B. subtilis* (Laurent, 1973) although no RNA primer has been demonstrated directly during initiation. The *dna*A and C genes may not be involved in this step since M. H. Hanna and P. L. Carl (personal communication) find that re-initiation after a period at the restrictive temperature is resistant to high concentrations of rifampicin and chloramphenicol in both the *dna*A and C mutants. This indicates that both initiator proteins and the alleged RNA primer were synthesized during the preceding period of restrictive temperature. The *dna*C gene product may facilitate RNA primer formation in phage ϕX174 DNA replication (Schekman *et al.*, 1972) but is not required for *oop* RNA primer formation during lambda DNA replication (Hayes and Szybalski, 1973a, b).

On the basis of the ColEl information, we have discussed a possible antirepressor function for *dna*A. ColEl is a small non-transmissible plasmid found in large numbers per cell and replicates for many hours in the presence of chloramphenicol (Clewell, 1972). Large plasmids (R-factors, ColV, Hly and F-factors) are present in only one or two copies per cell and initiate only once, or at most several times, before replication stops (Spratt, 1972). The replication of both of these plasmid classes was inhibited in *dna*C mutants at non-permissive temperatures (Goebel, 1974). Also chloramphenicol did not reverse the 43°C inhibition as in the *dna*A mutant. Multiple circular ColEl forms occurred at 43°C in the *dna*C mutants but new rounds of initiation were not demonstrated for the large transmissible Rl factor. Therefore Goebel suggested that the *dna*C gene product is an initiator protein for plasmid synthesis which must be synthesized *de novo* for the large Rl factor but can be re-used for the small ColEl factor. He further suggested that *dna*C might render membrane sites suitable for either replication or attachment of replicons. The small ColEl plasmids would repeatedly use the same *dna*C-mediated membrane site to generate many plasmid copies found in the cell. However the bacterial chromosome and large plasmids would only replicate at new *dna*C-mediated membrane sites.

To summarize, the *dna*C mutation may or may not be involved in RNA primer formation during chromosome initiation. The data of Hanna and Carl, which suggest the absence of a RNA primer role for *dna*C, conflict with those of Schekman *et al.* (1972) for phage ϕX174 SS–RF, which suggest a role in facilitating RNA primer formation. However these two processes are very different (bacterial chromosome initiation *in vivo* versus ϕ174 phage SS–RF RNA primer formation *in vitro*). Therefore, the specific biochemical function served by the *dna*C gene may be difficult to ascertain, since each process appears to be complex and composed of many separate biochemical functions. Although the idea of Goebel (1974), that *dna*C might act in a DNA-membrane attachment, is attractive, there is no conclusive evidence at present for the function of this protein.

(d) dna*E*: The *dna*E gene product, DNA polymerase III, is the first enzymic activity to be identified with a *dna* mutant. It has been discussed in detail in Section III, C.4 (p. 292).

(e) dna*F*: The *dna*F gene product appears to be one of the non-identical subunits of ribonucleotide reductase (Fuchs *et al.*, 1972). Since it blocks DNA replication by a known step in substrate formation, this mutation does not appear to be involved directly in the replication process, and will not be discussed further in this review.

(f) dna*G*: The *dna*G gene product has been purified in the same manner as the *dna*B and C(D) gene products (Wickner, S. *et al.*, 1973b). They were able to show temperature sensitivity of the mutant gene product, but no enzymic activity. The *dna*G protein would not incorporate dNTPs or NTPs and did not influence RNA or DNA synthesis catalysed by purified DNA polymerases I, II or III. There appeared to be no DNA-dependent or independent nucleoside triphosphatase, DNAse or RNAse H activities associated with the protein. The purified protein has a molecular weight of about 60,000 daltons, is insensitive to NEM and binds poorly to DNA. The *dna*G gene product did, however, stimulate *dna*A receptor activity in crude extracts. This suggests that *dna*G might function in some way similar to the *dna*A gene product.

The *dna*G gene product has also been partially purified using complementation of *dna*G cells concentrated on cellophane discs (Klein *et al.*, 1973). The cellophane disc method (see p. 278) was also utilized by Lark (1972b) to show that *dna*G regulates the "initiation" of synthesis of Okazaki pieces (see Section III, B1, p. 282). Lark (1972b) found that, in the *dna*G mutant, formation of Okazaki pieces was diminished after two minutes at a nonpermissive temperature (42°C), but the size of the pieces formed was larger. This suggested that gap closing could occur at the nonpermissive temperature between existing Okazaki pieces. If so, the *dna*G gene would appear to be involved in an earlier step of Okazaki-piece formation. The experiment was repeated in the cellophane *in vitro* system in the presence of NMN to inhibit ligase (Olivera *et al.*, 1968b; Olivera and Bonhoeffer, 1972b), and resulted in the accumulation of DNA pieces. Two distinct sizes of DNA were shown by alkaline sucrose-gradient analyses but, at 42°C, the peak of smaller fragments disappeared and only the more rapidly sedimenting peak remained. When the temperature was lowered to 25°C, the small peak re-appeared. This supports the concept that *dna*G facilitates "initiation" of Okazaki-piece formation in bacteria. The *dna*G gene product has also been implicated in RNA primer formation in phage ϕX174 DNA replication (Schekman *et al.*, 1972) and in *oop* RNA primer formation in phage lambda DNA replication (Hayes and Szybalski, 1973a, b).

To summarize the findings in the *dna* mutants of *E. coli*. The *dna*B, C(D), E, F and G gene products have been purified. However, only *dna*E and F have been associated with an enzymic activity, *dna*E being DNA polymerase III and *dna*F one of the subunits of ribonucleotide reductase. The *dna*B gene may have some associated ATPase activity but this protein may serve other functions as well. The functions of the other *dna* genes are far from clear at the moment. The possibility exists that an individual group of *dna* genes could be complex and multi-functional as indicated by the heterogeneity in the *dna*B group. And, of course, there may be undiscovered *dna* genes. Other complications might arise if some *dna* mutants could have a common function in both initiation and elongation. This was implied by the fact that RNA primer formation in the ϕX174 SS–RF system is facilitated not only by *dna*C(D) (an initiation mutant) but also by *dna*B and G (elongation mutants). Again, it is not known if RNA primer formation during phage synthesis can be equated with bacterial chromosome initiation or with Okazaki-piece "initiation". Since most of the purified proteins are on hand, the functions of the *dna* genes should be identified in the near future.

IV. Chromosome Replication During the Cell Cycle

Until 1960 there was considerable doubt regarding the nature of the chromatinic nuclear bodies in bacterial cells, whether these "nuclei" represented bacterial genomes, and whether they were replicated sequentially or in a parallel fashion. The observations of dichotomous replication by Sueoka and his coworkers (Section II, B, p. 265) showed that initiation of new rounds of replication did not require prior completion of earlier rounds, and also implied that the individual nuclear bodies, especially those found in increased number in rapidly growing cells, represented chromosomes that had not finished DNA replication. These observations of dichotomous replication, and evidence for a constant rate of synthesis at the replication fork (Yoshikawa *et al.*, 1964; Maaløe and Kjeldgaard, 1966), permitted a conceptual change from earlier interpretations and provided the foundation for the extended model later proposed by Cooper and Helmstetter (1968).

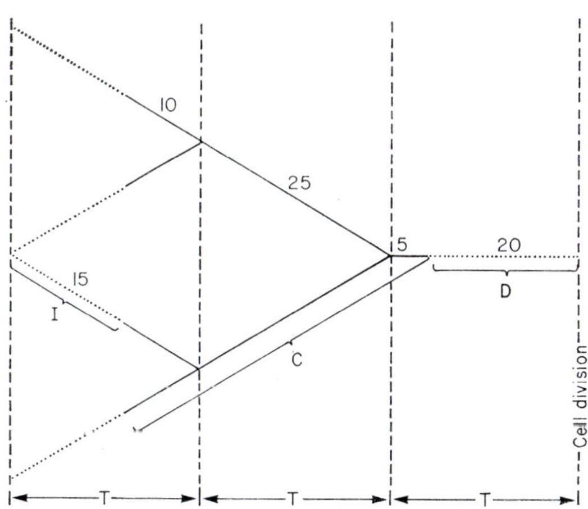

FIG. 14. Determination of chromosome configurations. The Cooper–Helmstetter model may be used to determine chromosome configurations at any cell age. One begins with the configuration in newborn cells, at time $t = 0$. One example is shown in the figure. For simplicity, nominal values of $C = 40$ minutes and $D = 20$ minutes are used. The full lines represent the chromosome, with numbered lengths in terms of times needed for replication of those segments. The dotted lines are imaginary extensions. The replication forks travel to the right as time passes. The dashed vertical lines indicate the positions of real or imaginary replication forks at time zero, and are separated by intervals of one generation time T. The final vertical line at the right represents the instant of cell division, the terminal event. For the generation time $T = 25$ minutes, as shown, the first replication fork is located 25 minutes to the left of cell division, and will complete replication of the chromosome terminus in 5 minutes. The second pair of replication forks are located 25 minutes to the left of the first. The next set of replication forks are imaginary and located at the position of the leftmost vertical line. After a period I, they will appear as real forks initiating replication at the chromosome origins. The structure of the chromosome at any later time during the cycle is determined by advancing all replication forks (real and imaginary) uniformly to the right for this time period. For example, at $t = 5$ minutes, there are three replication forks. Chromosome division occurs at 5 minutes, after which the cell contains two chromosomes, each with a single fork. At $t = 15$ minutes, replication is initiated at all new origins, and again each chromosome has three forks. At $t = 25$ minutes, the imaginary fork in the D period at the right has reached the final vertical line and cell division occurs, halving the number of chromosomes per cell. In general, $I + C + D = nT$, where $n = 1, 2, 3, \ldots$

A. Chromosome Replication in Rapidly Growing Cells

With the information in hand for dichotomous replication, the changes in rate of DNA synthesis in cells synchronized by membrane elution, as observed by Clark and Maaløe (1967) and by Helmstetter and his coworkers (Helmstetter, 1967; Helmstetter and Cooper, 1968; Cooper and Helmstetter, 1968; Helmstetter *et al.*, 1968), were interpreted as initiations of new rounds of replication. Taken together with the evidence that rapidly growing cells contain multiple replication forks (Section II, p. 265), these results were fitted with a model (Cooper and Helmstetter, 1968) which assumed: (1) that replication takes place at a constant rate at each replication fork, requiring a constant period (of duration C) to traverse the genome; and (2) that termination of a round of replication is followed by cell division after a second period (of duration D). The values for C and D (approximately 40 minutes and 20 minutes, respectively, at 37°C) are constant at all growth rates greater than one division per hour, and are independent of the medium in which the cells are grown. Where it has been tested, this model appears satisfactory for rapidly growing bacteria. It fits the results for *E. coli* and *S. typhimurium* obtained by several different methods in synchronous or exponentially increasing cultures (Cooper and Helmstetter, 1968; Helmstetter *et al.*, 1968; Kubitschek and Freedman, 1971; Cooper and Ruettinger, 1973).

The Cooper–Helmstetter model accounts for generation times shorter than 40 minutes by suggesting multiple initiations which lead to formation of branched chromosomes. For example, with a generation time of less than 20 minutes, cells would have seven replication points at birth. From the model one can predict DNA content and rate of synthesis throughout the cell cycle (Fig. 14). It should be noted, however, that this model assumes steady-state growth, such as occurs with exponentially growing cultures.

B. Chromosome Replication in Slowly Growing Cultures

In slowly growing synchronized cultures, observed replication patterns may depend upon the method of cell selection. With the membrane-elution technique, DNA replication consistently occurred over the first two-thirds of the cell cycle in *E. coli* at generation times longer than one hour (Cooper and Helmstetter, 1968; Helmstetter *et al.*, 1968). Similar results were later obtained for *S. typhimurium* with generation times longer than 48 minutes (Cooper and Ruettinger, 1973). On the other hand, when cells from chemostat cultures of *E. coli* THU were selected by velocity sedimentation in sucrose gradients, incorporation of ^{3}H-thymidine only occurred during the final part of the cell cycle

Kubitschek *et al.*, 1967). Very recently, however, Gudas and Pardee (1974), using the velocity-sedimentation method, obtained results similar to those with the elution technique. Autoradiographic results for slowly growing cultures of *E. coli* 15T$^-$ are also conflicting (C. Lark, 1966; Chai and Lark, 1970).

Because selection techniques might introduce bias, a method was devised for use with unselected cells from exponential-phase cultures; average amounts of DNA per cell were determined as a function of cell division rate. If replication occurs very early in the cycle, cells should have approximately two copies of the genome whereas, if replication occurs very late in the cycle, very few cells should contain more than a single chromosome. The latter result was obtained with very slowly growing cultures (Kubitschek and Freedman, 1971). In addition, the Cooper–Helmstetter model described above for rapidly growing cultures was found to fit the data at all growth rates.

Recent results for residual division of unselected cells, after exposure of slowly growing exponential-phase cultures to chloramphenicol or rifampicin, continue to support the conclusion that termination of DNA replication occurs very near the end of the cell cycle. Like DNA inhibitors, chloramphenicol and rifampicin prevent division of cells undergoing DNA synthesis but fail to block residual division in those cells that have completed a round of replication (Jones and Donachie, 1973; Dix and Helmstetter, 1973). When strains of *E. coli* or *S. typhimurium* were exposed to these inhibitors, residual division was of the magnitude expected for a constant value for the *D* period at all growth rates; the value for the *D* period was about 25 minutes for *E. coli* B/r, and about half that value for *S. typhimurium* (H. E. Kubitschek, unpublished results). Thus these results also extend the Cooper–Helmstetter model to include slowly growing exponential-phase cells.

This agreement between experiments with unselected cells suggests that the contrary results with selection techniques are due to perturbation of cell-cycle controls. In particular, uncoupling may occur by premature initiation of DNA replication very early in the cycle, and rapidly growing cells may be relatively immune to this premature initiation because DNA replication is already in process at birth of such cells.

In further support, DNA replication has recently been observed to occur during the terminal portion of the cycle in slowly growing cells of *E. coli* selected by velocity centrifugation in H_2O-D_2O isokinetic gradients (A. L. Koch and C. Blumberg, personal communication), and also in cultures synchronized by the Cutler–Evans (1966) technique (M. J. Kelly and C. J. Rupert, personal communication).

Moreover, Kelly and Rupert have obtained direct evidence for perturbation of the cell cycle by membrane filter attachment. They found that membrane filter-attached succinate-grown cells of *E. coli* B/r eluted during the first cell cycle were about 20–25 minutes older than their elution age and divided correspondingly earlier. In addition, when cells were synchronized by the elution technique and allowed to divide, and a portion of this control culture was put back onto the filter, the progeny of these membrane filter-bound cells were eluted some 20–25 minutes after the next division of the control culture. They conclude that there is a period between cell division and separation when cells will attach to the membrane filter. Perturbation of the cell division cycle can also be recognized by a change in the pattern of the distribution of cell sizes, or by the occurrence of hypersharp division steps in the synchronized culture (Kubitschek *et al.*, 1971). Hypersharp division steps of this kind were obtained in the synchrony reported by Gudas and Pardee (1974).

C. Termination of Chromosome Synthesis During the Cell Cycle

Synthesis of DNA in bacteria continues after starving them for required amino acids (Pardee and Prestidge, 1956; Cohen and Barner, 1954). This residual synthesis, amounting to approximately 40–50%, is due to the ability of cells to complete a round of replication even though protein synthesis is blocked (Maaløe and Hanawalt, 1961; Hanawalt *et al.*, 1961; see also reviews by Klein and Bonhoeffer, 1972; and by Smith, 1973). The requirement of synthesis of a "division protein" at or near termination of a normal round of DNA replication was shown by addition of rifampicin to block RNA synthesis, or chloramphenicol to block protein synthesis, in synchronized cultures. Division was blocked when the cells were exposed to either of these inhibitors during DNA replication, but residual division occurred if the inhibitor was added after termination of a round of DNA replication (Pierucci and Helmstetter, 1969; Jones and Donachie, 1973; Dix and Helmstetter, 1973).

A similar residual synthesis of DNA occurs at non-permissive temperatures in mutants that are temperature sensitive for initiation (see review by Gross, 1972). There is, nevertheless, a difference between the effects of these temperature shifts and those resulting from inhibition of protein synthesis. Temperature-sensitive initiation mutants complete a round of cell division even when they are shifted to the non-permissive temperature during DNA replication. In an elegant set of experiments, Marunouchi and Messer (1973) traced these cell-division differences to an additional requirement for protein synthesis in order to complete replication of a specific terminal segment of the chromosome.

Marunouchi and Messer (1973) first measured the rate of DNA replication in a temperature-sensitive *dna*A initiation mutant by incorporation of ^{3}H-thymidine during pulsed exposures. After starvation for amino acids in a culture that had been growing exponentially, the rate of replication dropped continually and the total amount of DNA increased by 48%. At this time the temperature was raised to 42°C to prevent further initiation, and the required amino acids were added back. In response, there was an immediate but brief burst in the rate of DNA synthesis. This was followed by division in those cells that previously had failed to divide because of the block in protein synthesis. These results indicated that DNA replication had not proceeded to chromosome termini in all cells, and that protein synthesis was required for synthesis of the terminal segment.

Experiments with synchronized cultures supported this explanation. Again these cultures were first starved of amino acids. After DNA replication in the absence of protein synthesis, for various periods of time, the required amino acids were added to permit protein synthesis, and nalidixic acid was added simultaneously to prevent further DNA replication. If protein synthesis must follow termination of a round of DNA replication for cell division to occur, this procedure should allow cell division. On the other hand, if protein synthesis is required instead for synthesis of a terminal chromosome segment, then this procedure should prevent cell division (assuming that termination of DNA synthesis is required for cell division). When protein synthesis was blocked early in the cell cycle there was no observable cell division. However, if amino-acid starvation was begun so late in the cell cycle that many of the cells had completed chromosome replication, some division did occur, but division was blocked during the following cycle. That is, restoring protein synthesis alone gave no division during the second cycle in the presence of nalidixic acid. It was concluded that the signal for cell division is not given by a newly initiated replication cycle, but rather that synthesis of a terminal segment of the chromosome is required for each cell division.

Since there is no initiation but continued elongation during amino-acid starvation, Marunouchi and Messer (1973) proposed that the entire chromosome should be synthesized except for the terminal segment which requires additional protein synthesis. They supported this concept by labelling the terminal segment with ^{3}H-thymidine in amino-acid starved *dna*A cells, by re-addition of amino acids at non-permissive temperature. The labelled segment appeared to be near the terminus since it replicated at a later time after another re-initiation cycle was instituted. Finally, Marunouchi and Messer (1973) showed that these terminal segments were not labelled when protein synthesis was inhibited. They again labelled terminal DNA segments with tritiated thymidine and permitted initiation of a second round of DNA replication in the presence of heavy isotopes, but prevented second-round terminal synthesis by amino-acid starvation. When this was done, the labelled DNA was found in a region of density intermediate between light reference DNA and hybrid-density DNA, and was consistent with chromosome replication up to, but not including, the terminal segment. Shearing decreased the density of the labelled fragments, again consistent with an unreplicated terminal segment. Shearing also allowed an estimate of the length of the terminal segment, which is about 0·5% of the total chromosome and has a molecular weight of about 10^7 daltons.

The requirement for protein synthesis to initiate the cell-division process has also been observed in filamentous cells of *E. coli* (Starka and Moravova, 1967; Nagai and Tamura, 1972) and in temperature-sensitive mutants of *B. subtilis* (Mendelson and Cole, 1972; Breakefield and Landman, 1973) as discussed in the review on cell division by Slater and Schaechter (1974). It may be that one or more proteins are required for disengagement of replication complexes from their terminal locations on the chromosome.

D. Initiation of Chromosome Replication

The proposal of Maaløe and Hanawalt (1961) that protein synthesis is required to initiate a new round of replication (see also Hanawalt *et al.*, 1961) led to intensive study of initiation, and proof of the model (Lark *et al.*, 1963). Many of the early results, however, were ambiguous because of assumptions that chromosome origins and termini were contiguous, or that the absence of protein synthesis inhibited only initiation. No considerations were given to a protein synthesis requirement for the terminal segment of the chromosome (Marunouchi and Messer, 1973). Early segments of the chromosome were labelled by first starving for amino acids to allow chromosomes to finish elongation. Then amino acids were added and the cells transferred first to medium containing tritiated thymidine for 5 minutes and then to medium containing BU. There was no radioactivity in the hybrid peak until 70% of the DNA had replicated, indicating both chromosome alignment and sequential replication. Labelled termini would have contributed only a minor fraction of the radioactivity in these experiments and were presumably not detected. The observation by Lark *et al.* (1963) that the rate of DNA synthesis appeared to be slower after addition of amino acids may be explained in part by the need for terminal segment synthesis before the next initiation can occur.

The studies of Lark and Renger (1969) may also provide, in retrospect, evidence for two independent initiation processes. The first step occurred soon after addition of amino acids and was resistant to 25μg of chloramphenicol/ml. The second step occurred some 15–20 minutes later, corresponding to a displacement of duration of the D period; at this time re-initiation became insensitive to 150 μg of chloramphenicol/ml. Again, in these experiments label incorporated into cells within an interval of 2–3 minutes appeared in cells that re-initiated over a period of 15–20 minutes, in agreement with incorporation in termini and origins separated in initiation by a corresponding D period. Evidence that two distinct steps are involved in initiation of chromosome replication was obtained by Lark and Lark (1964) and also by Messer (1972).

More direct evidence that protein synthesis is required for initiation of new rounds of DNA replication was obtained with a temperature-sensitive DNA initiation mutant of *E. coli* (Worcel, 1970). Exposure to the non-permissive temperature (40°C) for 10 minutes, or longer, induced asymmetric initiation, and repeated re-initiations were induced by recycling between the permissive and non-permissive temperatures. These induced re-initiations required protein synthesis; they were blocked if the temperature was shifted down in the presence of chloramphenicol for a period of time up to about 15 minutes following the shift, after which re-initiation and completion of a round of replication was again independent of protein synthesis (Schwartz and Worcel, 1971).

Synthesis of RNA may also be involved in the initiation of DNA replication (Lark, 1972a; Laurent, 1973). Both rifampicin and streptolydigin inhibited replication later in the cycle than did chloramphenicol, suggesting that initiation of DNA replication requires a transcriptive step in addition to an earlier period of protein synthesis. The requirement for RNA synthesis occurred about 10 minutes before initiation of DNA synthesis and about 5–10 minutes after the required protein synthesis. Lark concluded that the RNA product itself was required for initiation of replication, and not for further transcription. Because this RNA appeared stable, he suggested that the replication complex consists of proteins self-assembled around an RNA core.

There is an anomalous type of DNA replication which continues for long periods in the absence of protein synthesis (Kogoma and Lark, 1970; Lark, 1972a). This "stabilized" synthesis of DNA occurs when firstly DNA synthesis is inhibited and protein synthesis is allowed to continue for a sufficient period of time; then protein synthesis is blocked and the DNA inhibition is released. Lark (1972a) showed that transcription was required to initiate stabilized synthesis of DNA after thymine starvation and that, in any cycle, replication probably occurred via a single chromosome template. Subsequent replication cycles required neither RNA nor protein synthesis, suggesting that this RNA was stable and could be used again during successive rounds. The entire mode of DNA replication could be altered during stabilized synthesis, with normal replication replaced by a rolling circle, or anomalous mechanism, that continually conserves the replication complex.

At present the nature of the initiation mechanism of DNA replication is almost completely unknown. As discussed, initiation could involve membrane attachment, protein and RNA synthesis and the *dna*A and *dna*C gene products. Since a protein-synthesis requirement for termination is now known, and could have complicated earlier initiation studies, more progress on the mechanism of initiation should be forthcoming.

V. Discussion

We have attempted to categorize the study of bacterial DNA replication on three levels, namely the whole chromosome, intermediate fork formation and the biochemical events at the replication fork. This approach was taken because of the complexity of the bacterial replication process and the still existing gaps in knowledge. Difficulties have arisen because of two characteristics of replication. First, because replication has a stringent but structurally delicate dependency on the cell membrane, and on many replication proteins, an adequate reconstruction of the complete replication complex has never been achieved. In addition, many replication proteins are still unidentified. A second characteristic is that the process of chromosome replication is interrelated with other poorly understood processes such as cell division and recombination. Knowledge of some of these complex relationships may be important to understanding the DNA replication process itself.

We should like to summarize the progress that has been made since the review of Smith (1973), again categorizing the levels of discussion.

(a) *Whole chromosome.* (1) Although circularity has long been established in *E. coli*, the isolation of the *B. subtilis* chromosome as a circular molecule (Section II, A.1, p. 255) re-emphasizes the importance of this configuration in bacterial DNA replication. (2) The isolation of the *E. coli* chromosome in its folded form with associated membrane, RNA and protein represents a major step in understanding how the chromosome exists in the intact cell (Sections II, A2, p. 256, and A3, p. 258). (3) The requirement for protein synthesis for the terminal segment of the chromosomes indicates another locus upon which control proteins may be acting during DNA replication (Section IV, C, p. 312).

(b) *Intermediate* (*fork formation*). The evidence for bidirectionality of chromosome replication has been solidified and demonstrated in both *E. coli* and *B. subtilis* (Section II, B.1, p. 261).

(c) *Biochemistry*. (1) The roles of RNA synthesis in both initiation (Section IV, D, p. 314) and elongation (Section III, B, p. 282) have been established. The isolation of RNA primer has re-inforced the previous evidence that identified Okazaki DNA fragments as intermediates of DNA replication, and has established more clearly the discontinuous mode of elongation. (2) The discovery of the pol III* system in bacteriophage opens up a new realm of study in bacterial DNA replication (Sections III, A.3, p. 280 and III, C.4(b), p. 293). The future study of protein factor relationships to DNA polymerases should reveal much information on the mechanism of replication. (3) The purification of *dna* gene products should allow the identification of the *dna* functions, and expand the list of known essential replication proteins.

The areas where future research is needed are obvious. An adequately reconstructed *in vitro* system is still desirable, and perhaps the folded chromosome with purified membrane components will be the starting materials for a more well-defined bacterial *in vitro* DNA replication system (Sections II, A.2, p. 257 and 3, p. 258). Studies of the initiation and termination processes, at the biochemical level, have been few (Sections IV, C, p. 312 & D, p. 314). We do not know the mechanism of the RNA synthesis required for initiation, nor do we know the identity or functions of any initiation or termination proteins. A membrane-attachment role in initiation should be examined along with a clearer definition of what constitutes the structure of fork formation. Although three DNA polymerases have been purified, only DNA polymerase III has been established as an essential replication enzyme. All DNA polymerases should be further investigated in order to determine whether they serve specific functions or can substitute for each other during replication. A further search for non-enzymic control proteins similar to Copol III* (Section III, C.4(b), p. 293) is needed as well as the establishment of the functions of the *dna* gene products (Section III, C.8, p. 304). Another perplexing problem has been the role of ATP during DNA replication. Lastly, the relationship of DNA replication to cell division, DNA repair and genetic recombination should be studied. The identification of proteins serving common functions in two or more of these processes (such as DNA polymerase I) will aid our understanding of replication. The full or partial accomplishment of the above-mentioned areas of research should aid our understanding of bacterial DNA replication on both the physiological and molecular levels, and perhaps make the next review approachable at one instead of three levels of discussion.

VI. Acknowledgements

The authors wish to thank the many workers who have provided preprints of their recent research. We wish to thank particularly Dr. N. R. Cozzarelli for his discussion, helpful information and the detailed reading and criticism of the text, and Dr. P. C. Hanawalt for his reading and criticism of parts of the manuscript. The authors' work was supported by the U.S. Atomic Energy Commission.

References

Abe, M. and Tomizawa, J. (1967). *Proceedings of the National Academy of Sciences of the United States of America* **58**, 1911.

Alberts, B. (1971). *In* "Nucleic Acid-Protein Interactions and Nucleic Acid in Viral Infection", (D. W. Ribbons, J. F. Woessner and J. Schultz, eds.), p. 128, North Holland, Amsterdam.

Alberts, B. M., Amodio, F. J., Jenkins, M., Cutmann, E. D. and Ferris, F. L. (1968). *Cold Spring Harbor Symposium of Quantitative Biology* **33**, 289.

Alberts, B. M. and Frey, L. (1970). *Nature, London* **227**, 1313.

Arber, W. and Linn, S. (1969). *Annual Review of Biochemistry* **38**, 467.

Bak, A. L., Christiansen, C. and Stenderup, A. (1970). *Journal of General Microbiology*, **64**, 377.

Barbour, S. D. and Clark, A. J. (1970). *Proceedings of the National Academy of Sciences of the United States of America* **65**, 955.

Bazill, G. W. and Gross, J. D. (1972). *Nature New Biology* **240**, 82.

Bazill, G. W. and Gross, J. D. (1973). *Nature New Biology* **243**, 241.

Berg, P., Fancher, H. and Chamberlin, M. (1963). *In* "Symposium on Informational Macromolecules", (H. J. Vogel, V. Bryson and J. O. Lampen, eds.), p. 467, Academic Press, New York.

Berger, H. Jr. and Irwin, J. L. (1970). *Proceedings of the National Academy of Sciences of the United States of America* **65**, 152.

Beyersmann, D., Schlicht, M. and Schuster, H. (1971). *Molecular and General Genetics* **111**, 145.

Billen, D., Carreira, L. B., Hadden, C. T. and Silverstein, S. J. (1971b). *Journal of Bacteriology* **108**, 1250.

Billen, D., Carreira, L. B. and Silverstein, S. (1971a). *Biochemical and Biophysical Research Communications* **43**, 1150.

Bird, R. and Lark, K. G. (1965). *Journal of Molecular Biology*, **13**, 607.

Bird, R. and Lark, K. G. (1968). *Cold Spring Harbor Symposium of Quantitative Biology* **33**, 799.

Bird, R., Louran, J., Martuscelli, J. and Caro, L. (1972). *Journal of Molecular Biology* **70**, 549.

Blair, D. G., Sherratt, D. J., Clewell, D. B. and Helinski, D. R. (1972). *Proceedings of the National Academy of Sciences of the United States of America* **69**, 2518.

Bleecken, S., Strohbach, G. and Sarfert, E. (1966). *Zeitschrift für Allgemeine Mikrobiologie* **6**, 121.

Bode, H. R. and Morowitz, H. J. (1967). *Journal of Molecular Biology* **23**, 191.

Bott, K. F. and Wilson, G. A. (1967). *Journal of Bacteriology* **94**, 562.

Bowersock, D. and Moses, R. E. (1973). *Journal of Biological Chemistry* **248**, 7449.

Boyle, J. V., Goss, W. A. and Cook, T. M. (1967). *Journal of Bacteriology* **94**, 1664.

Breakefield, X. O. and Landman, O. E. (1973). *Journal of Bacteriology* **113**, 985.

Broker, T. R. and Lehman, I. R. (1971). *Journal of Molecular Biology* **60**, 131.

Brown, N. C. (1970). *Proceedings of the National Academy of Sciences of the United States of America* **67**, 1454.

Brown, N. C. (1971). *Journal of Molecular Biology* **59**, 1.

Brown, N. C. (1972a). *Federation Proceedings. Federation of American Societies for Experimental Biology* **31**, 442.

Brown, N. C. (1972b). *Biochimica et Biophysica Acta* **281**, 202.

Brown, N. C., Wisseman, C. L. III and Matsushita, T. (1972). *Nature New Biology* **237**, 72.

Brutlag, D. and Kornberg, A. (1972) *Journal of Biological Chemistry* **247**, 241.

Brutlag, D., Schekman, R. and Kornberg, A. (1971). *Proceedings of the National Academy of Sciences of the United States of America* **68**, 2826.

Burger, R. M. (1971). *Proceedings of the National Academy of Sciences of the United States of America* **68**, 2124.

Buttin, G. and Kornberg, A. (1966). *Journal of Biological Chemistry* **241**, 5419.

Buttin, G. and Wright, M. R. (1968). *Cold Spring Harbor Symposium of Quantitative Biology* **33**, 259.

Cairns, J. (1963a). *Journal of Molecular Biology* **6**, 208.

Cairns, J. (1963b). *Cold Spring Harbor Symposium of Quantitative Biology* **28**, 43.

Carl, P. L. (1970). *Molecular and General Genetics* **109**, 107.

Chai, N. S. and Lark, K. G. (1970). *Journal of Bacteriology* **104**, 401.

Clark, D. J. and Maaløe, O. (1967). *Journal of Molecular Biology* **23**, 99.

Clewell, D. B. (1972). *Journal of Bacteriology* **110**, 667.

Clewell, D. B., Evenchik, B. and Cranston, J. W. (1972). *Nature New Biology* **237**, 29.

Cohen, S. S. and Barner, H. D. (1954). *Proceedings of the National Academy of Sciences of the United States of America* **40**, 885.

Cooper, P. K. and Hanawalt, P. C. (1972a). *Journal of Molecular Biology* **67**, 1.

Cooper, P. K. and Hanawalt, P. C. (1972b). *Proceedings of the National Academy of Sciences of the United States of America* **69**, 1156.

Cooper, S. and Helmstetter, C. E. (1968). *Journal of Molecular Biology* **31**, 519.

Cooper, S. and Ruettinger, T. (1973). *Journal of Bacteriology* **114**, 966.

Coulter, C. L. and Cozzarelli, N. R. (1974). *Acta Crystallographica*, in the press.

Cozzarelli, N. R. and Low, R. L. (1973). *Biochemical and Biophysical Research Communications* **51**, 151.

Curtiss, R. III (1969). *Annual Review of Microbiology* **23**, 69.

Cutler, R. G. and Evans, J. E. (1966) *Journal of Bacteriology* **91**, 469.

Delius, H., Mantell, N. J. and Alberts, B. (1972). *Journal of Molecular Biology* **67**, 341.

Delius, H. and Worcel, A. (1974). *Journal of Molecular Biology* **82**, 107.

De Lucia, P. and Cairns, J. (1969). *Nature, London* **224**, 1164.

Dix, D. E. and Helmstetter, C. E. (1973). *Journal of Bacteriology* **115**, 786.

Dove, W. F., Hargrove, E., Ohashi, M., Haugli, F. and Guha, A. (1969). *Japanese Journal of Genetics* **44** (suppl. 1), 11.

Dubnau, D. (1970). *In* "Handbook of Biochemistry", 2nd ed., (H. A. Sober and R. A. Harte, eds.) pp. 139–145, Chemical Rubber Co., Ohio.

Durwald, H. and Hoffmann-Berling, H. (1971). *Journal of Molecular Biology* **58**, 755.

Earhart, C. F. (1970). *Virology* **42**, 429.

Earhart, C. F., Tremblay, G. Y., Daniels, M. J. and Schaechter, M. (1968). *Cold Spring Harbor Symposium of Quantitative Biology* **33**, 707.

Echols, H. (1971). *Annual Review of Biochemistry* **40**, 827.

Englund, P. T., Deutscher, M. P., Jovin, T. M., Kelly, R. B., Cozzarelli, N. R. and Kornberg, A. (1968). *Cold Spring Harbor Symposium of Quantatitive Biology* **33**, 1.

Epstein, R. H., Bolle, A., Steinberg, C. M., Kellenberger, E., Boy de la Tour, R., Chevallez, R., Edgar, R. S., Susman, M., Denhardt, G. H. and Lielausis, A. (1963). *Cold Spring Harbor Symposium of Quantitative Biology* **28**, 375.

Fleischman, R. A. and Richardson, C. C. (1971). *Proceedings of the National Academy of Sciences of the United States of America* **68**, 2527.

Fox, R. M., Mendelsohn, J., Barbosa, E. and Goulian, M. (1973). *Nature New Biology* **245**, 234.

Friedman, E. A. and Smith, H. O. (1973). *Journal of Biological Chemistry* **247**, 2846.

Fritsch, A. and Worcel, A. (1971). *Journal of Oklahoma State Dental Association* **59**, 207.

Fuchs, E. and Hanawalt, P. C. (1970). *Journal of Molecular Biology* **52**, 301.

Fuchs, J. A., Karlstrom, H. O., Warner, H. A. and Reichard, P. (1972). *Nature New Biology* **238**, 69.

Fujisawa, T. and Eisenstark, A. (1973). *Journal of Bacteriology* **115**, 168.

Fulton, C. (1965). *Genetics* **52**, 55.

Ganesan, A. T. (1971). *Proceedings of the National Academy of Sciences of the United States of America* **68**, 1296.

Ganesan, A. T. and Lederberg, J. (1965). *Biochemical and Biophysical Research Communications* **18**, 824.

Ganesan, A. T., Yehle, C. O. and Yu, C. C. (1973). *Biochemical and Biophysical Research Communications* **50**, 155.

Gass, K. B. and Cozzarelli, N. R. (1973a). *Journal of Biological Chemistry* **248**, 7688.

Gass, K. B. and Cozzarelli, N. R. (1973b). *In* "Methods in Enzymology", (L. Grossman and K. Moldave, eds.), Vol. 29, p. 27. Academic Press, New York.

Gass, K. B., Hill, T. C., Goulian, M., Strauss, B. S. and Cozzarelli, N. R. (1971). *Journal of Bacteriology* **108**, 364.

Gass, K. B., Low, R. L. and Cozzarelli, N. R. (1973). *Proceedings of the National Academy of Sciences of the United States of America* **70**, 103.

Gefter, M. L., Hirota, Y., Kornberg, T., Barnoux, C. and Wechsler, J. A. (1971). *Proceedings of the National Academy of Sciences of the United States of America* **68**, 3150.

Gellert, M. and Bullock, M. L. (1970). *Proceedings of the National Academy of Sciences of the United States of America* **67**, 1580.

Gilbert, W. and Dressler, D. (1968). *Cold Spring Harbor Symposium of Quantitative Biology* **33**, 473.

Goebel, W. (1970). *European Journal of Biochemistry* **15**, 311.

Goebel, W. (1973). *Biochemical and Biophysical Research Communications* **51**, 1000.

Goebel, W. (1974). *European Journal of Biochemistry* **41**, 51.

Goldmark, P. J. and Linn, S. (1970). *Proceedings of the National Academy of Sciences of the United States of America* **67**, 434.

Goldmark, P. J. and Linn, S. (1972). *Journal of Biological Chemistry* **247**, 1849.

Goulian, M. (1971). *Annual Review of Biochemistry* **40**, 855.

Greth, M. L. and Chevallier, M. R. (1973). *Biochemical and Biophysical Research Communications* **54**, 1.

Gross, J. D. (1972). *Current Topics in Microbiology and Immunology* **57**, 39.

Gudas, L. J. and Pardee, A. B. (1974). *Journal of Bacteriology* **117**, 1216.

Gyurasits, E. G. and Wake, R. B. (1973). *Journal of Molecular Biology* **73**, 55.

Habener, J. F., Bynum, B. S. and Shack, J. (1970). *Journal of Molecular Biology* **49**, 157.

Hall, F. W. and Lehman, I. R. (1969). *Journal of Biological Chemistry* **244**, 43.

Hanawalt, P. C., Maaløe, O., Cummings, D. J. and Schaechter, M. (1961). *Journal of Molecular Biology* **3**, 156.

Hara, H. and Yoshikawa, H. (1973). *Nature New Biology*, **244**, 200.

Hayes, S. and Szybalski, W. (1973a). *Molecular and General Genetics* **126**, 275.

Hayes, S. and Szybalski, W. (1973b). *In* "Molecular Cytogenetics", (B. A. Hamkalo and J. Papaconstantinou, eds.), p. 277. Plenum Press, New York.

Helinski, D. R. and Clewell, D. B. (1971). *Annual Review of Biochemistry* **40**, 899.

Helmstetter, C. E. (1967). *Journal of Molecular Biology* **24**, 417.

Helmstetter, C. E. (1974a). *Journal of Molecular Biology* **84**, 1.

Helmstetter, C. E. (1974b). *Journal of Molecular Biology* **84**, 21.

Helmstetter, C. E., and Cooper, S. (1968). *Journal of Molecular Biology* **31**, 507.

Helmstetter, C. E., Cooper, S., Pierucci, O. and Revelas, E. (1968). *Cold Spring Harbor Symposium of Quantitative Biology* **33**, 809.

Herrick, G. (1973). Ph.D. Thesis: Princeton University, Princeton, N.J.

Herrmann, R., Huf, J. and Bonhoeffer, F. (1972). *Nature New Biology* **240**, 237.

Hirose, S., Okazaki, R. and Tamanoi, F. (1973). *Journal of Molecular Biology* **77**, 501.

Hirota, Y., Mordoh, J. and Jacob, F. (1970). *Journal of Molecular Biology* **53**, 369.

Hirota, Y., Mordoh, J., Scheffler, I. and Jacob, F. (1972). *Federation Proceedings. Federation of American Societies for Experimental Biology* **31**, 1422.

Hohlfeld, R. and Vielmetter, W. (1973). *Nature New Biology* **242**, 130.

Hopwood, D. A. (1967). *Bacteriological Reviews* **31**, 373.

Huberman, J. A. and Riggs, A. D. (1968). *Journal of Molecular Biology* **32**, 327.

Huberman, J. A., Tsai, A. and Deich, R. A. (1973). *Nature, London* **241**, 32.

Hurwitz, J. and Wickner, S. (1974). *Proceedings of the National Academy of Sciences of the United States of America* **71**, 6.

Hurwitz, J., Wickner, S. and Wright, M. (1973). *Biochemical and Biophysical Research Communications* **51**, 257.

Jackson, R. W. and DeMoss, J. A. (1965). *Journal of Bacteriology* **90**, 1420.

Jacob, F., Brenner, S. and Cuzin, F. (1963). *Cold Spring Harbor Symposium of Quantitative Biology* **28**, 329.

Jacob, F. and Wollman, E. (1961). *Sexuality and the Genetics of Bacteria.* New York, Academic Press.

Jonasson, J. (1973). *Molecular and General Genetics* **120**, 69.

Jones, N. C. and Donachie, W. D. (1973). *Nature New Biology* **243**, 100.

Jovin, T. M., Englund, P. T. and Kornberg, A. (1969). *Journal of Biological Chemistry* **244**, 3009.

Kainuma, R. and Okazaki, R. (1970). *Journal of the Japanese Biochemical Society* **42**, 464.

Kallenbach, N. R. and Ma, R. (1968). *Journal of Bacteriology* **95**, 304.

Karkas, J. D. (1974) *Biochimica et Biophysica Acta* **340**, 152.

Karu, A. E. and Linn, S. (1972). *Proceedings of the National Academy of Sciences of the United States of America* **69**, 2855.

Karu, A. E., MacKay, V., Goldmark, P. J. and Linn, S. (1973). *Journal of Biological Chemistry* **248**, 4874.

Kasamatsu, H. and Vinograd, J. (1973). *Nature New Biology* **241**, 103.

Kelly, R. B., Atkinson, M. R., Huberman, J. A. and Kornberg, A. (1969). *Nature, London* **224**, 495.

Kelly, R. B., Cozzarelli, N. R., Deutscher, M. P., Lehman, I. R. and Kornberg, A. (1970). *Journal of Biological Chemistry* **245**, 39.

Klein, A. and Bonhoeffer, F. (1972). *Annual Review of Biochemistry* **41**, 301.

Klein, A., Nusslein, V., Otto, B. and Powling, A. (1973). *In* "DNA Synthesis *in Vitro*", (R. D. Wells and R. B. Inman, eds.), p. 185. University Park Press, Baltimore, Maryland.

Klotz, L. C. and Zimm, B. H. (1972). *Journal of Molecular Biology* **72**, 779.

Knippers, R. and Strätling, W. (1970). *Nature, London* **226**, 713.

Koerner, J. F. (1970). *Annual Review of Biochemistry* **39**, 291.

Kogoma, T. and Lark, K. G. (1970). *Journal of Molecular Biology* **52**, 143.

Konrad, E. B. and Lehman, I. R. (1974). *Proceedings of the National Academy of Sciences of the United States of America* **71**, 2048.

Konrad, E. B., Modrich, P. and Lehman, I. R. (1973). *Journal of Molecular Biology* **77**, 119.

Kornberg, A. (1961). *Enzymatic Synthesis of DNA*, Wiley, New York.

Kornberg, A. (1969). *Science* **163**, 1410.

Kornberg, T. and Gefter, M. L. (1971). *Proceedings of the National Academy of Sciences of the United States of America* **68**, 761.

Kozinski, A. and Felgenhauer, K. (1967). *Journal of Virology* **1**, 1193.

Kubitschek, H. E. and Freedman, M. L. (1971). *Journal of Bacteriology* **107**, 95.

Kubitschek, H. E., Bendigkeit, H. E. and Loken, M. R. (1967). *Proceedings of the National Academy of Sciences of the United States of America* **57**, 1611.

Kubitschek, H. E., Freedman, M. L. and Silver, S. (1971). *Biophysical Journal* **11**, 787.

Kuempel, P. L. and Veomett, G. E. (1970). *Biochemical and Biophysical Research Communications* **41**, 973.

Lange, D., Bujard, H., Wolff, B. and Russel, D. (1967). *Journal of Molecular Biology* **23**, 163.

Lark, C. (1966). *Biochimica et Biophysica Acta* **119**, 517.

Lark, C. and Lark, K. G. (1964). *Journal of Molecular Biology* **10**, 120.

Lark, K. G. (1966). *Bacteriological Reviews* **30**, 3.

Lark, K. G. (1969). *Annual Review of Biochemistry* **38**, 569.

Lark, K. G. (1972a). *Journal of Molecular Biology* **64**, 47.

Lark, K. G. (1972b). *Nature New Biology* **240**, 237.

Lark, K. G. and Renger, H. (1969). *Journal of Molecular Biology* **42**, 221.

Lark, K. G., Repko, T. and Hoffman, E. J. (1963). *Biochimica et Biophysica Acta* **76**, 9.

Laurent, S. (1973). *Journal of Bacteriology* **116**, 141.

Lehman, I. R. (1963). *Progress in Nucleic Acid Research* **2**, 83.

Lehman, I. R. (1967). *Annual Review of Biochemistry* **36**, 645.

Lehman, I. R. and Chien, J. R. (1973). *Journal of Biological Chemistry* **248**, 7717.

Lindahl, G., Hirota, Y. and Jacob, F. (1971). *Proceedings of the National Academy of Sciences of the United States of America* **68**, 2407.

Linn, S. and MacKay, V. (1974). *In* ICN/UCLA Winter Conferences on Molecular Biology, "Molecular Mechanisms for the Repair of DNA", Squaw Valley, Calif.

Little, J. W., Zimmerman, S. B., Oskinsky, C. K. and Gellert, M. (1967). *Proceedings of the National Academy of Sciences of the United States of America* **58**, 2004.

Low, R. L., Rashbaum, S. A. and Cozzarelli, N. R. (1974). *Proceedings of the National Academy of Sciences of the United States of America* **71**, 2973.

Lundquist, R., Manlapaz-Fernandez, P. and Olivera, B. M. (1974). *Journal of Molecular Biology* **83**, 541.

Maaløe, O. and Hanawalt, P. C. (1961). *Journal of Molecular Biology* **3**, 144.

Maaløe, O. and Kjeldgaard, N. O. (1966). "Control of Macromolecular Synthesis", W. A. Benjamin, New York.

Mackenzie, J. M., Neville, M. M., Wright, G. E. and Brown, N. C. (1973). *Proceedings of the National Academy of Sciences of the United States of America* **70**, 512.

Marunouchi, T. and Messer, W. (1973). *Journal of Molecular Biology* **78**, 211.

Marvin, D. A. (1968). *Nature, London* **219**, 485.

Masker, W. E. and Hanawalt, P. C. (1973). *Proceedings of the National Academy of Sciences of the United States of America* **70**, 129.

Masker, W. E. and Hanawalt, P. C. (1974). *Biochimica et Biophysica Acta* **340**, 229.

Masker, W. E., Hanawalt, P. C. and Shizuya, H. (1973). *Nature New Biology* **244**, 242.

Massie, H. R. and Zimm, B. H. (1965). *Proceedings of the National Academy of Sciences of the United States of America* **54**, 1636.

Masters, M. and Broda, P. (1971). *Nature New Biology* **232**, 137.

Masters, M. and Pardee, A. B. (1965). *Proceedings of the National Academy of Sciences of the United States of America* **54**, 64.

Matsushita, T. and Sueoka, N. (1974). *Journal of Bacteriology* **118**, 974.

Matsushita, T., White, K. P. and Sueoka, N. (1971). *Nature New Biology* **232**, 111.

McKenna, W. G. and Masters, M. (1972). *Nature, London* **240**, 536.

Mendelson, N. H. and Cole, R. M. (1972). *Journal of Bacteriology* **112**, 994.

Meselson, M. and Stahl, F. W. (1958). *Proceedings of the National Academy of Sciences of the United States of America* **44**, 671.

Messer, W. (1972). *Journal of Bacteriology* **112**, 7.

Modrich, P. and Lehman, I. R. (1971). *Proceedings of the National Academy of Sciences of the United States of America* **68**, 1002.

Moses, R. E. (1972). *Journal of Biological Chemistry* **247**, 6031.

Moses, R. E. and Richardson, C. C. (1970a). *Proceedings of the National Academy of Sciences of the United States of America* **67**, 674.

Moses, R. E. and Richardson, C. C. (1970b). *Biochemical and Biophysical Research Communications* **41**, 1557.

Moses, R. E. and Richardson, C. C. (1970c). *Biochemical and Biophysical Research Communications* **41**, 1565.

Nagai, K. and Tamura, G. (1972). *Journal of Bacteriology* **112**, 959.

Nagata, T. and Meselson, M. (1968). *Cold Spring Harbor Symposium of Quantitative Biology* **33**, 553.

Nath, K. and Hurwitz, J. (1974). *Journal of Biological Chemistry* **249**, 2605.

Neville, M. M. and Brown, N. C. (1972). *Nature New Biology* **240**, 80.

Nishimura, Y., Caro, L., Berg, C. M. and Hirota, Y. (1971). *Journal of Molecular Biology* **55**, 441.

Nishioka, Y. and Eisenstark, A. (1970). *Journal of Bacteriology* **102**, 320.

Nusslein, V., Otto, B., Bonhoeffer, F. and Schaller, H. (1971). *Nature New Biology* **234**, 285

Ohi, S., Bastia, D. and Sueoka, N. (1974). *Nature, London* **248**, 586.

Ohi, S. and Sueoka, N. (1973). *Journal of Biological Chemistry* **248**, 7336.

Ohki, M. and Tomizawa, J. (1968). *Cold Spring Harbor Symposium of Quantatitive Biology* **33**, 651.

Oishi, M. (1968). *Proceedings of the National Academy of Sciences of the United States of America* **60**, 329.

Oishi, M. (1969). *Proceedings of the National Academy of Sciences of the United States of America* **64**, 1292.

Oishi, M., Yoshikawa, H. and Sueoka, N. (1964). *Nature, London* **204**, 1069.

Okazaki, R., Arisawa, M. and Sugino, A. (1971). *Proceedings of the National Academy of Sciences of the United States of America* **68**, 2954.

Okazaki, T. and Kornberg, A. (1964). *Journal of Biological Chemistry* **239**, 259.

Okazaki, R., Okazaki, T., Sakabe, K., Sugimoto, K., Kainuma, R., Sugino, A. and Iwatsuki, N. T. (1968b). *Cold Spring Harbor Symposium of Quantitative Biology* **33**, 129.

Okazaki, R., Okazaki, T., Sakabe, K., Sugimoto, K. and Sugino, A. (1968a). *Proceedings of the National Academy of Sciences of the United States of America* **59**, 598.

Okazaki, R., Sugimoto, K., Okazaki, T., Imae, Y. and Sugino, A. (1970). *Nature, London* **228**, 223.

Okazaki, R., Sugino, A., Hirose, S., Okazaki, T., Imae, Y., Kainuma-Kuroda, R., Arisawa, M. and Kurosawa, Y. (1973). *In* "DNA Synthesis *in Vitro*", (R. D. Wells and R. B. Inman, eds.), p. 83. University Park Press, Baltimore, Maryland.

Olivera, B. M. and Bonhoeffer, F. (1972a). *Proceedings of the National Academy of Sciences of the United States of America* **69**, 25.

Olivera, B. M. and Bonhoeffer, F. (1972b). *Nature New Biology* **240**, 233.

Olivera, B. M., Hall, Z., Anraku, Y., Chien, J. and Lehman, I. R. (1968b). *Cold Spring Harbor Symposium of Quantitative Biology* **33**, 27.

Olivera, B. M., Hall, Z. W. and Lehman, I. R. (1968a). *Proceedings of the National Academy of Sciences of the United States of America* **61**, 237.

O'Sullivan, A. and Sueoka, N. (1967). *Journal of Molecular Biology* **27**, 349.

Otto, B., Bonhoeffer, F. and Schaller, H. (1973). *European Journal of Biochemistry* **34**, 440.

Painter, R. B. and Schaefer, A. (1969). *Nature, London,* **221**, 1215.

Pardee, A. B. and Prestidge, L. S. (1956). *Journal of Bacteriology* **71**, 677.

Pauling, C. and Hamm, L. (1968). *Proceedings of the National Academy of Sciences of the United States of America* **60**, 1495.

Pearson, G. D. and Hanawalt, P. C. (1971). *Journal of Molecular Biology* **62**, 65.

Pettijohn, D. E., Hecht, R. M., Stonington, O. G. and Stamato, T. D. (1973). *In* "DNA Synthesis *in Vitro*", (R. D. Wells and R. B. Inman, eds.), p. 145. University Park Press, Baltimore, Maryland.

Pierucci, O. and Helmstetter, C. E. (1969). *Federation Proceedings. Federation of American Societies of Experimental Biology* **28**, 1755.

Prescott, D. M. and Kuempel, P. L. (1972). *Proceedings of the National Academy of Sciences of the United States of America* **69**, 2842.

Pritchard, R. H. and Lark, K. G. (1964). *Journal of Molecular Biology* **9**, 288.

Quinn, W. G. and Sueoka, N. (1970). *Proceedings of the National Academy of Sciences of Washington* **67**, 717.

Radding, C. M. (1969). *Annual Review of Genetics* **3**, 363.

Rama Reddy, G. V., Goulian, M. and Hendler, S. S. (1971). *Nature New Biology* **234**, 286.

Richardson, C. C. (1969). *Annual Review of Biochemistry* **38**, 795.

Rodriguez, R. L., Dalbey, M. S. and Davern, C. I. (1973). *Journal of Molecular Biology* **74**, 599.

Rosenberg, B. H. and Cavalieri, L. F. (1968). *Cold Spring Symposium of Quantitative Biology* **33**, 65.

Rupp, W. D. and Ihler, G. (1968). *Cold Spring Harbor Symposium of Quantitative Biology* **33**, 647.

Ryter, A. and Jacob, F. (1963). *Compte rendu hebdomadaire des séances de l'Académie des sciences* **257**, 3060.

Sakabe, K. and Okazaki, R. (1966). *Biochimica et Biophysica Acta* **129**, 651.

Sanderson, K. E. (1967). *Bacteriological Reviews* **31**, 354.

Sanderson, K. E. and Demerec, M. (1964). *Microbial Genetics Bulletin* **20**, 11.

Sato, S., Tanaka, M. and Sugimura, T. (1970). *Biochimica et Biophysica Acta* **209**, 43.
Schaechter, M., Maaløe, O. and Kjeldgaard, N. O. (1958). *Journal of General Microbiology* **19**, 592.
Schaller, H., Otto, B., Nusslein, V., Huf, J., Herrmann, R. and Bonhoeffer, F. (1972). *Journal of Molecular Biology* **63**, 183.
Schandl, E. K. and Taylor, J. H. (1969). *Biochemical and Biophysical Research Communications* **34**, 291.
Schekman, R., Wickner, W., Westergaard, O., Brutlag, D., Geider, K., Bertsch, L. L. and Kornberg, A. (1972). *Proceedings of the National Academy of Sciences of the United States of America* **69**, 2691.
Schnös, M. and Inman R. B. (1970). *Journal of Molecular Biology* **51**, 61.
Schnös, M. and Inman, R. B. (1971). *Journal of Molecular Biology* **55**, 31.
Schwartz, M. and Worcel, A. (1971). *Journal of Molecular Biology* **61**, 329.
Siegel, P. J. and Schaechter, M. (1973). *Annual Review of Microbiology* **27**, 261.
Sigal, N., Delius, H., Kornberg, T., Gefter, M. L. and Alberts, B. (1972). *Proceedings of the National Academy of Sciences of the United States of America* **69**, 3537.
Signer, E. (1971). *In* "The Bacteriophage Lambda", (A. D. Hershey, ed.), p. 139. Cold Spring Harbor Laboratory, New York.
Silverstein, S. and Billen, D. (1971). *Biochimica et Biophysica Acta* **247**, 383.
Slater, M. and Schaechter, M. (1974). *Bacteriological Reviews* **39** 119.
Smith, D. W. (1973), *Progress in Biophysical Molecular Biology* **26**, 323.
Smith, D. W. and Hanawalt, P. C. (1967). *Biochimica et Biophysica Acta* **149**, 519.
Smith, D. W., Schaller, H. E. and Bonhoeffer, F. J. (1970). *Nature, London* **226**, 711.
Spratt, B. G. (1972). *Biochemical and Biophysical Research Communications* **48**, 496.
Starka, J. and Moravova, J. (1967). *Folia Microbiologica, Praha* **12**, 240.
Stein, G. and Hanawalt, P. (1969). *Journal of Molecular Biology* **46**, 135.
Stein, G. H. and Hanawalt, P. (1972). *Journal of Molecular Biology* **64**, 393.
Stonington, O. G. and Pettijohn, D. E. (1971). *Proceedings of the National Academy of Sciences of the United States of America* **68**, 6.
Sueoka, N. (1971). *Genetics* **68**, 349.
Sueoka, N., Matsushita, T., Ohi, S., O'Sullivan, M. A. and White, K. P. (1973). *In* "DNA Synthesis *in Vitro*", (R. D. Wells and R. B. Inman, eds.), p. 385. University Park Press, Baltimore, Maryland.
Sueoka, N. and Quinn, W. G. (1968). *Cold Spring Harbor Symposium of Quantitative Biology* **33**, 695.
Sueoka, N. and Yoshikawa, H. (1963). *Cold Spring Harbor Symposium of Quantitative Biology* **28**, 47.
Sugino, A., Hirose, S. and Okazaki, R. (1972). *Proceedings of the National Academy of Sciences of the United States of America* **69**, 1863.
Sugino, A. and Okazaki, R. (1972). *Journal of Molecular Biology* **64**, 61.
Sugino, A. and Okazaki, R. (1973). *Proceedings of the National Academy of Sciences of the United States of America* **70**, 88.
Tait, R. C. and Smith, D. W. (1974). *Nature, London* **249**, 116.
Taylor, A. L. and Trotter, C. D. (1972). *Bacteriological Reviews* **36**, 504.
Thomas, R. and Mousset, S. (1970). *Journal of Molecular Biology* **47**, 179.
Tomizawa, J., Anraku, N. and Iwama, Y. (1966). *Journal of Molecular Biology* **21**, 247.
Tsai, R. L. and Green, H. (1973). *Journal of Molecular Biology* **73**, 307.
Van der Vliet, P. C. and Levine, A. J. (1973). *Nature New Biology* **246**, 170.
Vapnek, D. and Rupp, W. D. (1970). *Journal of Molecular Biology* **53**, 287.

Vosberg, H. P. and Hoffman-Berling, H. (1971). *Journal of Molecular Biology* **58**, 739.

Vovis, G. F. and Buttin, G. (1970). *Biochimica et Biophysica Acta* **224**, 29.

Wake, R. G. (1972). *Journal of Molecular Biology* **68**, 501.

Wake, R. G. (1973). *Journal of Molecular Biology* **77**, 569.

Wang, J. C. (1971). *Journal of Molecular Biology* **55**, 523.

Wang, H. F. and Sternglanz, R. (1974). *Nature, London* **248**, 147.

Ward, C. B. and Glaser, D. A. (1969). *Proceedings of the National Academy of Sciences of the United States of America* **63**, 800.

Watson, J. D. (1965). *In* "Molecular Biology of the Gene", pp. 261–273. W. A. Benjamin, New York.

Watson, J. D. (1972). *Nature New Biology* **239**, 197.

Watson, J. D. and Crick, F. H. C. (1953). *Nature, London* **171**, 737.

Wechsler, J. A. (1973). *In* "DNA Synthesis *in Vitro*", (R. D. Wells and R. B. Inman, eds.), p. 375. University Park Press, Baltimore, Maryland.

Weintraub, H. (1972). *Nature New Biology* **236**, 195.

Werner, R. (1971). *Nature, London* **230**, 570

White, K. and Sueoka, N. (1973). *Genetics* **73**, 185.

Wickner, R. B., Ginsberg, B., Berkower, I. and Hurwitz, J. (1972a). *Journal of Biological Chemistry* **247**, 489.

Wickner, R. B. and Hurwitz, J. (1972). *Biochemical and Biophysical Research Communications* **47**, 202.

Wickner, R. B., Wright, M., Wickner, S. and Hurwitz, J. (1972b). *Proceedings of the National Academy of Sciences of the United States of America* **69**, 3233.

Wickner, S., Berkower, I., Wright, M. and Hurwitz, J. (1973a). *Proceedings of the National Academy of Sciences of the United States of America* **70**, 2369.

Wickner, S., Hurwitz, J., Nath, K. and Yarbrough, L. (1972). *Biochemical and Biophysical Research Communications* **48**, 619.

Wickner, S., Wright, M. and Hurwitz, J. (1973b). *Proceedings of the National Academy of Sciences of the United States of America* **70**, 1613.

Wickner, S., Wright, M. and Hurwitz, J. (1974). *Proceedings of the National Academy of Sciences of the United States of America* **71**, 783.

Wickner, W., Brutlag, D., Schekman, R. and Kornberg, A. (1972). *Proceedings of the National Academy of Sciences of the United States of America* **69**, 965.

Wickner, W. and Kornberg, A. (1973). *Proceedings of the National Academy of Sciences of the United States of America* **70**, 3679.

Wickner, W., Schekman, R., Geider, K. and Kornberg, A. (1973). *Proceedings of the National Academy of Sciences of the United States of America* **70**, 1764.

Wolf, B., Newman, A. and Glaser, D. A. (1968). *Journal of Molecular Biology* **32**, 611.

Worcel, A. (1970). *Journal of Molecular Biology* **52**, 371.

Worcel, A. and Burgi, E. (1972). *Journal of Molecular Biology* **71**, 127.

Worcel, A. and Burgi, E. (1974). *Journal of Molecular Biology* **82**, 91.

Wright, M. and Buttin, G. (1969). *Bulletin de la Société de Chimie Biologique* **51**, 1373.

Wright, M., Buttin, G. and Hurwitz, J. (1971). *Journal of Biological Chemistry* **246**, 6543.

Wright, M., Wickner, S. and Hurwitz, J. (1973). *Proceedings of the National Academy of Sciences of the United States of America* **70**, 3120.

Yamaguchi, K. and Yoshikawa, H. (1973). *Nature New Biology* **244**, 204.

Yoshikawa, H. and Haas, M. (1968). *Cold Spring Harbor Symposium of Quantitative Biology* **33**, 843.

Yoshikawa, H., O'Sullivan, A. and Sueoka, N. (1964). *Proceedings of the National Academy of Sciences of the United States of America* **52**, 973.

Yoshikawa, H. and Sueoka, N. (1963a). *Proceedings of the National Academy of Sciences of the United States of America* **49**, 559.

Yoshikawa, H. and Sueoka, N. (1963b). *Proceedings of the National Academy of Sciences of the United States of America* **49**, 806.

Young, F. E. and Wilson, G. A. (1972). *Spores* V, 77 (H. O. Halvorson, R. Hanson, and L. L. Campbell, eds.), American Society for Microbiology, Washington D.C.

Youngs, D. A. and Smith, K. C. (1973). *Nature New Biology* **244**, 240.

AUTHOR INDEX

Numbers in italics refer to the pages on which references are listed at the end of each article.

A

B

C

D

E

F

G

H

I

J

K

L

M

N

O

P

T

Y

Z

SUBJECT INDEX

A

B

C

D

E

F

G

H

I

K

L

M

N

O

P

R

S

T

U